Lecture Notes in Mathematics

Edited by A. Dold and B. Eckmann

541

Measure Theory

Proceedings of the Conference Held at Oberwolfach, 15–21 June, 1975

Edited by A. Bellow (formerly A. Ionescu Tulcea) and D. Kölzow

Springer-Verlag
Berlin · Heidelberg · New York 1976

Editors
Alexandra Bellow
(formerly A. Ionescu Tulcea)
Department of Mathematics
Northwestern University
Evanston, Illinois 60201/USA

Dietrich Kölzow
Mathematisches Institut der
Universität Erlangen-Nürnberg
Bismarckstraße 1 1/2
8520 Erlangen/BRD

Library of Congress Cataloging in Publication Data

Conference on Measure Theory, Oberwolfach, Ger., 1975.
Measure theory.

(Lecture notes in mathematics ; 541)
1. Measure theory--Congresses. I. Bellow, Alexandra, 1935- II. Kölzow, Dietrich. III. Title. IV. Series: Lecture notes in mathematics (Berlin) ; 541.
QA3.L28 no. 541 [QA312] 510'.8s [515'.42] 76-40183

AMS Subject Classifivations (1970): 28-02, 46A40, 46GXX, 60-02

ISBN 3-540-07861-4 Springer-Verlag Berlin · Heidelberg · New York
ISBN 0-387-07861-4 Springer-Verlag New York · Heidelberg · Berlin

Printed in Germany
Printing and binding: Beltz Offsetdruck, Hemsbach/Bergstr.

FOREWORD

This volume contains the contributions to the Conference on Measure Theory, held at the Research Institute for Mathematics at Oberwolfach, from June 15 to June 21, 1975.

The organizers and the editors would like to express their gratitude to the participants for their contributions, to the administration of the Oberwolfach Institute for making the Conference possible, and, last not least, to the Springer-Verlag for its readiness to publish these Proceedings.

A. Bellow D. Kölzow

PARTICIPANTS

G. Aumann
Mathematisches Institut, Technische Universität München, Barerstr. 23, 8000 München 2, Deutschland

A. Bellow (formerly Ionescu-Tulcea)
Department of Mathematics, College of Arts and Sciences, Northwestern University, Evanston, IL 60201, USA

K. Bichteler
Department of Mathematics, University of Texas at Austin, Austin, TX 78712, USA

J. Bliedtner
Fakultät für Mathematik, Universität Bielefeld, Kurt-Schumacher-Str. 6, 4800 Bielefeld, Deutschland

V. M. Bogdan (formerly W. M. Bogdanowicz)
Department of Mathematics, Catholic University of America, Washington DC 20017, USA

S. D. Chatterji
Ecole Polytechnique Fédérale de Lausanne, Département de mathématiques, 33 Ave de Cour, 1007 Lausanne, Suisse

G. Y. H. Chi
Department of Mathematics, University of Pittsburgh, Pittsburgh, PA 15260, USA

J. R. Choksi
Department of Mathematics, McGill University, Box 6070, Station A, Montreal (101), Quebec, Canada, H3C 3G1

J. P. R. Christensen
Københavns Universitets Matematiske Institut, Universitetsparken 5, 2100 København Ø, Denmark

J. Diestel
Department of Mathematics, Kent State University, Kent OH 44242, USA

L. E. Dubins
Department of Mathematics, University of California at Berkeley, Berkeley, CA 94720, USA

B. Eifrig
Institut für Angewandte Mathematik, Universität Heidelberg, Im Neuenheimer Feld 5, 6900 Heidelberg, Deutschland

Th. Eisele
Institut für Angewandte Mathematik, Universität Heidelberg, Im Neuenheimer Feld 5 , 6900 Heidelberg, Deutschland

G. Fichera
Istituto Matematico della Università di Roma, 00185 Roma, Italia

D. H. Fremlin
Department of Mathematics, University of Essex, Colchester, Essex, Great Britain

Z. Frolík
Matematický ústav ČSAV, Žitná 25, 11567 Praha 1, Czechoslovakia

B. Fuchssteiner
Gesamthochschule Paderborn, Fachbereich Mathematik, Pohlweg, 4790 Paderborn, Deutschland

J. Gapaillard
Université de Nantes, Institut de Mathématiques et d'Informatique, 38, Bd. Michelet, B. P. 1044, 44037 Nantes Cedex, France

M. Gattinger
Institut für Angewandte Mathematik, Universität Erlangen, Egerlandstr. 3, 8520 Erlangen, Deutschland

M. Gattinger (formerly Pehmler)
Mathematisches Institut, Universität Erlangen, Bismarckstr. 1 1/2, 8520 Erlangen, Deutschland

P. Georgiou
Department of Mathematics, University of Athens, Athens, Greece

V. Goodman
Department of Mathematics, Indiana University, Swain Hall East, Bloomington, IN 47401, USA

S. Graf
Mathematisches Institut, Universität Erlangen, Bismarckstr. 1 1/2, 8520 Erlangen, Deutschland

L. Gross
Department of Mathematics, White Hall, Cornell University, Ithaca, NY 14850, USA

M. de Guzman
Universidad de Madrid, Departamento de Ecuaciones Diferenciales, Facultad de Ciencias, C. U., Madrid 3, España

W. Hackenbroch
Fachbereich Mathematik, Universität Regensburg, Universitätsstr. 31, 8400 Regensburg, Deutschland

D. Hoffmann
Fachbereich Mathematik, Universität Konstanz, Postfach 7733, 7750 Konstanz, Deutschland

R. E. Huff
Department of Mathematics, Pennsylvania State University, 230 Mc Allister Bldg., University Park, PA 16802, USA

D. Kahnert
Mathematisches Institut, Universität Stuttgart, Pfaffenwaldring 57, 7000 Stuttgart 80, Deutschland

S. Kakutani
Department of Mathematics, Yale University, New Haven, CT 06520, USA

D. A. Kappos
Lykabetton 29, Athens 135, Greece

D. Kölzow
Mathematisches Institut, Universität Erlangen, Bismarckstr. 1 1/2, 8520 Erlangen, Deutschland

G. Knowles
Institut für Angewandte Mathematik und Informatik, Universität Bonn, Wegelerstr. 6, 5300 Bonn, Deutschland

W. A. J. Luxemburg
California Institute of Technology, Pasadena, CA 91125, USA

G. Mägerl
Mathematisches Institut, Universität Erlangen, Bismarckstr. 1 1/2, 8520 Erlangen, Deutschland

D. Maharam-Stone
Department of Mathematics, University of Rochester, Rochester, NY 14627, USA

P. R. Masani
Department of Mathematics, University of Pittsburgh, Pittsburgh, PA 15260, USA

K. Musiał
Institut Matematyczny PAN, Pl. Grunwaldzki 2/4, 50-384 Wroclaw, Poland

Z. R. Pop-Stojanovic
Department of Mathematics, University of Florida, 205 Walker Hall, Gainesville, FL 32601, USA

E. Rauch
FB 6, Mathematik-Naturwissenschaften, Lehrstuhl für Mathematik IV, Gesamthochschule Siegen, Hölderlinstr. 3, 5900 Siegen-Weidenau, Deutschland

P. Ressel
Institut für Mathematische Stochastik, Universität Freiburg, Hermann-Herder-Str. 10, 7800 Freiburg i.Br., Deutschland

M. Sion
Department of Mathematics, University of British Columbia, Vancouver (8), Br. Columbia, Canada

D. Sondermann
Lehrstuhl f. Theoretische Volkswirtschaftslehre, Universität Hamburg, von Melle Park 15, 2000 Hamburg 13, Deutschland

T. P. Srinivasan
Department of Mathematics, University of Kansas, Lawrence, KS 66045, USA

A. H. Stone
Department of Mathematics, University of Rochester, Rochester, NY 14627, USA

W. Strauss
Mathematisches Institut A, Universität Stuttgart, Pfaffenwaldring 57, 7000 Stuttgart 80, Deutschland

G. E. F. Thomas
Rijksuniversiteit Groningen, Mathematisch Instituut, Hoogbouw WSN, Universiteitscomplex Paddepool, Postbus 800, Groningen, Nederland

F. Topsøe
Københavns Universitets Matematiske Institut, Universitetsparken 5, 2100 København Ø, Denmark

T. Traynor
Department of Mathematics, University of Windsor, Windsor, Ontario, Canada, N9B 3P4

H. von Weizsäcker
Mathematisches Institut, Universität München, Theresienstr. 39, 8000 München 2, Deutschland

J. D. M. Wright
90A Bulmershe Rd., Berkshire, Reading, England

A. C. Zaanen
Mathematisch Instituut, Rijksuniversiteit te Leiden, Wassenaarseweg
80, Leiden, Nederland

CONTRIBUTIONS

SET AND POINT TRANSFORMATIONS ON HOMOGENEOUS SPACES

J.R. Choksi[+] and R.R. Simha

(McGill Univ. and Tata Inst. of Fund. Res.)

If (X,M,μ) is a measure space and (E,μ) is its measure algebra, one can ask whether every automorphism of (E,μ) is induced by an invertible measurable point map of X. In general the answer is no, it was shown by von Neumann [8] that the answer is yes when X is a Polish space and μ is a σ-finite measure (necessarily regular) on the Borel subsets of X. This was generalized by Maharam [7] to the direct product measure on an uncountable product of Polish spaces each with a normalized measure, and by the first author [1],[2] to an arbitrary σ-finite measure on the product σ-algebra of such a product. One can ask the question for Baire or Borel measures on an arbitrary compact, Hausdorff space: one notes that a regular Borel measure and its Baire contraction have isomorphic measure algebras. The answer to both questions is no in general, see [9] or [3], sec. 2; hence it is no, even on a product of unit intervals, except for completion regular measures such as the product measure, for which Maharam's result shows it is true. The result in [2] shows that it is true for all Baire measures on such a product of intervals. Subsequently the first author [3] showed that the same result holds for all Baire measures on an arbitrary compact group. The proof did not fully use the algebraic properties of the group but rather only the homogeneity, but the expected generalisation to homogeneous spaces at first ran into difficulties. We here announce that the result does generalise: specifically it holds for all Baire measures on an arbitrary locally compact σ-compact homogeneous space, under the action of a locally compact σ-compact group G. Further, if the space is of the form G/L, and μ is taken to be the essential Baire measure, then the result holds without any assumption of σ-compactness. These results are stated below with an indication of the idea of the proof. Complete proofs will appear in a paper by us [4], to appear in Advances in Mathematics.

+ Presented by the first author, who acknowledges support of the National Research Council of Canada in the preparation of this work, and a travel grant from McGill University to attend the Conference.

For completion regular measures such as the direct product measure, it has been asked whether the point transformation inducing the automorphism can be chosen to be Lusin measurable. It was shown by D.H. Fremlin, at this conference, that this is false; his counter-example is included in these proceedings. This has some interesting consequences: e.g. it follows that on an uncountable product of unit intervals with product Lebesgue measure, there exists a measure preserving automorphism of the measure algebra, which cannot be approximated, in the metric topology, by an automorphism induced by a homeomorphism of the product space.

We now state the main results.

<u>THEOREM</u> 1. Let M be a locally compact homogeneous space acted upon by a locally compact, σ-compact group G, so that M is σ-compact and homeomorphic to G/L for some closed subgroup L of G. Let μ be a σ-finite Baire measure on M, which, since it is finite on compact sets, is necessarily the restriction to the Baire sets $\mathcal{B}^o$ (of M or G/L) of a Radon measure $\bar{\mu}$ on M. Then every automorphism φ of the measure algebra E of $(M,\mathcal{B}^o,\mu)$ (or $(G/L,\mathcal{B}^o,\mu)$) is induced by an invertible, completion Baire measurable point transformation T of M.

<u>THEOREM</u> 2. Let M be a locally compact homogeneous space, acted upon by a locally compact, σ-compact group G, μ a σ-finite Baire measure on M, necessarily the restriction of a Radon measure $\bar{\mu}$ to the Baire sets $\mathcal{B}^o$. Let U be an invertible isometry of $L^p(M,\mathcal{B}^o,\mu)$, $1 \leq p < \infty$, $p \neq 2$, or a positive invertible isometry of $L^2(M,\mathcal{B}^o,\mu)$. [Note that $L^p(M,\mathcal{B}^o,\mu)$ coincides with $L^p(M,\mathcal{B},\bar{\mu})$.] Then there exists an invertible, completion Baire measurable, point transformation T of M such that $(Uf)(x) = f(T^{-1}x)\,\alpha(x)$, with $|\alpha(x)|^p = \omega_T(x)$ [for $p = 2$, $\alpha(x) = \omega_T^{\frac{1}{2}}(x)$], where $\omega_T(x)$ is defined, for all $X \in \mathcal{B}^o$, by

$$\mu(T^{-1}X) = \int_X \omega_T(x)\,\mu(dx)\,.$$

<u>THEOREM</u> 3. Let G be a locally compact group, L a closed subgroup, $\bar{\mu}$ a Radon essential measure on the locally compact homogeneous space G/L, μ its restriction to the Baire sets $\mathcal{B}^o$ of G/L. Then every automorphism φ of the measure algebra E of $(G/L,\mathcal{B}^o,\mu)$ is induced by an invertible, completion Baire measurable point transformation T of G/L.

<u>COROLLARY</u> 1. If $L = \{e\}$, then $G/L = G$, so the theorem holds for groups.

<u>COROLLARY</u> 2. If $G/L = G$, and if $\bar{\mu}$ is Haar measure, then T, T^{-1} are $\mathcal{B}_{\bar{\mu}}$ measurable.

Our method of proof in the σ-compact case is as follows. We first note that M is then necessarily of the form G/L. We then express G as a projective limit of G/H where H is compact normal in G and G/H is metrisable. To do this we can no longer use the Peter-Weyl Theorem (this is what was done for compact groups in [3]), but a theorem of Kakutani and Kodaira [5] is adequate for our needs. From this we get that G/L is the projective limit of G/LH. We now attempt to follow the argument used for compact groups in [3]: we consider the set of point realizations T_K of φ restricted to the measure algebra of G/LK, where each K is compact normal in G and is 'invariant' under φ; we partially order these in the natural way and apply Zorn's lemma. While many refinements are needed to tackle the homogeneous space case, the critical problem is, as in the case of a compact group, to extend the point transformation from G/LK, K compact normal, to $G/L(K \cap H)$ where H is compact normal with G/H metrisable. If $L(K \cap H)$ were equal to $LK \cap LH$, a fairly straightforward genealization of the argument in [3] would give the result. But this is not in general so, and this proved the main stumbling block in our attempts to prove Theorem 1. We were finally able to prove two group theoretic propositions which showed that for each K, there were enough groups H with this property (i.e. such groups H were cofinal in a sufficiently strong way), and we were then able to obtain the desired extension of the point transformation, and so complete the proof of Theorem 1. Theorem 2 follows trivially from Theorem 1, using the theorem of Lamperti ([6], Theorem 3.1). Finally to prove Theorem 3, we show that G/L is the disjoint union of open σ-compact spaces and that these may further be replaced by larger ones which are, in addition, invariant under φ. We show that each of these is homogeneous under the action of a locally compact σ-compact group, and then use Theorem 1. The rest of the proof is then straightforward. There is a trivial generalization of Theorem 2 to the non σ-compact case. Full details, including statements and proofs of the group theoretic propositions and other subsidiary results (the proof of Theorem 1 has nine lemmas) are given in our paper [4]. We have attempted here only to give the basic ideas of the proof.

REFERENCES

1. J. R. Choksi, Automorphisms of Baire measures on generalized cubes, Z. Wahrscheinlichkeitstheorie und Verw. Geb. 22 (1972), 195-204.

2. ________, Automorphisms of Baire measures on generalized cubes II, Z. Wahrscheinlichkeitstheorie und Verw. Geb. 23 (1972), 97-102.

3. ________, Measurable transformations on compact groups, Trans. American Math. Soc. 184 (1973), 101-124.

4. J.R. Choksi and R.R. Simha, Measurable transformations on homogeneous spaces, Advances in Math. (to appear).

5. S. Kakutani and K. Kodaira, Über das Haarsche Mass in der lokal bikompakten Gruppe, Proc. Imperial Acad. Tokyo, 20 (1944), 444-450.

6. J. Lamperti, On the isometries of certain function spaces, Pacific J. Math. 8 (1958), 459-466.

7. D. Maharam, Automorphisms of products of measure spaces, Proc. American Math. Soc. 9 (1958), 702-707.

8. J. von Neumann, Einige Sätze über die messbare Abbildungen, Ann. of Math. (2) 33 (1932), 574-586.

9. R. Panzone and C. Segovia, Measurable transformations on compact spaces and o.n. systems on compact groups, Rev. Un. Mat. Argentina 22 (1964), 83-102.

Dept. of Mathematics
McGill University
Box 6070, Station A
Montreal, Quebec
Canada, H3C 3G1

and

School of Mathematics
Tata Institute of
Fundamental Research
Homi Bhabha Road
Bombay 400 005, India

ON THE UNIQUENESS OF PREIMAGES OF MEASURES

by

K.-Th. Eisele, Heidelberg

Given the following situation:

Let E and F be compact spaces and $f: E \to F$ a continuous surjective mapping from E onto F. Let ν be a positive regular Borel measure on F with $\nu(F) = 1$.

The problem is, how much the uniqueness of a positive regular Borel measure μ on E, such that the image $\mu \cdot\cdot f$ of μ with respect to f is ν:

$$\mu \cdot\cdot f = \nu \quad (\text{i.e. } \mu(f^{-1}(B)) = \nu(B) \text{ for all Borel subsets B of F}),$$

is determined by the ν-measure of the set

$$Y = \{ y \in F, \quad \exists x_o, x_1 \in E, x_o \neq x_1 \text{ and } f(x_o) = f(x_1) = y \}.$$

The main result is stated in the theorem below.

This problem has been put to me by Prof. Böge, to whom I would like to express my thanks for his support troughout. I know of no other literature on this problem.

For short, by a Borel measure we mean here always a positive regular Borel measure.

By the Riesz - representation theorem the measure ν is uniquely determined by his values $\nu(g) = \int g \, d\nu$ for $g \in \mathfrak{C}(F)$.

But $\mu \cdot\cdot f = \nu$ is equivalent to

$$\mu(g \cdot f) = \nu(g) \qquad \text{for all } g \in \mathfrak{C}(F),$$

So we may define the Banach space

$$\mathfrak{D} = \{ h \in \mathfrak{C}(E);\ \exists g \in \mathfrak{C}(F),\ h = g \cdot f \}$$

and on $\mathfrak{D}$ the positive linear operator I by

$$I(h) = \nu(g) \qquad \text{if } h = g \cdot f \in \mathfrak{D}.$$

I is well-defined. By this and the Riesz - representation theorem for μ, our problem leads to an investigation of all possible extensions

of I to positive linear operators on $\mathfrak{C}(E)$.
Let $\|\cdot\|$ be the supremum norm on $\mathfrak{D}$, $\mathfrak{C}(E)$ or $\mathfrak{C}(F)$ respectively.
By $\nu(F) = 1$ we have

$$|I(h)| = |\nu(g)| \leq \|g\| = \|h\|$$

for $h = g\circ f \in \mathfrak{C}(E)$. The Hahn - Banach theorem yields the existence of a positive linear operator K on $\mathfrak{C}(E)$ with

$$K(h) = I(h) \text{ for } h \in \mathfrak{D} \text{ and } |K(h)| \leq \|h\| \text{ for } h \in \mathfrak{C}(E).$$

So we have in any way the existence of a Borel measure μ satisfying

$$\mu \bullet\bullet f = \nu .$$

A sufficient condition for the uniqueness of μ is given by the following

Lemma

If for the outer measure ν^* we have $\nu^*(Y) = 0$, then there exists at most one Borel measure μ on E with $\mu \bullet\bullet f = \nu$.

Proof: Since $\nu^*(Y) = \inf\{\nu(U), U \supseteq Y, U \text{ open}\} = 0$, there exists for $\varepsilon > 0$ an open set $U \supseteq Y$ with $\nu(U) = \eta < \varepsilon$. $f^{-1}(U)$ is open in E. So $E \setminus f^{-1}(U)$ is compact and $f \upharpoonright E \setminus f^{-1}(U)$ is a continous bijection, i.e. a homeomorphism, between $E \setminus f^{-1}(U)$ and $F \setminus U$. Hence, there is exactly one Borel measure μ_U on $E \setminus f^{-1}(U)$ such that

$$\mu_U \bullet\bullet (f \upharpoonright E \setminus f^{-1}(U)) = \nu \upharpoonright F \setminus U.$$

In particular

$$\mu_U(E \setminus f^{-1}(U)) = 1-\eta > 1-\varepsilon .$$

Suppose now, that μ' and μ'' are two preimages of ν with respect to f.

We have $\mu' \upharpoonright E - f^{-1}(U) = \mu_U = \mu'' \upharpoonright E - f^{-1}(U)$

and $\|\mu' \upharpoonright f^{-1}(U)\| = \|\mu'' \upharpoonright f^{-1}(U)\| = \mu(f^{-1}(U)) = \eta < \varepsilon.$

Hence $\|\mu' - \mu''\| = \|\mu' \upharpoonright f^{-1}(U) - \mu'' \upharpoonright f^{-1}(U)\| \leq 2\eta < 2\varepsilon.$

This shows $\mu' = \mu''$.

However, the converse of the lemma does not hold without any further hypothesis. This is elucidated by the following example, which Prof. D.H. Fremlin kindly told me of, when we met in Oberwolfach.

Example:

Let $E = [0,1] \times \{1,2\}$ be the "split interval".
We order E by

$$\langle\alpha,i\rangle < \langle\beta,j\rangle \quad \text{iff} \quad \alpha<\beta \text{ or } (\alpha=\beta \text{ and } i=1 < 2=j).$$

Let E be endowed by the order topology induced by $<$, for which the open intervals with respect to $<$ form a basis.
The following points state some facts about E.

For $i = 1,2$ let $\pi_i : [0,1] \to E$ be defined by $\pi_i(\alpha) = (\alpha,i)$.
For a subset $A \subseteq E$ let $\rho(A) = \pi_1^{-1}(A) \cap \pi_2^{-1}(A)$.

1. If (x',x'') is an open interval of E, then $\rho((x',x''))$ is an open interval of $[0,1]$ and it is $(x',x'') = \emptyset$ iff $\rho((x',x'')) = \emptyset$.

2. Every open subset of E is a countable union of open intervals.

Proof: Since the open intervals form a basis, every open subset U of E has the form $U = \bigcup_{\lambda\in\Lambda} (x'_\lambda, x''_\lambda)$. E being linearly ordered by $<$, we may suppose, that the representation $U = \bigcup_{\lambda\in\Lambda} (x'_\lambda, x''_\lambda)$ uses only maximal non-empty intervals, i.e.

$$(x'_\lambda, x''_\lambda) \neq (x'_{\lambda'}, x''_{\lambda'}) \quad \text{implies} \quad x''_\lambda \leq x'_{\lambda'} \quad \text{or} \quad x''_{\lambda'} \leq x'_\lambda.$$

Therefore with $x'_\lambda = (\alpha_\lambda, i_\lambda)$ and $x''_\lambda = (\beta_\lambda, j_\lambda)$ $(\lambda\in\Lambda)$ we have

$$\alpha_\lambda, \beta_\lambda \notin \rho(U) \quad \text{and} \quad \rho(U) = \bigcup_{\lambda\in\Lambda} \rho((x'_\lambda, x''_\lambda)) = \bigcup_{\lambda\in\Lambda} (\alpha_\lambda, \beta_\lambda),$$

where $\beta_\lambda \leq \alpha_{\lambda'}$ or $\beta_{\lambda'} \leq \alpha_\lambda$ for all $\lambda, \lambda' \in \Lambda$, $(x'_\lambda, x''_\lambda) \neq (x'_{\lambda'}, x''_{\lambda'})$.
But in $[0,1]$ every set of open, disjoint and non-empty intervals is countable. So Λ may be supposed to be countable.

3. For $A \subseteq E$ set $\sigma(A) = \pi_1^{-1}(A) \,\Delta\, \pi_2^{-1}(A) =$

$$= (\pi_1^{-1}(A) \setminus \pi_2^{-1}(A)) \cup (\pi_2^{-1}(A) \setminus \pi_1^{-1}(A)).$$

For all Borel subsets B of E we have: $\sigma(B)$ is at most countable.

Proof: The system of sets, for which the assertion holds, is closed under countable unions and complements. Since $\overline{\overline{\sigma((x',x''))}} \leq 2$ for all open intervals (x',x'') of E, by 2. the assertion is true for all open subsets of E and, therefore for all Borel subsets.

4. E is compact.

Proof: Let A be an infinite subset of E and let us suppose, without loss of generality, that $A_1 = A \cap ([0,1] \times \{1\})$ is infinite. But in $[0,1]$ there exists a monotonous sequence $(\alpha_n)_{n<\omega} \subseteq A_1$

and a point $\gamma \in [0,1]$ such that $\gamma = \lim_n \alpha_n$.

If $\alpha_n \leq \gamma$ for all n, then $\langle \alpha_n, 1 \rangle \in A$ converges to $\langle \gamma, 1 \rangle$ in E.

If $\alpha_n > \gamma$ for all n, then $\langle \alpha_n, 1 \rangle \in A$ converges to $\langle \gamma, 2 \rangle$.

Therefore, each infinite subset A of E has a culmination point in E.

5. E is not metrizable.

Proof: If, otherwise, E would be a compact metrizable space, there exists a sequence $((x_n', x_n''))_{n<\omega}$ of open intervals forming a basis for E. Let $x_n' = \langle \alpha_n, i_n \rangle$, $x_n'' = \langle \beta_n, j_n \rangle$ and $\gamma \in (0,1] \setminus (\{\alpha_n, n<\omega\} \cup \{\beta_n, n<\omega\})$. But then

$$U_0 := (\langle 0,1 \rangle, \langle \gamma, 2 \rangle) \neq \bigcup \{ (x_n', x_n''), (x_n', x_n'') \subseteq U_0 \} =: U_1$$

since $\langle \gamma, 1 \rangle \in U_0$; but on the other hand $\beta_n < \gamma$ for all n with $(x_n', x_n'') \subseteq U_0$ implies $\langle \gamma, 1 \rangle \notin U_1$. So $((x_n', x_n''))_{n<\omega}$ is not a basis for E; contradiction!

We are now going to construct a counter-example to the converse direction of the foregoing lemma.

Set $F = [0,1]$ and define $f : E \to F$ by

$$f(\langle \alpha, j \rangle) = \alpha \qquad \text{for } \langle \alpha, j \rangle \in E.$$

f is continuous and surjective and $Y = \{ y \in F; \overline{\overline{f^{-1}\{y\}}} \geq 2 \} = F$.

Let ν denote the Lebesque measure on $F = [0,1]$, and μ be any Borel measure on E. We want to show that μ is uniquely determined by the relation $\mu \circ\circ f = \nu$.

First, for any countable subset $\mathfrak{C}$ of E we have

$$0 \leq \mu(\mathfrak{C}) \leq \mu(f^{-1}(f(\mathfrak{C}))) = \nu(f(\mathfrak{C})) = 0,$$

since $f(\mathfrak{C})$ is countable. For any subset A of E it holds that

$$\rho(A) \times \{1,2\} \subseteq A \subseteq (\rho(A) \cup \sigma(A)) \times \{1,2\} = \rho(A) \times \{1,2\} \cup \sigma(A) \times \{1,2\}$$

and $\rho(A) \times \{1,2\} = f^{-1}(\rho(A))$.

By 3. it follows therefore that

$$\mu(B) = \mu(f^{-1}(\varphi(B))) = \nu(\varphi(B))$$

for every Borel subset B of E.
Hence, there exists exactly one Borel measure μ on E with $\mu^{\bullet\bullet} f = \nu$, though $\nu(Y) = \nu([0,1]) = 1$.

However, if we restrict ourselves to compact metrizable spaces E, we get the following

Theorem

Let E be a compact metrizable space, F a compact space, and $f : E \to F$ a continuous, surjective mapping. Then it holds that

(i) $Y = \{y \in F, \ \overline{\overline{f^{-1}(y)}} \geq 2\}$ is universally measurable in F

and (ii) for any Borel measure ν on F with $\nu(F) = 1$ we have $\nu(Y) = 0$ if and only if there exists exactly one Borel measure μ on E with $\mu^{\bullet\bullet} f = \nu$.

Proof:

(i) The mapping $f \times f : E \times E \to F \times F$ is continuous and surjective. Let $\Delta E = \{(x,x), x \in E\}$ and $\Delta F = \{(y,y), y \in F\}$ be the diagonal embeddings of E and F respectively. Since the compact metrizable space $E \times E$ has a countable basis, the compact sets ΔE and $(f \times f)^{-1}(\Delta F)$ are measurable with respect to the product algebra generated by $\mathcal{B}(E) \otimes \mathcal{B}(E)$, hence so is $(f \times f)^{-1}(\Delta F) \setminus \Delta E$.
Let π denote the projection from $E \times E$ to the first component. By the continuity of f and π, $f \circ \pi((f \times f)^{-1}(\Delta F) \setminus \Delta E)$ is an analytical and therefore, with regard to Choquet's theorem, a universally measurable subset of F. We claim

$$Y = f \circ \pi((f \times f)^{-1}(\Delta F) \setminus \Delta E);$$

for $y \in Y$, iff $\exists\, x_1, x_2 \in E,\ x_1 \neq x_2$ and $f(x_1) = f(x_2) = y$

iff $\exists\, (x_1, x_2) \in (f \times f)^{-1}(\Delta F) \setminus \Delta E,\ f(x_1) = y$

iff $y \in f \circ \pi((f \times f)^{-1}(\Delta F) \setminus \Delta E)$.

This proves (i).

(ii) The implication "$\curvearrowright$" has already been shown by the introductory remarks and the lemma. For the converse, we suppose $\nu(Y) = \eta > 0$, and we have to show that there are at least two different Borel measures μ_1 and μ_2, $\mu_1 \neq \mu_2$, on E with $\mu_1 \cdot\cdot f = \mu_2 \cdot\cdot f = \nu$, or that there are two different positive linear operators K_1 and K_2 on $\mathcal{C}(E)$ extending I.

On E we have the equivalence relation R_f defined by

$$R_f\,(x',x") \quad \text{iff} \quad f(x') = f(x") .$$

By a known theorem (cf. [1] §6, N$^{\underline{o}}$ 8, théorème 4 et remarque) there is a Borel set B_o of E, such that each equivalence class of R_f meets B_o in exactly one point. In other words, f is a bijection between B_o and F.

Now we can easily define a first Borel measure μ_1 on E by setting

$$\mu_1(E \setminus B_o) = 0$$

and $\mu_1(B) = \nu(f(B))$ for all Borel sets $B \subseteq B_o$,

where $f(B)$ as an analytical set in F is universally measurable.

Since $\mu_1(f^{-1}(D)) = \mu_1(f^{-1}(D) \cap B_o) = \nu(f(f^{-1}(D) \cap B_o)) = \nu(D)$ for all Borel subsets D of F, we have $\mu_1 \cdot\cdot f = \nu$.

Let $B_1 = \pi(\,(f \times f)^{-1}(\Delta F) \setminus \Delta E\,)$. By the first part of the proof B_1 is universally measurable and

$$B_1 = f^{-1}(Y) = \{x_1 \in E,\ \exists\, x_2 \in E,\ x_1 \neq x_2 \text{ and } f(x_1) = f(x_2)\} .$$

Then also the sets $B_2 = B_1 \cap B_o$ and $B_3 = B_1 \setminus B_o$ are universally measurable. By definition we have

$$Y = f(B_2) = f(B_3)$$

and $\mu_1(B_2) = \nu(Y) = \eta > 0$, whereas $\mu_1(B_3) = 0$.

μ_1 being regular, we find an open set $U \supseteq B_3$ with $\mu_1(U) = \eta/3$.

Since E has a countable basis and the characteristic function 1_U is lower semicontinuous, there exists an increasing sequence $(h_n)_{n<\omega}$ of positive continuous functions on E with $1_U = \sup_n h_n$.

For any function $h : E \to \overline{\mathbb{R}}^+$ we define $f h : F \to \overline{\mathbb{R}}^+$ by

$$fh(y) := \sup \{h(x);\ f(x) = y\} \qquad \text{for } y \in F .$$

One easily verifies that $f\, 1_U = 1_{f(U)}$ and

$$f\, 1_U = f\,(\sup_n h_n) = \sup_n f\, h_n .$$

If $h \in \mathcal{C}^+(E)$, then $f\,h$ is positive and bounded, and for all $\alpha > 0$ the set $\{y : f\,h\,(y) > \alpha\} = f(\{x \in E,\ h(x) > \alpha\})$ is universally measurable. Therefore $f\,h$ is integrable with respect to every Borel measure on E. Further, if $h = g \circ f \in \mathfrak{D}$ with $g \in \mathcal{C}^+(F)$, then $f\,h = f(g \circ f) = g$, since $f\,h(y) = \sup \{g \circ f(x),\ f(x) = y\} = g(y)$.

Collecting the above results, we receive by Lebesgue's theorem

$0 < \eta = \nu(Y) = \nu(f(B_3)) = \nu(f\,1_U) = \nu(\sup_n fh_n) = \sup_n \nu(fh_n)$.

So let n_o be a natural number with $\nu(fh_{n_o}) \geq \frac{2}{3}\eta$.

On the other hand we have $\mu_1(h_{n_o}) \leq \mu_1(1_U) \leq \frac{1}{3}\eta$.

The relation $\mu_1 \circ\circ f = \nu$ shows that μ_1, regarded as a positive linear operator on $\mathcal{C}(E)$, satisfies $\mu_1 \geq I$. This gives us the following inequalities:

$$\sup \{I(h),\ h \in \mathfrak{D},\ h \leq h_{n_o}\} \leq \mu(h_{n_o}) \leq \tfrac{1}{3}\eta < \tfrac{2}{3}\eta \leq \nu(fh_{n_o}) \leq$$

$$\leq \inf \{I(h),\ h \in \mathfrak{D},\ h \geq h_{n_o}\},$$

where the last inequality follows from the implications

$$h \in \mathfrak{D},\ h \geq h_{n_o} \curvearrowright \exists g \in \mathcal{C}(F),\ h = g \circ f \geq h_{n_o} \curvearrowright f(g \circ f) = g \geq fh_{n_o}$$

$$\curvearrowright I(h) = \nu(g) \geq \nu(fh_{n_o}) .$$

This proves that $h_{n_o} \notin \mathfrak{D}$, because otherwise the ends of the above chain of inequalities would be equal to $I(h_{n_o})$.

On the Banach subspace of $\mathcal{C}(E)$, generated by $\mathfrak{D} \cup \{h_{n_o}\}$, we define the positive linear operator K' by

$$K'(\alpha h + \beta h_{n_o}) = \alpha I(h) + \beta \tfrac{2}{3}\eta \quad \text{for } \alpha, \beta \in \mathbb{R} \text{ and } h \in \mathfrak{D}.$$

The Hahn - Banach theorem now yields a positive linear operator K on $\mathcal{C}(E)$ with $K \geq K' \geq I$, but $K(h_{n_0}) = \frac{2}{3}\eta \neq \mu_1(h_{n_0})$ shows $K \neq \mu_1$.

q.e.d.

References

[1] N. Bourbaki: Topologie Générale, chap. 9, Hermann Paris 1958

Addendum: After this paper has been written, G. Mägerl, Erlangen, pointed out to me, that part (ii) of the above proof can considerably be shorten, if we replace the section theorem by a proposition, which asserts the existence of measurable functions g_0, g_1 from F to E, disjoint on Y and with $f \circ g_0 = f \circ g_1 = id_F$.

HAAR-MASS UND HAUSDORFF-MASS

Von D. Kahnert, Universität in Stuttgart, BRD

Es sei G eine separable lokalkompakte Gruppe ohne isolierte Punkte. Wie Birkhoff [1] und Kakutani [5] gezeigt haben, gibt es eine linksinvariante Metrik d auf G, die mit der Gruppentopologie verträglich ist. Ist $h: [0,\infty) \to [0,\infty)$ eine Hausdorff-Funktion, d.h. gilt $h(t_1) \leqq h(t_2)$ falls $t_1 \leqq t_2$ ist und

$$\lim_{t\to 0} h(t) = h(0) = 0,$$

so wird durch

$$L_h(X) = \liminf_{q\to 0} \left\{ \sum_{i=1}^{\infty} h(d(X_i)) : \bigcup_{i=1}^{\infty} X_i \supset X;\ d(X_i) \leqq q \text{ für } i\in \mathbb{N} \right\} \quad (X\subset G)$$

das zugehörige äußere Hausdorff-Maß erklärt. Bekanntlich ist L_h ein metrisches äußeres Maß, was zur Folge hat, daß die Einschränkung von L_h auf das System $\mathfrak{B}$ der Borelmengen von G ein Maß ist.
Da L_h linksinvariant ist, d.h.

$$L_h(x+X) = L_h(X) \quad \text{für } x\in G \text{ und } X\in \mathfrak{B},$$

stellt sich folgendes Problem:

<u>Frage:</u> Gibt es eine Hausdorff-Funktion h, so daß L_h ein Haar-Maß auf G ist?

Vermutlich ist diese Frage positiv zu beantworten. Jedoch sind bisher nur Sonderfälle bekannt. Sicher ist L_h ein Haar-Maß auf G, wenn es eine kompakte Nullumgebung U in G gibt mit $0<L_h(U)<\infty$. Nun gibt es zwei einfache Methoden, um Abschätzungen für $L_h(U)$ zu gewinnen. Einmal gilt

$$L_h(U) \leqq \liminf_{q\to 0} N_q(U)h(q),$$

wobei für $q>0$

$$N_q(U) = \min\{k \mid \text{es existieren } U_i : \bigcup_{i=1}^{k} U_i = U,\ d(U_i) \leqq q \text{ für } i=1,2,\dots,k\}$$

ist.

Gibt es andererseits eine subadditive reellwertige Mengenfunktion λ auf $\mathfrak{B}$ mit $\lambda(U)>0$ und positive reelle Zahlen a und b, so daß für alle $X\in\mathfrak{X}$ mit $d(X)\leqq a$ die Abschätzung

$$\lambda(X) \leqq b\cdot h(d(X))$$

gilt, dann folgt

$$L_h(U) \geqq \lambda(U)/b>0.$$

Ist μ ein festes Haar-Maß auf G, so lassen sich mit Hilfe dieser Methoden leicht Hausdorff-Funktionen h_1 und h_2 angeben mit

$$L_{h_1} \leqq \mu \leqq L_{h_2} .$$

Sei nämlich $U = K(0,r) = \{x\in G : d(x,0)\leqq r\}$ eine kompakte Nullumgebung in G,

$$h_1(t) = \begin{cases} \mu(U)/N_t(U), & \text{für } t>0 \\ 0, & \text{für } t=0 \end{cases}$$

und

$$h_2(t) = \begin{cases} \sup\{\mu(X): X\subset U,\ X\in\mathfrak{X},\ d(X)\leqq t\}, & \text{falls } t\leqq d(U) \\ \mu(U), & \text{falls } t>d(U). \end{cases}$$

Ist $L_{h_1}(U)>0$, so ist L_{h_1} ein Haar-Maß und $L_{h_1}\leqq\mu$. Im Falle $L_{h_1}(U)=0$ folgt $L_{h_1}(G)=0$, also $L_{h_1}<\mu$. Es ist nicht bekannt, ob der Fall $L_{h_1}(U)=0$ eintreten kann. Diese Frage steht im Zusammenhang mit einem Problem von Mycielski ([9], Problem P).

Es gibt eine positive reelle Zahl a, so daß für alle $X\in\mathfrak{X}$ mit $d(X)\leqq a$ die Ungleichung

$$\mu(X) \leqq h_2(d(X))$$

gilt. Damit ist $\mu \leqq L_{h_2}$. Es ist nicht bekannt, ob der Fall $L_{h_2}(U)= \infty$ eintreten kann.

I. Hinreichende Bedingungen und Beispiele.

Im folgenden sei U eine kompakte Nullumgebung von G, μ ein Haar-Maß auf G und

$$f(q) = \sup\{\mu(X): X\in\mathfrak{X},\ d(X)\leqq q\} \quad \text{für } q>0.$$

Bedingung (1):

$$M = \limsup_{q\to 0} N_q(U)\, f(q) < \infty \qquad \text{und}$$
$$h(q) = \mu(U)/N_q(U) \qquad \text{für } q>0.$$

Behauptung: $(1) \Rightarrow (\mu(U)/M)\cdot\mu \leqq L_h \leqq \mu.$

Beweis: Es ist $L_h(U) \leqq \mu(U)$ und $M \geqq \mu(U)$. Zu jedem $\varepsilon>0$ gibt es ein $a>0$, so daß für alle $X \in \mathfrak{G}$ mit $0<d(x)\leqq a$

$$\mu(X) \leqq (M+\varepsilon)/N_{d(X)}(U) = (M+\varepsilon)h(d(X))/\mu(U)$$

ist. Daraus folgt

$$L_h(X) \geqq (\mu(U)/M)\ \mu(X) \text{ für alle } X \in \mathfrak{G}.$$

Auf den folgenden Spezialfall der Bedingung (1) beziehen sich unsere Beispiele 1, 2 und 3.

Bedingung (2):

Es gibt ein $a>0$ und zu jedem $q\in(0,a]$ eine ausgezeichnete Borelmenge I_q mit folgenden Eigenschaften (i) und (ii).

(i) Zu jeder Teilmenge X von G mit $0<d(X)=q\leqq a$ gibt es ein $x\in G$:

$$X \subset x+I_q,\ d(X) = d(I_q).$$

(ii) Es gibt Elemente $x_1^q,\dots,x_{n_q}^q \in G$ mit

$$U \subset \bigcup_{i=1}^{n_q} x_i^q + I_q,$$

so daß

$$\lim_{q\to 0} n_q \mu(I_q) = \mu(U)$$

ist.

Behauptung: $(2) \Rightarrow M = \limsup_{q\to 0} N_q(U) f(q) \leqq \mu(U)$, also $L_h=\mu$.

Beweis: Sei $q\in(0,a]$, $X\in\mathfrak{G}$ und $0<d(X)=r\leqq q$.

Dann ist

$$N_q(U)\mu(X) \leq N_r(U)\mu(X) \leq N_r(U)\mu(I_r)$$
$$\leq n_r \, \mu(I_r),$$

also $\limsup_{q \to 0} N_q(U)f(q) \leq \mu(U)$.

<u>Beispiel 1.</u> Sei $G = \mathbb{R}_n$ metrisiert durch

$$d((x_i),(y_i)) = \max_{1 \leq i \leq n} |x_i - y_i|,$$

$a=1$, $U = [-1,1]^n$, $I_q = [-q/2,q/2]^n$ für $q \in (0,1]$.

Dann ist die Bedingung (2) erfüllt, wenn das Lebesgue-Maß μ auf übliche Weise definiert wird (s. auch [12], Th.30).

<u>Beispiel 2.</u> Sei $G = U = T = \mathbb{R}/\mathbb{Z} = \{x \in \mathbb{C} : |x| = 1\}$ mit der natürlichen Metrik p versehen, so daß $p(T_1)=1$ ist, $a = 1/2$ und $I_q = K(0,q/2)$ für $q \in (0,a]$.

<u>Beispiel 3.</u> Sei d eine nichtarchimedische Metrik auf G, d.h.

$$d(x,y) \leq \max \{d(x,z),d(y,z)\} \text{ für } x,y,z \in G.$$

Wir wählen $a>0$, so daß $K(0,a)$ kompakt ist und definieren

$$I_q = \cup\{K(0,r): d(K(0,r)) \leq q,\ r>0\} \text{ für } q \in (0,a].$$

Dann ist I_q von der Form $K(0,r_q)$, und es gibt eine natürliche Zahl n_q mit $n_q \cdot \mu(I_q) = \mu(U)$.

<u>Beispiel 4.</u> Das Quadrupel (G_i,U_i,μ_i,d_i) erfülle die Bedingung (1) für $1 \leq i \leq n$.

Sei $G = G_1 \times G_2 \times \ldots \times G_n$, $U = U_1 \times U_2 \times \ldots \times U_n$, $\mu = \mu_1 \times \mu_2 \times \ldots \times \mu_n$

und

$$d((x_i),(y_i)) = \max_{1 \leq i \leq n} d_i(x_i,y_i).$$

Dann erfüllt auch (G,U,μ,d) die Bedingung (1) mit

$$M \leq M_1 M_2 \cdots M_n.$$

<u>Beispiel 5.</u> Sei $G=U=T_\infty = \prod_{i=1}^{\infty} T_i$ mit $T_i=T$ für $i \in \mathbb{N}$, μ das Haar-Maß auf G mit $\mu(G) = 1$ und

$$d((x_i),(y_i)) = \sup \{p(x_i,y_i)/2^i : i=1,2,..\}.$$

Sei $X \in \mathfrak{L}$ und $d(X)>0$. Es gibt eine natürliche Zahl n mit

$$\frac{1}{2^{n+1}} < d(X) = r \leqq \frac{1}{2^n}$$

und zusammenhängende Mengen $X_1,\dots,X_n$ in T_1:

$$p(X_i) = r\cdot 2^i \qquad \text{für } 1\leqq i \leqq n,$$

$$X \subset X_1 \times X_2 \times \dots \times X_n \times T \times T \times T \times \dots \quad = Y.$$

Dann ist

$$d(X) = d(Y),$$

$$\mu(X) \leqq \mu(Y) = \prod_{i=1}^{n} \frac{p(X_i)}{2}$$

und

$$N_r(G) \leqq \prod_{i=1}^{n} \left(1 + \frac{2}{p(X_i)}\right).$$

Man erhält

$$M \leqq \prod_{n=1}^{\infty} \left(1 + \frac{1}{2^n}\right) < \infty \text{ bei Bedingung (1).}$$

II. Endlich-dimensionale Gruppen.

Wir setzen in diesem Abschnitt über die Gruppe G zusätzlich voraus, daß sie von endlicher topologischer Dimension ist (in Zeichen: $\dim G<\infty$; s. [4]).
Der folgende Satz verallgemeinert ein Ergebnis von Goetz [2], das besagt, daß es auf jeder Liegruppe G eine linksinvariante Metrik gibt, so daß jedes Haar-Maß auf G ein Hausdorff-Maß ist. Ferner machen wir eine weitere Aussage über den metrisierbaren Raum G, die motiviert ist durch folgendes Resultat von Pontryagin und Schnirelmann [10]:

Für jeden kompakten metrisierbaren Raum X gilt

$$\dim X = \inf_{d} \liminf_{q \to 0} \frac{\log N_q((X,d))}{\log 1/q},$$

wobei für d alle Metriken zugelassen sind, die mit der Topologie von X verträglich sind.

<u>Satz:</u> Es gibt eine linksinvariante Metrik d auf G, die die Topologie von G induziert, und eine stetige Hausdorff-Funktion h:

(a) L_h ist ein Haar-Maß auf (G,d),

(b) $$\limsup_{q \to 0} \frac{\log N_q(U)}{\log (1/q)} = \dim G$$

für jede kompakte Nullumgebung U in G.

Zunächst wenden wir uns dem Fall dim G = 0 zu.

<u>Hilfssatz 1:</u> Ist G eine lokalkompakte [kompakte] metrisierbare Gruppe ohne isolierte Punkte und dim G=0, dann gibt es eine linksinvariante [invariante] nichtarchimedische Metrik d auf G und eine stetige Hausdorff-Funktion h:

(a') L_h ist ein Haar-Maß auf G [$L_h(G) = 1$],

(b') $$\lim_{q \to 0} N_q(U)q^s = 0 \text{ für alle } s>0 \text{ und}$$

für alle kompakten Nullumgebungen U in G,

(c') $h(2t)/h(t) \leqq 2$ für alle $t>0$.

<u>Beweis von Hilfssatz 1:</u> Es gibt eine offene kompakte Untergruppe U_1 von G. Ist G kompakt, so sei U_1=G. Ferner existiert eine Folge (U_n) offener kompakter Normalteiler von U_1 (s. [3], S.62 und S.70):

$$U_{n+1} \subset U_n,\ U_{n+1} \neq U_n\ (n \in \mathbb{N}) \text{ und } \bigcap_{n=1}^{\infty} U_n = \{0\}.$$

Sei i_n = ind $(U_1 : U_n)$ für $n \in \mathbb{N}$. Ist (q_n) eine beliebige strikt fallende Nullfolge reeller Zahlen, so definieren wir eine zugehörige Hausdorff-Funktion h wie folgt:

$$h(x) = \begin{cases} \frac{1}{i_1}, & \text{falls } x \geqq q_1 \\ \frac{1}{i_n}, & \text{für } x = q_n\ (n \in \mathbb{N}) \\ 0, & \text{für } x=0 \\ \text{linear}, & \text{sonst.} \end{cases}$$

Wir wählen nun (q_n) speziell so, daß

$$\lim_{n \to \infty} i_{n+1} q_n^s = 0 \qquad \text{für alle } s>0$$

und (c') gilt. Für Elemente $x,y \in G$ sei dann

$$d(x,y) = \sup \{q_n : -y + x \notin U_{n+1},\ n \in \mathbb{N}\} \quad (\sup \emptyset = 0).$$

Man kontrolliert ohne Mühe, daß d eine linksinvariante Metrik auf G ist, die der verschärften Dreiecksungleichung

$$\max \{d(x,z), d(y,z)\} \geqq d(x,y) \quad \text{für alle } x,y,z \in G$$

genügt.

Die Einschränkung von d auf U_1 liefert sogar eine invariante Metrik, denn für Elemente $x,y,z \in U_1$ und $n \in \mathbb{N}$ gilt

$$-(y+z) + (x+z) = -z + (-y+x) + z \in U_n \Leftrightarrow -y + x \in U_n,$$

da U_n ein Normalteiler von U_1 ist.

Die von der Metrik d erzeugte Topologie ist mit der ursprünglichen Gruppentopologie identisch, da in beiden Fällen (U_n) eine Basis für den Umgebungsfilter der Null bildet.

Für $n \in \mathbb{N}$ ist $d(U_n)=q_n$. Denn für Elemente $x,y \in U_n$ gilt $-y+x \in U_n$, also $d(x,y) \leqq q_n$. Für $x \in U_n \setminus U_{n+1}$ und $y \in U_{n+1}$ ist $-y+x \notin U_{n+1}$, also $d(x,y) \geqq q_n$.

Nach Beispiel 3 folgt $L_h(U_1)=1$, also (a'). Ist $q \in (0,q_1)$, dann gibt es ein $n \in \mathbb{N}$ mit $q_{n+1} \leqq q < q_n$.

Man erhält

$$N_q(U_1) = N_{q_{n+1}}(U_1) = i_{n+1}$$

und

$$N_q(U_1)q^s \leqq i_{n+1}\, q_n^s \quad \text{für alle } s>0.$$

Somit ist

$$\lim_{q \to 0} N_q(U_1)q^s \leqq \lim_{n \to \infty} i_{n+1}\, q_n^s = 0 \text{ für alle } s>0,$$

woraus sich (b') ergibt.

<u>Hilfssatz 2:</u> Sei H eine topologische Gruppe und V eine symmetrische Umgebung der Null in G. Auf V existiere eine Metrik p, die mit der Topologie von V verträglich ist, so daß

$$z+x\in V \wedge z+y\in V \Rightarrow p(z+x,z+y) = p(x,y)$$

für $z\in G$ und $x,y\in V$ gilt (p ist "lokal-linksinvariant").
Dann gibt es eine linksinvariante Metrik d auf H sowie positive reelle Zahlen α,β und γ, so daß

$$\alpha d(x,y) \leq p(x,y) \leq \beta d(x,y)$$

für alle $x,y\in V$ mit $p(x,y)<\gamma$ ist.

Ein Beweis kann leicht mit Hilfe des bekannten Metrisationssatzes von Birkhoff und Kakutani ([3], S.68) geführt werden. Möglicherweise läßt sich ohne dieses Hilfsmittel direkt zeigen, daß $\alpha=\beta=1$ gewählt werden kann. Wäre dies der Fall, so könnte die Bedingung (c') in Hilfssatz 1 im weiteren außer Betracht gelassen werden.

Beweis des Hauptsatzes: Da der Fall dim G=0 mit Hilfssatz 1 erledigt ist, sei nun dim $G=n\in\mathbb{N}$. Aufgrund eines Struktursatzes von Montgomery und Zippin ([8], S.184 und S.237) gibt es eine kompakte Gruppe G_1 mit dim $G_1=0$ und eine n-gliedrige lokale Liegruppe G_2, so daß $G_1\times G_2$ isomorph (topologisch und algebraisch) zu einer Umgebung der Null von G ist. Nach Ergebnissen von Goetz [2] gibt es eine symmetrische Umgebung U der Null in G_2, eine Umgebung W in $\mathbb{R}_n$, eine lokal-linksinvariante Metrik d_2 auf U sowie eine Abbildung T von W auf U mit zugehörigen positiven Konstanten r und s, so daß

$$r|x-y| \leqq d_2\ (T(x),T(y))\leqq s|x-y|$$

ist. Wir können W als beschränkt voraussetzen.
Ist $g(t)=t^n$ für $t\geqq 0$, so folgt

$$0 < L_g(U)<\infty \quad \text{und} \quad \limsup_{q\to 0} N_q(U)q^n<\infty \quad .$$

Besteht G_1 aus endlich vielen Elementen, so sei d_1 die diskrete Metrik auf G_1. Anderenfalls gibt es nach Hilfssatz 1 eine linksinvariante Metrik d_1 und eine stetige Hausdorff-Funktion h_1, so daß

$$L_{h_1}\ ((G_1,d_1)) = 1,\ h_1(2t)/h_1(t)\leqq 2 \quad \text{für } t>0$$

und

$$\lim_{q\to 0} N_q((G_1,d_1))q^s = 0 \qquad \text{für alle } s>0$$

ist. In beiden Fällen betrachten wir folgende Metrik $d_1\times d_2$ auf $G_1\times U$

$$d_1 \times d_2 \;((x_1,y_1),(x_2,y_2)) = \max \;\{d_1(x_1,x_2), d_2(y_1,y_2)\}.$$

Ist G_1 endlich, so ist L_g ein lokal-linksinvariantes Borelmaß auf $G_1 \times U$,

$$0 < L_g(G_1 \times U) < \infty$$

und

$$\limsup_{q \to 0} N_q(G_1 \times U) q^n < \infty .$$

In diesem Fall sei h=g.

Ist G_1 nicht diskret, so folgt nach Ergebnissen von Wegmann

$$0 < L_{h_1 g} \;(G_1 \times U) \qquad ([14], \text{ Satz } 2)$$

und

$$L_{h_1 g} \;(G_1 \times U) < \infty \quad ([14], \text{ Beweismethode von Satz } 7).$$

Ferner erhält man

$$\limsup_{q \to 0} N_q(G_1 \times U) q^{n+s}$$

$$\leq (\lim_{q \to 0} N_q(G_1) q^s)(\limsup_{q \to 0} N_q(U) q^n) = 0.$$

Wir setzen $h = h_1 g$.

Insgesamt erhält man durch Einbettung von $G_1 \times U$ in G eine symmetrische Umgebung V der Null in G, eine lokal-linksinvariante Metrik p auf V, so daß

$$0 < L_h((V,p)) < \infty$$

und

$$\limsup_{q \to 0} N_q(V) q^{n+s} = 0$$

ist.

Sei nun d eine linksinvariante Metrik auf G, wie sie in Hilfssatz 2 angegeben wird. Da wegen (c')

$$\limsup_{t \to 0} h(\lambda t)/h(t) < \infty \qquad \text{für alle } \lambda \geq 1$$

wird, folgen die Aussagen (a) und (b) des Satzes.

<u>Schlußbemerkungen</u> 1) Die Aussage (b) des Satzes läßt sich wie folgt verschärfen. Ist f eine feste Hausdorff-Funktion und

$\lim_{t\to 0} f(t)/t^n = 0$ $(n\in\mathbb{N}\cup\{0\})$, so kann statt (b)

$$\lim_{q\to 0} N_q(U)f(q) = 0$$

erreicht werden.

2) Ist $\dim G=n\in\mathbb{N}$ und $g(t)=t^n$ für $t\geq 0$, dann gibt es nicht in jedem Fall eine zulässige linksinvariante Metrik d auf G, so daß L_g ein Haar-Maß auf G wird. Sei etwa $\dim G_1=n$ und G_2 eine überabzählbare lokalkompakte Gruppe mit $\dim G_2=0$. Ist d eine beliebige Metrik auf $G=G_1\times G_2$ $(\dim G=n)$, so gilt nach Szpilrajn [13]

$$L_g(G_1\times\{0\}) > 0.$$

Damit ist das L_g-Maß von G nicht σ-endlich, also L_g kein Haar-Maß auf G.

3) Die im Literaturverzeichnis erwähnten Arbeiten [6], [7] und [11] werden hier nicht benötigt. Sie stehen jedoch in engerem Zusammenhang mit unserem Thema.

Literatur

[1] G.Birkhoff, A note on topological groups, Compositio Math. 3 (1936), 427-430.

[2] A.Goetz, Bemerkungen über Hausdorffsche Maße und Hausdorffsche Dimensionen in Lieschen Gruppen, Colloq.Math. 5 (1958), 55-65.

[3] E.Hewitt and K.A.Ross, Abstract harmonic analysis.I. New York 1963.

[4] W.Hurewicz and H.Wallmann, Dimension theory. Princeton University Press 1941.

[5] S.Kakutani, Über die Metrisation der topologischen Gruppen, Proc. Imp. Acad. Tokyo 12 (1936), 82-84.

[6] V.Klee, Polyhedral sections of convex bodies, Acta Math. 103 (1960), 243-267.

[7] L.H.Loomis, The intrinsic measure theory of Riemannian and Eudidian metric spaces, Ann. of Math. (2) 45 (1944), 367-374.

[8] D.Montgomery and L.Zippin, Topological transformation groups. New York 1955.

[9] J.Mycielski, Remarks on invariant measures in metric spaces, Colloq. Math. 32 (1974), 105-112.

[10] L.Pontryagin et L.Schnirelmann, Sur une propriêté de la dimension, Ann. of Math. (2) 33 (1932), 156-162.

[11] B.Riečan, On some properties of Haar Measure, Mat. Časop. 17 (1967), 59-63.

[12] C.A.Rogers, Hausdorff measures. Cambridge University Press 1970.

[13] E.Szpilrajn, La dimension et la mesure, Fund. Math. 28 (1937), 81-89.

[14] H.Wegmann, Die Hausdorff-Dimension von kartesischen Produkten metrischer Räume, J. reine angew. Math. 246 (1971), 46-75.

A UNIFIED PROOF OF FUBINI THEOREM FOR BAIRE AND BOREL MEASURES

J. L. Kelley
University of California

T. P. Srinivasan
University of Kansas

We like the following unified proof of Fubini theorem for Baire and for Borel measures in a locally compact Hausdorff space (Baire = δ-ring generated by compact G_δ's; Borel = δ-ring generated by all compact sets). The unification results from a reformulation of the theorem in terms of initial families and their Daniell extensions. The reformulation emerged during a discussion with Klaus J. Bichteler.

A pre-integral is a positive linear function I on a vector lattice L of real valued functions with truncation ($f \in L \Longrightarrow f \wedge 1 \in L$) such that: if $\{f_n\}_n$ is an increasing sequence of members of L with $\sup_n f_n \in L$ then $\sup_n I(f_n) = I(\sup_n f_n)$. A pre-integral I on L is <u>super continuous</u> iff for each increasing net $\{f_\alpha\}_\alpha$ of members of L whose pointwise supremum is a member of L, we have $\sup_\alpha I(f_\alpha) = I(\sup_\alpha f_\alpha)$. Each pre-integral can be extended to an integral by the method of Daniell extension. Recall that in the case of an arbitrary pre-integral this extension is through approximation by functions which are suprema of increasing sequences of non-negative members of L; in the case of a super continuous pre-integral the above extension is through approximation by functions which are suprema of increasing nets of non-negative members of L. To distinguish between the two, we call the later extension the <u>Daniell super extension.</u> In either case we denote the extension of a pre-integral (I, L) by (I^1, L^1); I^1 is the Daniell integral extending I and L^1 is the family of I^1-integrable functions.

We need an approximation lemma; we omit its proof. Denote by L_u (respectively L_τ) the family: $\{f : f$ is real valued and is the pointwise limit of an increasing sequence (respectively net) $\{f_\alpha\}_\alpha$ of members of L with $\sup_\alpha I(f_\alpha) < \infty \}$. Let $L_{u\ell}$ (respectively $L_{\tau\ell}$) be the family: $\{g : g$ is real valued and is the pointwise limit of a decreasing sequence $\{h_n\}_n$ of members of L_u (respectively L_τ) with $\lim_n I^1(h_n) > -\infty$.

APPROXIMATION LEMMA. Let (I^1, L^1) be the Daniell extension (respectively the Daniell super extension) of a pre-integral (respectively a super continuous pre-integral) I on L and let f be a bounded real valued function. Then $f \in L^1$ iff there is a bounded g in $L_{u\ell}$ (respectively $L_{\tau\ell}$) such that $g \geq f$ and g-f is dominated by some bounded I^1-null member of $L_{u\ell}$ (respectively $L_{\tau\ell}$).

THEOREM (FUBINI). Let I be a pre-integral on M, a space of real valued functions on a set X, J a pre-integral on N, a space of real valued functions on a set Y and Q, a pre-integral on V, a space of real valued functions on $X \times Y$. If I and J are super continuous, assume that Q is also super continuous. Let (I^1, M^1), (J^1, N^1) and (Q^1, V^1) denote respectively their Daniell extensions or in the super continuous case, their Daniell super extensions. Suppose that for each member f in V and each member x in X, the function $y \to f(x, y)$ is a member of N and the function $x \to J_y f(x, y)$ is a member of M (the subscript y in J_y denoting integration with respect to y) and suppose that $I_x J_y f(x, y) = Q(f(x, y)$. Then for each Q^1-integrable function f and each x, I^1 a.e. in X, the function $y \to f(x, y)$ is J^1-integrable, the function $x \to J^1_y(fx, y)$ defined I^1 a.e. is I^1-integrable and $I^1_x J^1_y f(x, y) = Q^1 f(x, y)$.

PROOF. We give the proof for Daniell extensions. Exactly the same argument applies to the case of Daniell super extensions as well.

The proof is a direct application of the Approximation Lemma. Recall the definitions of the families L_u and $L_{u\ell}$ corresponding to a pre-integral (I, L), which preceded the statement of the lemma. Denote by M_u, $M_{u\ell}$; N_u, $N_{u\ell}$ and V_u, $V_{u\ell}$ the corresponding families for the pre-integrals (I, M), (J, N) and (Q, V). By the Approximation Lemma, each bounded Q^1-integrable function f is a difference g-h where g is a bounded member of $V_{u\ell}$ and h is a non-negative function dominated by a bounded null member of $V_{u\ell}$. If the conclusion of the theorem holds for the functions g and h in place of f, clearly then it will hold for f as well. It is fairly straight forward to check that the conclusion holds for each member of V_u and then for each member of $V_{u\ell}$ also, whence it holds for g. Let k be a nonnegative null member of $V_{u\ell}$ dominating h. Then

$$I_x^1 \; J_y^1 \; k(x,y) = Q^1 \; k(x,y) = 0 \; .$$

Consequently $J_y^1 k(x,y) = 0$ for x, I^1 a.e. and in turn for each such x, $k(x,y) = 0$ for y, J^1 a.e. Since $0 \leq h(x,y) \leq k(x,y)$ for all (x, y) it follows that $y \to h(x,y)$ is J^1-integrable and null for x, I^1 a.e. and consequently $x \to J_y^1 h(x,y) = 0$ for x, I^1 a.e. is I^1-integrable and null, whence $I_x^1 J^1 h(x,y) = 0 = Q^1(h)$. Thus the conclusion of the theorem holds for h just like for g and then it holds for each bounded Q^1-integrable function f. The conclusion then extends to an arbitrary Q^1-integrable function f since each such f is the pointwise limit of the increasing sequence $\{f \wedge n\}_n$ of bounded integrable functions.

Let μ_0, ν_0 denote finite valued measures on the Baire δ-rings $\mathcal{B}_X^0$, $\mathcal{B}_Y^0$ where X, Y are locally compact Hausdorff spaces. Take for M the family of $\mathcal{B}_x^0$-simple functions and for I, the restriction of the integral I_{μ_0} to M. The Daniell extension (I^1, M^1) of (I, M) can be easily identified. The measure corresponding to I^1 is just the completion of μ_0 and I_{μ_0} coincides with the restriction of I^1 to the family of μ_0 integrable functions. Introduce the pair (J, N) similarly, in terms of ν_0 and $\mathcal{B}_y^0$. Define the pair (Q, V) in the product space $X \times Y$ as usual: V is the family of finite linear combinations of characteristic functions of rectangles with sides in $\mathcal{B}_X^0$, $\mathcal{B}_Y^0$ respectively and Q is the obvious functional on V. The δ-ring generated by the family of Baire rectangles coincides with the Baire δ-ring $\mathcal{B}_{X \times Y}^0$. Consequently the restriction of the Daniell extension (Q^1, V^1) of (Q, V) to the Baire sets in $X \times Y$ gives a Baire measure in $X \times Y$, the product of the Baire measures μ_0 and ν_0, denoted by $\mu_0 \otimes \nu_0$. The preceding theorem specialized to the triple (I, M), (J, N) and (Q, V) yeilds the following Corollary:

COROLLARY. Fubini Theorem holds for the Baire measures μ_0, ν_0 and $\mu_0 \otimes \nu_0$.

The continuous functions with compact supports are clearly Baire integrable. Denote the families of such functions on $X \times Y$ by $C_c(X \times Y)$. If $f \in C_c(X \times Y)$ then by the preceding corollary:

$$\int f\, d(\mu_0 \times \mu_0) = Q(f) = I_x J_y \; f = \int \left(\int f d\nu_0 \right) d\mu_0 .$$

Clearly for each $x \in X$ the function $x \to f(x, y)$ belongs to $C_c(Y)$ and the function $x \to \int f(x, y) d\nu_0(y)$ belongs to $C_c(X)$. Consequently, if I, J, Q denote the restrictions of I_{μ_0}, I_{ν_0} and $I_{\mu_0 \otimes \nu_0}$ to $C_c(X)$, $C_c(Y)$ and $C_c(X \times Y)$, the choice of the triple $(I, C_c(X))$, $(J, C_c(Y))$, $(Q, C_c(X \times Y))$ to take the place of the pre-integrals (I, L), (J, M) and (Q, V) in the statement of our main theorem is admissible. But now all these pre-integrals are super continuous and so the conclusion of the theorem applies to their Daniell super extensions I^1, J^1 and Q^1. This gives us almost immediately the Fubini theorem for products of Borel measures as follows.

Let μ, ν be finite valued regular measures on the Borel δ-rings $\mathcal{B}_X$, $\mathcal{B}_Y$ in X and Y respectively; let μ_0, ν_0 denote their Baire restrictions. Then $\mu_0 \otimes \nu_0$ is is a Biare measure in $X \times Y$. Its unique regular Borel extension is defined to be the Borel product of μ and ν. We denote the Borel product by $\mu \otimes' \nu$. Define the triples $(I, C_c(X))$, $(J, C_c(Y))$ and $(Q, C_c(X \times Y))$ as in the preceding paragraph and let I^1, J^1, Q^1 denote their Daniell super extensions. It is easy to see that the Borel measures μ, ν and $\mu \otimes' \nu$ coincide with the measures given by the integrals I^1, J^1 and Q^1 on the corresponding families of Borel sets and further, the σ-compactly supported integrable functions for I^1, J^1 and Q^1 are actually integrable with respect to these Borel restrictions (up to Borel null sets). We have therefore the following corollary to our main theorem.

COROLLARY (FUBINI THEOREM FOR BOREL PRODUCTS). Let μ and ν be Borel measures in X and Y and let $\mu \otimes' \nu$ be their Borel product. Let f be any $\mu\otimes'\nu$ integrable function. Then

i) for μ a.e. x, the function $y \to f(x,y)$ defined ν a.e. is equal ν a.e. to a member of $L^1(\nu)$ and

ii) the function defined μ a.e. by $x \to \int f(x,y)d\nu(y)$ agrees μ a.e. with a member of $L^1(\mu)$ and

$$\int\left(\int f(x,y)\,d\nu(y)\right)d\mu(x) = \int f(x,y)\,d\mu\otimes'\nu\,.$$

INHERITNESS OF COMPACTNESS AND PERFECTNESS OF MEASURES BY THICK SUBSETS

by Kazimierz Musiał
Wrocław University
and
Institute of Mathematics, Polish Academy of Sciences

Introduction.

Let $(X,\mathcal{C},\mu)$ be a finite positive measure space and let Z be a thick subset of X (that means, that $\mu^*(Z) = \mu(X)$, where μ^* is the outer measure induced by μ).

It is well known that if ν is the restriction of μ to Z, then ν is a measure. It is natural to ask what other properties of μ (besides the countable additivity) hold for ν.

In this paper two properties of μ are examined from that point of view, namely the perfectness and the compactness (for definitions and properties of compact and perfect measures see [2] and [3] respectively).

The perfectness of ν was examined by Sazonov [4], his results are given here, in a suitable form, for completeness (Theorem 1 and 2).

The compactness of ν was examined, for some collection of measures, by Vinokurov [6], who has obtained a necessary and sufficient condition for the compactness of ν in terms of Stone's representation of the algebra Z

In this paper, using a modification of the method applied in [6], we get a necessary and sufficient condition for the compactness of ν in a simpler form.

Having the above condition it is easy to construct examples of perfect but non-compact measures. The first such example has been published by Mahkamov and Vinokurov in [1] (the existence of such an example has been announced by Vinokurov in [6]).

Throught the paper we assume that all measures under consideration are countably additive and finite. For a collection $\mathcal{F}$ of sets we denote by $\sigma(\mathcal{F})$ the σ-algebra generated by $\mathcal{F}$; if Z is a set then we write $Z \cap \mathcal{F}$ instead of $\{Z \cap E : E \in \mathcal{F}\}$; if $\mathcal{A}$ is a σ-algebra then by an $\mathcal{A}$-atom we mean any set $e \in \mathcal{A}$ with the property that $A \in \mathcal{A}$ and $A \subset e$ imply $A = \emptyset$ or $A = e$; if μ is a measure on a σ-algebra $\mathcal{A}$ and $\mathcal{B} \subset \mathcal{A}$, then we write $\mathcal{B}^+$ in order to denote the family of all elements of $\mathcal{B}$ of positive measure. $\mathcal{B}_R$ denotes the Borel sets of the real line R.

A measure space $(X, \mathcal{C}, \mu)$ is said to be compact (perfect) if and only if μ is compact (perfect).

1. Perfect measures.

We begin with an easy observation:

Proposition. Let $(X, \mathcal{A}, \mu)$ be a measure space with a countably generated $\mathcal{A}$. Then μ is perfect if and only if for some sequence $\{A_n\}_{n=1}^{\infty}$ of sets generating $\mathcal{A}$, there exists a set $B \in \mathcal{B}_R$ such that $B \subset h(X)$ and $\mu(X - h^{-1}(B)) = 0$, where $h: X \to R$ is the Marczewski function [5] defined by

$$h(x) = 2 \sum_{n=1}^{\infty} I_{A_n}(x)\, 3^{-n} .$$

Proof. Let h and B be as above. Since the measure μh^{-1} defined on $\mathcal{B}_R$ is perfect [4] and $h^{-1}: \mathcal{B}_R \cap B \to h^{-1}(B) \cap \mathcal{A}$ is a σ-isomorphism, the measure μ considered on $\mathcal{A} \cap h^{-1}(B)$ is perfect [3]. Hence μ is perfect as well.

THEOREM 1. Let $\mathcal{A}$ be a countably generated σ-algebra on X, and let μ be a perfect measure on $\mathcal{A}$. If Z is a thick subset of X, then the restriction of μ to Z is perfect measure if and only if there exists a μ-null set $N \in \mathcal{A}$ with the property $e \cap Z \neq \emptyset$ for every $\mathcal{A}$-atom e outside of N.

Proof. Necessity. Let ν be the restriction of μ to Z and let $f: X \to R$ be such that $f^{-1}(\mathcal{B}_R) = \mathcal{A}$. Since $f|Z : Z \to R$ is $(\mathcal{A} \cap Z, \mathcal{B}_R)$-measurable, therefore there exists $B \in \mathcal{B}_R$, such that $B \subset f(Z)$ and

$$\nu(Z - f^{-1}(B)) = 0$$

Hence, setting $N = X - f^{-1}(B)$, we have $e \cap Z \neq \emptyset$ for every $\mathcal{A}$-atom $e \subset f^{-1}(B) = X - N$.

Sufficiency. Let $\{A_n\}_{n=1}^{\infty}$ be a sequence generating $\mathcal{A}$ and let $D_n = A_n \cap Z$, $n = 1, \ldots$. If $h: X \to R$ is the Marczewski function of $\{A_n\}$ then $h|Z$ is the Marczewski function of the sequence $\{D_n\}$.

Setting $g(x) = h(x)$ for $x \in X - N$ and zero whenever $x \in N$, we obtain an $(\mathcal{A}, \mathcal{B}_R)$-measurable function, and so in view of the perfectness of μ there exists $B \in \mathcal{B}_R$ such that $B \subset g(X)$ and $\mu(g^{-1}(B)) = \mu(X)$.

Clearly $A = (h|Z)^{-1}(B - \{0\}) = h^{-1}(B - \{0\}) \cap Z \in Z \cap \mathcal{A}$ and $\nu(Z - A) = 0$.

Thus, ν is perfect in view of the Proposition.

Since any measure space $(X, \mathcal{E}, \mu)$ is perfect if and only if all measure spaces $(X, \mathcal{A}, \mu|\mathcal{A})$, ($\mathcal{A} \subset \mathcal{E}$ is countably generated), are perfect [3], we get the following generalization of the above theorem.

THEOREM 2. Let $(X, \mathcal{E}, \mu)$ be a perfect measure space and let Z be a thick subset of X. Then the restriction of μ to Z is perfect if and only if for every countably generated sub-σ-algebra $\mathcal{A} \subset \mathcal{E}$ there exists a μ-null set $N_{\mathcal{A}} \in \mathcal{A}$ such that

$$e \cap Z \neq \emptyset \quad \text{for every } \mathcal{A}\text{-atom } e \subset X - N_{\mathcal{A}}.$$

2. Compact measures.

Our considerations are based on the following lemma, essentially due to Vinokurov [6].

LEMMA. Let $(X, \mathcal{E}, \mu)$ be a compact measure space and let $\mathcal{S}$ be a countably multiplicative compact class aproximating $\mathcal{E}$ with respect to μ. Then, for each countable family $\mathcal{K} \subset \mathcal{E}$, there exists a countable algebra $\mathcal{K}_0 \supset \mathcal{K}$ and a set $N \in \sigma(\mathcal{K}_0)$ such that $\mathcal{K}_0 \subset \mathcal{E}$ and all $\sigma(\mathcal{K}_0)$-atoms outside of N are elements of $\mathcal{S}$.

Proof. We shall construct a sequence of countable algebras

$$\mathcal{R}_1 \subset \mathcal{R}_2 \subset \ldots \subset \mathcal{E}$$

with the following property:
for every $A \in \mathcal{R}^+_{n-1}$ and for every $\varepsilon > 0$ there exists a set $B \in \mathcal{R}^+_n$ and a set $S \in \mathcal{S}$ such that $B \subset S \subset A$ and $\mu(A - B) < \varepsilon$.

Let $\mathcal{R}_1$ be the algebra generated by $\mathcal{K}$. Having constructed the algebras $\mathcal{R}_1, \ldots, \mathcal{R}_n$, take for every $A \in \mathcal{R}^+_n$ a sequence of sets $S^n_A \in \mathcal{S}$, $n = 1, \ldots$ and a sequence of sets $B^n_A \in \mathcal{E}^+$ sach that

$$B^n_A \subset S^n_A \subset A \quad \text{and} \quad \mu(A - B^n_A) < 1/n .$$

Let $\mathcal{R}_{n+1}$ be the algebra generated by $\mathcal{R}_n$ and the all sets B^n_A, $n = 1, \ldots$, $A \in \mathcal{R}_n$.

We shall show that the countable collection $\mathcal{K}_0 = \bigcup_{n=1}^{\infty} \mathcal{R}_n$ has the desired properties. In order to do it take an $\varepsilon > 0$ and a sequence ε_{nk}, $n, k = 1, 2, \ldots$ of positive numbers such that $\sum_{k=1}^{\infty} \varepsilon_{nk} < \varepsilon_n$ and $\sum_{n=1}^{\infty} \varepsilon_n < \varepsilon$, and assume that $\mathcal{K}_0 = \{A_k\}_{k=1}^{\infty}$. Then take for every n and every $A_k \in \mathcal{K}^+_0$ a set $A_{1_k} \in \mathcal{K}^+_0$ such that $A_{1_k} \subset A_k$, $\mu(A_k - A_{1_k}) < \varepsilon_{nk}$ and there exists an $S \in \mathcal{S}$ with the property $A_{1_k} \subset S \subset A_k$.
Now, if

$$N_{\varepsilon_n} = M \cup \left(\bigcup_{k=1}^{\infty} (A_k - A_{1_k}) \right),$$

where M is the union of all μ-null sets from $\mathcal{K}_0$, and $e \subset X - N_{\varepsilon_n}$ is a $\sigma(\mathcal{K}_0)$-atom, then $e \subset A_k$ for some k. On the other hand the properties of N_{ε_n} follows that $e \subset A_{1_k}$.

Thus, if $e \subset X - N_{\varepsilon_n}$ is a $\sigma(\mathcal{K}_0)$-atom, then for every $A \in \mathcal{K}_0$ and containing e there exist $B \in \mathcal{K}^+_0$ and $S \in \mathcal{S}$ such that $e \subset B \subset S \subset A$. Hence e is a countable intersection of elements of the compact class $\mathcal{S}$ and therefore $e \in \mathcal{S}$.

Setting $N = \bigcap_{n=1}^{\infty} N_{\varepsilon_n}$, we have $N \in \sigma(\mathcal{K}_0)$, $\mu(N) = 0$ and all $\sigma(\mathcal{K}_0)$-atoms $e \subset X - N$, are in $\mathcal{S}$.

From now on, we denote by Λ the set of all limit ordinals less than ω_1 and by Γ the remaining ordinals less than ω_1.

THEOREM 3. Let $(X, \mathcal{E}, \mu)$ be a compact measure space with $\mathcal{E}$ generated by $\aleph_1$ sets, and let Z be a thick subset of X. Then the measure space $(Z, Z \cap \mathcal{E}, \mu|Z = \nu)$ is compact iff there exists a net

<u>of countably generated</u> σ-<u>algebras</u> $\mathcal{C}_\lambda \subset \mathcal{C}$, <u>ordered by ordinals</u> $\lambda < \omega_1$. <u>and a net of</u> μ-<u>null sets</u> $N_\lambda \in \mathcal{C}_\lambda$ <u>with the following properties</u>:

(i) $\mathcal{C}_0 \subset \mathcal{C}_1 \subset \ldots \subset \mathcal{C}_\lambda \subset \ldots$ <u>and</u> $\mathcal{C} = \bigcup_{\lambda<\omega_1} \mathcal{C}_\lambda$

(ii) $N_0 \subset N_1 \subset \ldots \subset N_\lambda \subset \ldots$;

(iii) $N_\lambda = \bigcup_{\gamma<\lambda} N_\gamma$ <u>and</u> $\sigma(\bigcup_{\gamma<\lambda} \mathcal{C}_\gamma) = \mathcal{C}_\lambda$

<u>for every</u> $\lambda \in \Lambda$;

(iv) $e \cap Z \neq \emptyset$ <u>for every</u> $\mathcal{C}_\lambda$-<u>atom</u> $e \subset X - N_\lambda$, $\lambda < \omega_1$.

<u>Proof</u>. Necessity. Assume that $\nu = \mu|Z$ is compact and let $\mathcal{S} \subset Z \cap \mathcal{C}$ be a countably multiplicative compact class approximating $Z \cap \mathcal{C}$ with respect to ν .

If Σ is an arbitrary countable family of sets from $Z \cap \mathcal{C}$, then denote by $\mathcal{S}(\Sigma)$ a countable algebra contained in $Z \cap \mathcal{C}$ such that $\Sigma \subset \mathcal{S}(\Sigma)$ and all $\sigma(\mathcal{S}(\Sigma))$-atoms outside fo some ν-null set from $\sigma(\mathcal{S}(\Sigma))$ are elements of $\mathcal{S}$ (in view of Lemma, such an algebra $\mathcal{S}(\Sigma)$ always exists).

Then take an arbitrary net of countable algebras $\mathcal{R}_\gamma \subset \mathcal{C}$, $\gamma \in \Gamma$ such that $\mathcal{R}_0 \subset \mathcal{R}_1 \subset \ldots \mathcal{R}_\gamma \subset \ldots$ and $\mathcal{C} = \sigma(\bigcup_{\gamma<\omega_1} \mathcal{R}_\gamma)$, put $\mathcal{P}_\gamma = Z \cap \mathcal{R}_\gamma$ and

$$\mathcal{B}_0 = \mathcal{S}(\mathcal{P}_0), \ldots , \quad \mathcal{B}_\xi = \mathcal{S}(\mathcal{P}_\xi \cup (\bigcup_{\gamma<\xi} \mathcal{B}_\gamma)), \ldots$$

where $\xi \in \Gamma$.

Then take a countable algebras $\mathcal{E}_\gamma \subset \mathcal{C}$ such that

$$Z \cap \mathcal{E}_\gamma = \mathcal{B}_\gamma , \ \mathcal{R}_\gamma \subset \mathcal{E}_\gamma \text{ and } \mathcal{E}_1 \subset \mathcal{E}_2 \subset \ldots \subset \mathcal{E}_\gamma \subset \ldots$$

and put $\mathcal{E}_\lambda = \bigcup_{\gamma<\lambda} \mathcal{E}_\gamma$, whenever $\lambda \in \Lambda$ and $\mathcal{C}_\xi = \sigma(\mathcal{E}_\xi)$ for all. $\xi<\omega_1$.

Clearly the condition (ii) is satisfied.

From the properties of $\mathcal{B}_\gamma$ follows, that for every $\gamma \in \Gamma$ there exists $M_\gamma \in \mathcal{C}_\gamma$ such that $\mu(M_\gamma) = 0$ and $e \cap Z \in \mathcal{S}$ for every $\mathcal{C}_\gamma$-atom $e \subset X - M_\gamma$ with the non-empty intersection $e \cap Z$.

Since the restrictions of μ and ν to $\mathcal{C}_\gamma$ and $\sigma(\mathcal{B}_\gamma)$ respectively are perfect measures and because ν is the restriction of μ to Z , there exists (in view of theprem 1) for every $\gamma \in \Gamma$ a set

$K_\gamma \in \mathcal{E}_\gamma$ such that $\mu(K_\gamma) = 0$ and $e \cap Z \neq \emptyset$ for every $\mathcal{E}_\gamma$-atom $e \subset X - K_\gamma$.

Put $N_0 = M_0 \cup K_0$ and $N_\gamma = M_\gamma \cup K_\gamma \cup \bigcup_{\alpha<\gamma} N_\alpha$, whenever $\gamma, \alpha \in \Gamma$ and $N_\lambda = \bigcup_{\gamma<\lambda} N_\gamma$ if $\lambda \in \Lambda$.

Clearly $N_0 \subset N_1 \subset \ldots$ and $N_\xi \in \mathcal{E}_\xi$, for every $\xi < \omega_1$.

We shall prove, that the system $(\mathcal{E}_\lambda, N_\lambda)_{\lambda<\omega_1}$ satisfies the conditions (i) → (iv).

Since (i) - (iii) are satisfied, it if sufficient to prove (iv).

It is easy to see that (iv) is satisfied for non-limit ordinals, therefore let us assume that $\lambda \in \Lambda$ and let $e \subset X - N_\lambda$ be a $\mathcal{E}_\lambda$-atom.

In view of (ii) we have $e \subset X - N_\gamma$, for every $\gamma < \lambda$, and so for every $\gamma < \lambda$ there exists a $\mathcal{E}_\gamma$-atom $e_\gamma \supset e$ such that $e_\gamma \subset X - N_\gamma$, because of $N_\gamma \in \mathcal{E}_\gamma$.

Hence

$$e = \bigcap_{\gamma<\lambda} e_\gamma$$

and then

$$e = \bigcap_{\gamma_n<\lambda} e_{\gamma_n} ,$$

for some sequence $\gamma_0 < \gamma_1 < \ldots$, $\gamma_n \in \Gamma$, cofinal with λ.

Since $\gamma_n \in \Gamma$ we have $e_{\gamma_n} \cap Z \neq \emptyset$ (by the application of (iv) to non limit ordinals) and so $e_{\gamma_n} \cap Z \in \mathcal{S}$ because of the construction of N_{γ_n}, $n = 0,1,\ldots$.

The compactness of $\mathcal{S}$ implies

$$e \cap Z = \bigcap_{n=0}^{\infty} (e_{\gamma_n} \cap Z) \neq \emptyset .$$

Sufficiency. It is sufficienz to assume here that N_λ is a union of $\mathcal{E}_\lambda$-atoms and belongs to $\overline{\mathcal{E}}$ (the completion of $\mathcal{E}$ with respect to μ).

Let $\mathcal{S} \subset \mathcal{E}$ be a countably multiplicative compact class aproximating $\mathcal{E}$ with respect to μ.

Then, let

$$\mathcal{S}_\lambda = \{ S \in \mathcal{S} : \text{is a union of } \mathcal{E}_\lambda\text{-atoms and } S \subset X - N_\lambda \}$$

and

$$\varphi^{\lambda} = Z \cap \mathcal{S}_{\lambda} , \quad \lambda < \omega_1$$

We shall prove the compactness of φ^{λ} , $\lambda < \omega_1$.

In order to do it let $S_n \in \varphi^{\lambda}$, $n = 1,\dots$ be an arbitrary sequence such that $\bigcap_{k=1}^{n} S_k \neq \emptyset$, $k = 1,\dots$ and let $\overline{S_n} \in \mathcal{S}_{\lambda}$ be such that $S_n = \overline{S_n} \cap Z$.

Since $\bigcap_{k=1}^{n} \overline{S_k} \neq \emptyset$ for every n , we have $\bigcap_{n=1}^{\infty} \overline{S_n} \neq \emptyset$, because of the compactness of $\mathcal{S}_{\lambda}$.

Moreover $\bigcap_{n=1}^{\infty} \overline{S_n} \subset X - N_{\lambda}$, since $\overline{S_n} \subset X - N_{\lambda}$, $n = 1,\dots$.

Hence,

$$\bigcap_{n=1}^{\infty} S_n = Z \cap \bigcap_{n=1}^{\infty} \overline{S_n} \neq \emptyset$$

in view of the condition (iv) and so φ^{λ} is compact for every $\lambda < \omega_1$.

Let

$$\mathcal{T} = \left\{ E \subset Z : E = \bigcap_{n=1}^{\infty} E_{\lambda_n} , E_{\lambda_n} \in \varphi^{\lambda_n} , \lambda_n < \omega_1 , n = 1,\dots \right\}$$

We shall prove, that $\mathcal{T}$ is a compact class aproximating $\mathcal{E} \cap Z$ with respect to $\nu = \mu | Z$.

Take $F_n \in \mathcal{T}$, $n = 1,\dots$ with the property $\bigcap_{k=1}^{n} F_k \neq \emptyset$ for every n . If $F_n \in \varphi^{\lambda_n}$, then $F_n = \overline{F_n} \cap Z$ for some $\overline{F_n} \in \mathcal{S}_{\lambda_n}$ $n = 1,\dots$.

Clearly $\bigcap_{k=1}^{n} F_k \neq \emptyset$ for every n and so $\bigcap_{n=1}^{\infty} \overline{F_n} \neq \emptyset$.

Let λ be the first ordinal with the property $\lambda_n \leq \lambda$ for every n .

Since $\overline{F_n}$ is a union of $\mathcal{E}_{\lambda}$-atoms and $\overline{F_n} \subset X - N_{\lambda_n}$, therefore $\bigcap_{n=1}^{\infty} \overline{F_n}$ is a union of $\mathcal{E}_{\lambda}$-atoms and $\bigcap_{n=1}^{\infty} \overline{F_n} \subset X - N_{\lambda}$ in view of (iii).

Now, (iv) implies the relation $\bigcap_{n=1}^{\infty} \overline{F_n} \cap Z \neq \emptyset$, and hence

$\bigcap_{n=1}^{\infty} F_n \neq \emptyset$.

Thus the compactness of $\mathcal{T}$ is proved.

Since $\mathcal{E} = \bigcup_{\gamma<\omega_1} \mathcal{E}_\gamma$, so in order to prove that $\mathcal{T}$ approximates $\mathcal{E} \cap Z$ with respect to ν, it is sufficient to show, that for every $\varepsilon > 0$, $\gamma < \omega_1$ and $E \in \mathcal{E}_\gamma^+ \cap Z$ there exists $F \in \mathcal{T}$ such that $E \supset F$ and $\nu(E - F) < \varepsilon$.

So take $\gamma < \omega_1$, $E \in \mathcal{E}_\gamma^+ \cap Z$ and $\varepsilon > 0$.

Since $N_\gamma \in \mathcal{E}$, therefore there exists $\gamma_0 \geq \gamma$ and $M_0 \in \mathcal{E}_{\gamma_0}$ such that $\mu(M_0) = 0$ and $N \subset M_0$.

Let $\overline{E} \in \mathcal{E}_{\gamma_0}$ be such that $E = \overline{E} \cap Z$. Clearly $\nu(E) = \mu(\overline{E}) = \mu(\overline{E} - M_0)$ and by the compactness of μ and because of the inclusion $\mathcal{S} \subset \mathcal{E}$, there exists $S_0 \in \mathcal{S}$ such that

$$X - N_\gamma \supset \overline{E} - M_0 \supset S_0 \quad \text{and} \quad \mu[(\overline{E} - M_0) - S_0] < \varepsilon/2$$

Let $\gamma_1 \geq \gamma_0$ be such that $S_0 \in \mathcal{E}_{\gamma_1}$ and there exists a null set $M_1 \in \mathcal{E}_{\gamma_1}$ containing N_{γ_0}. Clearly $\mu(S_0) = \mu(S_0 - M_1)$.

Since μ is compact and $\mathcal{S} \subset \mathcal{E}$, there exists $S_1 \in \mathcal{S}$ such that

$$X - N_{\gamma_0} \supset S_0 - M_1 \supset S_1 \quad \text{and} \quad \mu[(S_0 - M_1) - S_1] < \varepsilon/4$$

Going on in this way, we get a sequence

$$\overline{E} \supset S_0 \supset S_1 \supset \dots \supset S_n \supset$$

with the following properties:

$$\overline{\overline{E}} \in \mathcal{E}_{\gamma_0},$$

$$S_n \in \mathcal{E}_{\gamma_{n+1}} \quad \text{and} \quad S_n \in \mathcal{S}, \quad n = 0,1,\dots, \quad \gamma_1 < \gamma_2 < \dots$$

$$S_{n+1} \subset X - N_{\gamma_n}, \qquad n = 0,1,\dots$$

$$\mu(S_n - S_{n+1}) < \varepsilon/2^{n+1}, \qquad n = 0,1,\dots$$

Hence $S = \bigcap_{n=1}^{\infty} S_n \neq \emptyset$, $S \subset \bigcap_{n=0}^{\infty} (X - N_{\gamma_n}) = X - N_\lambda$, and $S \in \mathcal{E}_\lambda$, where λ is the first ordinal with the property $\lambda_n \leq \lambda$, $n = 1,\dots$.

Setting $F = S \cap Z$, we have

$$F \in \mathcal{S}^{\lambda} \subset \mathcal{T}$$

and
$$\nu(E - F) = \mu(\overline{E} - S) < \varepsilon ,$$

and that completes the proof of the theorem.

Having proved the above theorem, it is easy to construct an example of a perfect measure, which is not compact.

Example 1. Let $T = \{t_0 < t_1 < \ldots < t_\gamma < \ldots\}$ be a set ordered by the ordinals $\gamma \in \Gamma$.

Then let $X_t = \{0,1\}$, let μ_t be a uniform probability measure on $\mathcal{X}_t = \{\emptyset,\{0\},\{1\},X_t\}$, and let

$$(X,\mathcal{C},\mu) = \prod_{t \in T} (X_t, \mathcal{X}_t,\mu_t) ,$$

that means, $X = \prod_{t\in T} X_t$, $\mathcal{C}$ is the σ-algebra generated by all finite dimensional cylinders, and μ is the direct product of the measures μ_t , $t \in T$.

Moreover, let

$$\mathcal{B}_\gamma = \sigma\left(\prod_{\alpha\leq\gamma} \mathcal{X}_{t_\alpha}\right) \times \prod_{\alpha>\gamma} X_{t_\alpha} , \qquad \alpha,\gamma \in \Gamma .$$

Choose for each $\lambda \in \Lambda$, a sequence $\gamma_0 < \gamma_1 < \ldots$ cofinal with λ and put

$$M^\lambda = \left\{\{x_{t_\gamma}\}_{\gamma<\lambda} : x_{t_{\gamma_n}} = 0 , n=1,\ldots\right\}$$

and $M_\lambda = M^\lambda \times \prod_{\gamma>\lambda} X_{t_\gamma}$

Clearly $\mu(M_\lambda) = 0$.

We shall prove that the measure space $(Z, \mathcal{C}\cap Z, \nu) = (X - \bigcup_{\lambda<\omega_1} M_\lambda, Z \cap \mathcal{C} , \mu|Z)$ is perfect but it is not compact.

We shall divide the proof into three steps.

1. ν is a measure.

In order to show it, it is sufficient to prove that $\mu^*(Z) = 1$ and the last equality holds if and only if $(\mu|\mathcal{B}_\gamma)^*(Z) = 1$, for every $\gamma \in \Gamma$.

So take $\gamma \in \Gamma$ and a point

$$x \in \prod_{\alpha \leq \gamma} X_{t_\alpha} - \bigcup_{\lambda < \gamma} M^\lambda \quad , \quad \alpha \in \Gamma \quad , \quad \Lambda \in \Lambda$$

Since $\{x\} \times \prod_{\alpha > \gamma} \{1\} \notin M_\lambda$, whenever $\lambda > \gamma$ and $\mu(X - \bigcup_{\lambda < \gamma} M_\lambda) = 1)$, we have $(\mu|\mathcal{B}_\gamma)^*(Z) = 1$ and hence $\mu^*(Z) = 1$.

2. ν is perfect.

Since $\mathcal{C} \cap Z = \bigcup_{\gamma \in \Gamma} (\mathcal{B}_\gamma \cap Z)$ and $\mathcal{B}_0 \subset \mathcal{B}_1 \subset \dots$, for the perfectness of ν it is sufficient to show the perfectness of ν restricted to $Z \cap \mathcal{B}_\gamma$.

But $\nu|\mathcal{B}_\gamma$ is σ-isomorphic with the restriction of μ, to

$$\mathcal{B}_\gamma \cap (X - \bigcup_{\lambda < \gamma} M_\lambda) \quad ,$$

and the last measure is perfect.

3. ν is not compact.

Suppose, that ν is compact with $\mathcal{C}_\gamma$, and N_γ as in the Theorem 3, and take an arbitrary $\gamma_0 \in \Gamma$.

Since $\mathcal{B}_\gamma$, $\gamma \in \Gamma$ are countably generated, therefore for each $\gamma_n \in \Gamma$ there exists $\gamma_{n+1} \in \Gamma$ such that $\gamma_{n+1} > \gamma_n$ and

$$\mathcal{B}_{\gamma_n} \subset \mathcal{C}_{\gamma_{n+1}} \quad , \quad n = 0,1,\dots$$

Let λ be the first ordinal greater that γ_n, $n = 0,1,\dots$.

Clearly, we have

$$\bigcup_{n=0}^{\infty} \mathcal{B}_{\gamma_n} \subset \bigcup_{n=1}^{\infty} \mathcal{C}_{\gamma_n} \subset \mathcal{C}_\lambda \, ,$$

and hence

$$\mathcal{B}_\lambda = \sigma\Big(\bigcup_{n=0}^{\infty} \mathcal{B}_{\gamma_n}\Big) \subset \mathcal{C}_\lambda \quad .$$

M_λ is a union of $\mathcal{B}_\lambda$-atoms, and therefore it is also a union of $\mathcal{E}_\lambda$-atoms ; N_λ is a union of $\mathcal{E}_\lambda$-atoms by assumption.

If $e \subset M_\lambda$ is a $\mathcal{E}_\lambda$-atom, then from the construction of Z follows that $e \cap Z = \emptyset$ and so $e \subset N_\lambda$, by (iv) of the Theorem 2.

Hence

$$M_\lambda \subset N_\lambda = \bigcup_{\gamma<\lambda} N_\gamma .$$

This is however impossible, since any set M_λ is contained in a countable union of μ-null sets A_γ , such that A_γ is a union of $\mathcal{B}_\gamma$-atoms, where $\gamma < \lambda$.

<u>Example</u> 2. Let T , $(X,\mathcal{E},\mu)$ and M_λ be as in the Example 1.

Put

$$Y_\lambda = \left\{ \left\{x_t\right\}_{t\in T} \in X : \; x_{t_\alpha} = 1 \text{ whenever } \alpha > \lambda \right\}$$

for every $\lambda \in \Lambda$, and

$$Y = \bigcup_{\lambda<\omega_1} Y_\lambda$$

Setting $Z = Y - \bigcup_{\lambda<\omega_1} M_\lambda$, we get the perfect measure space $(Z, \mathcal{E}\cap Z, \mu|Z)$ which is not compact, and for which the cardinality of Z is equal to the continuum.

The proof is similar to that given in the example 1, it is only sufficient to observe that if $\mathcal{G} \subset \mathcal{E}$ approximates $\mathcal{E}$ with respect to μ , then $Y\cap\mathcal{G}$ approximates $Y\cap\mathcal{E}$ with respect to $\mu|Y$ and so $\mu|Y$ is compact.

We now mention two qestions for which we do not know the answers:

Q1. It the restriction of a compact meaure to any sub-σ-algebra compact? (The answer to this questionis unknown even in the case of Lebesgue measure on Borel sets of the unit interval).

Q2. Let $(X,\mathcal{A})$ and $(Y,\mathcal{B})$ be measurable spaces and let μ be a compact measure on the σ-algebra $\sigma(\mathcal{A}\times\mathcal{B})$. Are the marginals of μ compact ?

References.

[1] Mahkamov B.M., Vinokurov V.G. , On the spaces with perfect but non-compact measures (in Russian), Naucn. zap. Taskent, Institut Nar. Hoz., 71(1973) p. 97-103 .

[2] Marczewski E. , On compact measures, Fund.Math. 40(1953) p.113-124 ,

[3] Ryll-Nardzewski C. , On quasi-compact measures, ibidem, p. 125-130.

[4] Saznonov V.V. , On perfect measures (in Russian), Inz. Acad. Nauk SSSR , 26(1962), p. 391-414.

[5] Szpilrajn E.(Marczewski) , The charcteristic function of a sequence of sets and some of its applications, Fund. Math. 31(1938), p. 207-229 .

[6] Vinokurov V.G. , Compact measures and products of Lebesgue spaces (in Russian), Mat. Sbornik, 74(1967), p. 434-472.

TOPOLOGY AND MEASURE THEORY

A. H. Stone

This talk was concerned with three problems involving both topology and measure theory. I am grateful to D. Fremlin and P. Masani for subsequent discussion and information about the first of the problems, which provided considerable improvements (incorporated here).

1. Suppose a measurable space $(X, \mathfrak{B})$ and a metrizable topological space Y are given. Define measurability of a map $f: X \to Y$ naively, by requiring $f^{-1}(G) \in \mathfrak{B}$ for every open G in Y (and hence for every Borel G in Y). Now let Y be a Banach space. Problem: If f, g are two such measurable maps of X in Y, need their sum $f + g$ be measurable in this sense ?

If Y is separable, the answer is well known to be "yes", by an elementary argument using the fact that Y has a countable base of open sets. But it was shown by J. Nedoma [4] that the answer is in general "no". However, if we require that X be a <u>measure space</u> $(X, \mathfrak{B}, m)$, where m is a (finite or σ-finite) countably additive measure, the situation is more complicated. The answer is still "no" in general, as Masani has observed. But <u>if m is complete (with respect to null sets), then the answer "yes" is consistent with standard set theory</u>. Roughly this is because f and g induce measures on Y which (by a theorem of Marczewski and Sikorski [3]) live on separable subsets of Y, under the (relatively consistent) assumption that there are no measurable cardinals (for real-valued measures). The completeness of m then enables one to reduce the situation to the case in which Y is separable. (See [7] for details, generalizations and references.) Conversely, Fremlin observed that if such a measurable cardinal exists, say on a set Y, then by taking $X = Y \times Y$ (with completed product measure) and $E = \ell^1(Y)$ one can construct measurable $f, g : X \to E$ for which $f + g$ is not measurable. Thus the problem for complete measure spaces is <u>equivalent</u> to the question of existence of measurable (real-valued) cardinals.

The problem can be reformulated in apparently more general terms, by thinking of $f + g$ as the composition of the map $(f, g): X \to Y \times Y$ with the "addition map" of $Y \times Y$ into Y. Since addition is continuous, the measurability of $f + g$ will follow from that of (f, g). In a sense, the converse is also true, because of the following theorem of Fremlin:

For a measurable space $(X, \mathcal{B})$, the following are equivalent:

(1) Whenever Y is a discrete space and f, g are measurable maps of X into Y, the map $(f, g): X \to Y \times Y$ is measurable (with respect to the discrete topology on $Y \times Y$).

(2) Whenever Y, Z are metrizable spaces and the maps $f: X \to Y$, $g: X \to Z$ are measurable, then $(f, g): X \to Y \times Z$ is measurable (with respect to the product topology).

(3) Whenever E is a Banach space and f, g are measurable maps of X into E, then $f + g: X \to E$ is measurable.

A brief indication of the proof is as follows. That (2) implies (3) has been pointed out above. To see that (3) implies (1), apply (3) to the Banach space $M(Y \times Y)$. Finally, in proving (1) implies (2), we may assume $Y = Z$ in (2) (replace them by their topological sum); then the use of a σ-discrete base for Y allows Y to be replaced, in effect, by countably many discrete spaces.

In the case of greatest topological interest, X is an absolutely analytic metric space (that is, a Souslin-F subset of a complete metric space, not necessarily separable) and $\mathcal{B}$ is its family of Borel sets. We have the theorem: <u>For such</u> $(X, \mathcal{B})$, <u>the above equivalent statements all hold</u>. For, in (3) above, the maps f and g must be of bounded Borel class, by a recent result of

D. Preiss [6]; hence, by a theorem of R. Hansell [2], f and g are limits of pointwise convergent sequences of maps of smaller Borel classes, and the result follows by transfinite induction from the case (class 0) in which f and g are continuous.

2. The second problem arises from work of J. Choksi [1] concerned with transformations on locally compact groups; a troublesome invariant null set had to be dealt with. Without essential loss of generality we may assume that the underlying space is the unit square $I \times I$, that the "bad" set B is a Borel subset of $I \times I$ meeting each "vertical" $\{x\} \times I$ in an uncountable set, and that a Borel isomorphism T of I onto I is given. Choksi observed [1, p.115] that his argument could be greatly simplified if there exists a Borel isomorphism T^* of B onto B that sends verticals to verticals and induces T on the first co-ordinate -- that is, the diagram

$$\begin{array}{ccc} B & \xrightarrow{T^*} & B \\ \pi \downarrow & & \downarrow \pi \\ I & \xrightarrow{T} & I \end{array}$$

commutes (where $\pi(x, y) = x$). This would follow at once if (for instance) there exists a Borel isomorphism θ of B onto a standard set, say $I \times I$, such that θ preserves verticals. Unfortunately such a θ need not in general exist (as J. E. Jayne pointed out to me in a letter); one can get a counterexample by exploiting the existence of a plane G_δ set whose projection is not Borel. However, it would suffice for Choksi's purpose if, instead of requiring θ and θ^{-1} to preserve Borel sets, we merely require them to be absolutely measurable. Whether such a θ always exists, under the above hypotheses on B , I do not know. But if more is assumed about B , the existence of a suitable θ can be shown. Using P to denote the space of irrational numbers, we have:

Theorem If B is a G_δ subset of $P \times P$, meeting each "vertical" $\{x\} \times P$ $(x \in P)$ in a dense set, then a suitable vertical-preserving bijection Θ of B onto $P \times P$ exists.

The idea of the proof is as follows. Take a complete metric ρ for B ; and, for $i = 1,2,\ldots$, cover B by a sequence of pairwise disjoint "rectangles" A^i_j , of diameters less than $1/i$, open-closed in $P \times P$, and which moreover meet B in sets of ρ-diameter less than $1/i$. Arrange further that the ith covering refines the $(i-1)^{st}$, and that each vertical segment of a rectangle in the $(i-1)^{st}$ covering is split into infinitely many pieces by the ith . Using fixed enumerations of these rectangles, we assign to each $(x, y) \in B$ a sequence of positive integers $n_1, n_2,\ldots$; roughly speaking, n_i indicates which of the ith stage rectangles meeting $\{x\} \times P$ and contained in the $(i-1)^{st}$ stage rectangle containing (x, y) contains (x, y). Then define

$$\Theta(x, y) = (x, n_1, n_2,\ldots) \in P \times N^{\aleph_0} = P \times P .$$

The absolute measurability of Θ and Θ^{-1} comes from the fact that they are measurable with respect to the Borel field generated by the analytic sets.

The replacement of I by P , and the requirement that B meets every vertical in a dense set (rather than in an uncountable set, which for G_δ sets B would be weaker), do not interfere with the intended application to Choksi's situation, though a more general result would, of course, be desirable. But the restriction to G_δ sets B is severe (in Choksi's situation the "bad" set can be taken to be a $G_{\delta\sigma}$, but not in general a G_δ), and relaxation of it is badly needed. One would hope that the theorem would be true for arbitrary Borel B .

3. The third problem concerns the classification of topological

measure spaces under measure-preserving homeomorphisms. To be specific: which compact metrizable measure spaces $(X, \mathcal{T}, m)$, say with regular Borel measures, are equivalent, under a measure-preserving homeomorphism, to subspaces of $I^{\aleph_0}$ (with product topology and product Lebesgue measure) ? It is trivially necessary that m vanishes on singletons, and that $m(X) \leq 1$. The case $m(X) = 1$ is probably difficult, as it would require an extension of the Oxtoby-Ulam theorem characterizing Lebesgue measure on finite-dimensional cubes [5]. But the case $m(X) < 1$ should be accessible; so we ask: If $(X, \mathcal{T}, m)$ is a compact metrizable measure space, with a regular Borel measure m, vanishing on singletons and such that $m(X) < 1$, is X necessarily equivalent to a subspace of $I^{\aleph_0}$ (by a measure-preserving homeomorphism) ? The answer can be shown to be "yes" if X is topologically a subspace of the plane, and the method (successive approximations) ought to work at any rate for finite-dimensional X. Since a measure-preserving variant of Urysohn's lemma is available, one might even hope that a direct method (without approximation) might work. However, in the finite-dimensional case, the imbedding of X in $I^{\aleph_0}$ would have to be rather pathological. The special case in which $X = I$ topologically can be dealt with by threading an arc through an increasingly dense sequence of disjoint Cantor sets of positive measure.

References

[1] J. R. Choksi: Measurable transformations on compact groups, Trans. Amer. Math. Soc. 184 (1973), 101-124.

[2] R. Hansell: On Borel mappings and Baire functions, Trans. Amer. Math. Soc. 194 (1974), 195-211.

[3] E. Marczewski and R. Sikorski: Measures on nonseparable metric spaces, Colloq. Math. 1 (1948), 133-139.

[4] J. Nedoma: Note on generalized random variables, Trans. First Prague Conference on Information Theory. Statistical Decision Functions and Random Processes, 1956; Czechoslovak Academy of Sciences, Prague, 1957; 139-141.

[5] J. Oxtoby and S. Ulam: Measure-preserving homeomorphisms and metrical transitivity, Annals of Math. 43 (1941), 874-920.

[6] D. Preiss: Completely additive disjoint system of Baire sets is of bounded class, Comm. Math. Univ. Carolinae 15 (1975), 341-344.

[7] A. H. Stone: Some problems of measurability, Topology Conference, V.P.S.U. 1973 (Lecture Notes in Mathematics No. 375, Springer Verlag 1974), 242-248.

University of Rochester
Rochester, New York, U.S.A.

SUBMEASURES AND THE PROBLEM ON THE EXISTENCE OF CONTROL MEASURES

Jens Peter Reus Christensen.

University of Copenhagen

In the first part of the present note we survey without proofs the main results of a joint paper with Wojchiech Herer [1]. In the second part we discuss some open problems, the most interesting of which is the problem on the existence of control measures for invariant Maharam submeasures.

Let $\mathcal{A}$ be a Boolean algebra of subsets of some set X. A submeasure φ defined on $\mathcal{A}$ is a function fulfilling the following axioms

1) $\varphi(\emptyset) = 0$; $A \subseteq B \Rightarrow \varphi(A) \leq \varphi(B)$

2) $\varphi(A \cup B) \leq \varphi(A) + \varphi(B)$ for all $A, B \in \mathcal{A}$.

Many set functions arising naturally in the study of group valued measures and covering problems are submeasures. A submeasure φ is called pathological if $\varphi(X) \neq 0$ and φ does not dominate a non negative finitely additive real valued measure. If $\varphi(X) = 1$ the submeasure φ is normalized; we denote by $\Phi(\mathcal{A})$ the set of normalized submeasures defined on $\mathcal{A}$. An important result is

Theorem 1: Let $e > 0$ be an arbitrary positive number. Then there exist a finite set X and a normalized submeasure φ defined on the Boolean algebra $\mathcal{P}(X)$ of subsets of X such that any non negative measure u defined on $\mathcal{P}(X)$ and dominated by φ satisfies $u(X) \leq e$.

For the proof of this result see [1] Th. 1. The constructed submeasure is defined on the subsets of a finite abelian group and is invariant under translations.

Using the above result it is easy to show that if the Boolean algebra $\mathcal{A}$ of subsets of X is atomless, then the set of pathological submeasures is a dense G_δ subset of $\Phi(\mathcal{A})$ (with respect to the natural compact topology of setwise convergense). In particular there exist on every atomless $\mathcal{A}$ a pathological submeasure.

If the Boolean algebra $\mathcal{A}$ of subsets of X is a σ-field and φ is a countably subadditive submeasure on $\mathcal{A}$ then there exist not any non negative countably additive measure u on $\mathcal{A}$ such that $u(A) = 0$ implies $\varphi(A) = 0$ and there is also no non trivial countably additive non negative measure v such that $\varphi(A) = 0$ implies

$v(A) = 0$. For the proof of the above important remarks see [1] Th. 2.

Theorem 2: Let φ be a pathological submeasure defined on the Boolean algebra $\mathcal{A}$ of sebsets of the set X. Then there is no non trivial finitely additive non negative measure u on $\mathcal{A}$ such that u is φ-continuous (i.e. $\forall e > 0 \exists \delta > 0 \forall A \in \mathcal{A} : \varphi(A) \leq \delta \Rightarrow u(A) \leq e$).

Proof: Let Ω be the Stone space of the Boolean algebra . The algebra $\mathcal{A}$ may be identified in the usual way with the clopen sets of Ω. By compactness the set function ψ defined by

$$\psi(B) = \inf.\{\sum_{k=1}^{\infty} \varphi(A_k) \mid B \subseteq \bigcup_{k=1}^{\infty} A_k\}$$

is a countably subadditive extension of φ to all subsets of Ω. Any measure u on $\mathcal{A}$ has a unique extension to a countably additive Radon measure defined on the Borel sets of Ω.

If u were φ-continuous then for compact subsets of Ω we have $\varphi(A) = 0$ implies $u(A) = 0$. By regularity of u this is also true for Borel sets and this contradiction with the preceding remarks finishes the proof.

The preceding investigations were motivated by the following problem (which is still open); "Let $\mathcal{B}$ be a σ-field of subsets of the set X and m a countably additive measure defined on with values in an abelian Polish group. Does there exist a countably additive probability measure with the same zero sets as m?".

A Maharam submeasure φ defined on $\mathcal{B}$ is by definition a submeasure which is sequentially point continuous (i.e. for every sequence $A_n \in \mathcal{B}$ whose characteristic functions tends pointwise to the characteristic function of $A \in \mathcal{B}$ we have

$$\varphi(A_n) \to \varphi(A) \qquad).$$

If $\mathcal{B}$ is a separable σ-field the above problem is equivalent to the problem whether or not every Maharam submeasure admits a probability measure with the same zero sets. In this form the question was raised by D.Maharam (see [2]).

Our results above possibly indicates that the answer is in the negative since we are able to construct a countably subadditive submeasure defined on all subsets of a compact metrizable space (say the Cantor space) which is parhological and therefore does not have a probability measure with the same zero sets according to the above remarks (for the construction of such a submeasure see [1]). There seems however to be no way to obtain this submeasure with the strong continuity property of D.Maharam.

Let us denote by $\Phi_m(\mathcal{A})$ the set of normalized Maharam submeasures defined on the Boolean algebra $\mathcal{A}$ of subsets of X. If we try to adapt the Baire category argument of [1] (which shows the existence of pathological submeasures) a main difficulty turns out to be that $\Phi_m(\mathcal{A})$ is apparently not a Baire space with the topology of setwise convergence (in any case we cannot show it). Then we should possibly try the much stronger topology induced by the metric

$$d(\varphi,\psi) = \operatorname{sub}\{|\varphi(A) - \psi(A)| \mid A \in \mathcal{A}\}.$$

It is easily seen that $\Phi_m(\mathcal{A})$ is indeed a complete metric space with this metric and therefore has the Baire property. Also it can easily be shown that the pathological submeasures in $\Phi_m(\mathcal{A})$ is a G_δ with respect to this topology. Unfortunately we run into serious difficulties if we try to show the density in a way analogous to [1].

If φ is a Maharam submeasure defined on the σ-field of subsets of X then a necessary condition for the existence of a control measure is that a Fubini type theorem is valid for the product of X with any measure space. Its not unlikely that this condition is also sufficient; in any case if the Maharam submeasure is defined on the Borel field of a compact metrizable abelian group and is invariant then a Fubini type theorem is sufficient for the Haar measure to be control measure. We do not know whether or not every Maharam submeasure has the property that a Fubini type theorem is valid for the product with any measure space.

An interesting problem left open in [1] is the asymptotic behavior of the number sequence a_n where a_n is defined as the maximum of all positive numbers e with the property that for all normalized submeasures defined on the subsets of a finite set with cardinality n one can always find a positive dominated by that submeasure and with total mass bigger than e or equal to e .

REFERENCES

1) J.P.R.Christensen and Wojchiech Herer, On the existence of pathological submeasures and the construction of exotic topological groups, Math. Ann. 213, 203-310 (1975).

2) Maharam, D. ; An algebraic characterization of measure algebras, Ann.Math. 48, 154-167 (1947).

ON DISINTEGRATIONS AND CONDITIONAL PROBABILITIES*

by

Lester E. Dubins

University of California, Berkeley

Without denying the importance of measures which are countably additive, Bruno de Finetti has long advocated that countable additivity not be taken as an axiom for measure theory. One reason is this.

Though its finite additivity can always be preserved under extension, the countable additivity of a probability measure sometimes cannot be. An interesting example is obtained by letting $P(A)$ be 0 or 1 according as A is a countable set, or a cocountable set, of ordinals. This countably additive P cannot be extended to the set of all sets of countable ordinals so as to be countably additive, as is implied by a result of Ulam [5]. From this follows Banach and Kuratowski's result that it is consistent with Cantor's continuum hypothesis that Lebesgue measure on the unit interval cannot be extended to all subsets so as to preserve countable additivity.

In contrast, if, as in this talk, countable additivity is not taken as an axiom, then, not only does every P extend to all events, but every P can even be extended to be a full conditional probability [Krauss, 1968]. To explain this, and for other purposes, it is useful to borrow, and minimally modify, some useful terminological conventions of [de Finetti, 1972, 1974].

A random number on a nonempty set, Ω, is a real-valued function defined on Ω. Since there is some simplification in

*This research was prepared with the support of National Science Foundation, Grant No. MPS 75-09459.

considering only those that are bounded, it is explicitly assumed now, and will be implicitly assumed hereafter, that random numbers are always bounded. A <u>prevision supported by</u> Ω is an order preserving, linear functional, P, defined on a linear space of random numbers on Ω, including the constants, which satisfied $P(c)=c$. (Since every probability measure on Ω extends to a prevision on Ω, it should cause no confusion if the letter P designates both previsions and probabilities). If P is defined for all random numbers, then P is <u>full</u>. A <u>full conditional prevision</u> on a nonempty set, Ω, is a mapping Q which assigns to each nonempty subset A of Ω a full prevision, Q_A, supported by A, and which satisfies:

(1) $$Q_A(BX) = Q_A(B)\cdot Q_{AB}(X)$$

for all random numbers X and all events A and B provided only that AB is nonempty. (As in [1] and [2], the letter, 'B', in [1] designates an event as well as the indicator of that event.)

Krauss' result can be formulated, thus: <u>For every prevision P on Ω, there is a full, conditional prevision, Q, on Ω such that</u> $Q_\Omega=P$.

Suppose that π is a partition of Ω, that is, a collection of nonempty, pairwise-disjoint, sets whose set-theoretic union is Ω. A π-<u>strategy</u> is a mapping Q which assigns to each $A \in \pi$, a prevision, Q_A, which is supported by A.

The definition of π-stretegy would be unsatisfactory if it did not permit this not-at-all surprising theorem.

<u>Theorem 1</u>. Every π-strategy can be extended so as to be a full conditional prevision.

As formulated here, Theorem 1 is somewhat more general than [3, Theorem 5], but since their proofs are essentially the same, and since the theorem will not be logically relied on in what follows, no proof need be given here.

For simplicity of exposition, assume henceforth that π-strategies, Q, are always full, that is, that each Q_A is full.

For each $\omega \in \Omega$ and partition π, let $\pi(\omega)$ be the unique $A \in \pi$ which contains ω. Let X range over the random numbers and define

$$\sigma(X|\omega) = Q_{\pi(\omega)}(X \cdot \pi(\omega)). \tag{2}$$

Of course, for each ω, $\sigma(\cdot|\omega)$ is a prevision supported by $\pi(\omega)$; moreover, for any ω and ω' which belong to the same element of π, $\sigma(\cdot|\omega)$ is identical with $\sigma(\cdot|\omega')$. Plainly, any σ which satisfies these two properties determines the π-strategy Q, so it is justifible to call such a σ a π-strategy, too. If, for some π-strategy, σ, and prevision, P',

$$P(X) = \int \sigma(X|\omega)\, dP'(\omega) \tag{3}$$

for all X in the domain of P, then P is π-disintegrable, σ is a π-disintegration of P, and (P,π) is disintegrable. In (3), it is implicitly assumed that for all X in the domain of P, $\sigma(X|\cdot)$ is in the domain of P', that is, that $\sigma(X|\cdot)$ is P'-integrable.

As is trivial, if P is discrete, then, for every π, P is π-disintegrable. Nevertheless, there do exist (P,π) which are not disintegrable. For a simple example, see [de Finetti, 1972, p.205]. More surprisingly, if $\mathcal{D}(\pi)$ denotes the previsions that are π-disintegrable, then there exist π and P such that P is orthogonal to every member of $\mathcal{D}(\pi)$. Such a P is plainly outside the norm-closure of $\mathcal{D}(\pi)$, that is, P cannot be approximated in the total variation norm by previsions that are π-disintegrable. The example given in [3] of this phenomenon is purely finitely additive. It would be of interest to characterize $\mathcal{D}(\pi)$ as well as its orthogonal complement $\mathcal{D}^{\perp}(\pi)$. It would also be good to know whether there is a π and a countably additive P for which P is in $\mathcal{D}^{\perp}(\pi)$. I suspect that even the Legesgue integral on the unit interval, Ω, may fail to be disintegrable if π is suitably chosen, for example, thus: Two points of Ω belong to the same element of π if, and only if, they differ by a rational number.

For a closely related example, let Ω be the set of all infinite sequences of zeroes and ones and let C be the set of all continuous, real-valued, functions defined on Ω when Ω is endowed with the usual product topology. If $d\omega$ represents ordinary, fair, coin-tossing measure on Ω and P is defined on C by

$$P(X) = \int X(\omega)\, d\omega \qquad X \in C, \tag{4}$$

then P is the <u>elementary prevision</u>. If π is the set of atoms of the tail sigma field, and σ is a π-strategy such that, for each $X \in C$, $\sigma(X|\cdot)$ is measurable with respect to the completion of $d\omega$, and

$$P(X) = \int \sigma(X|\omega)\, d\omega, \tag{5}$$

then σ is a <u>measurable</u>, <u>tail-disintegration</u> of the elementary prevision, P. If, for almost all ω, $\sigma(\cdot|\omega)$ is purely finitely additive, then σ will be said to be purely finitely additive. Though the elementary prevision is obviously countably additive, one can easily establish:

<u>Theorem 2</u>. Every measurable, tail-disintegration of the elementary prevision, P, is purely finitely additive.

I expect to publish details elsewhere, but the essentials of the proof are given here, thus. For each $\omega \in \Omega$, the mapping defined on C by $X \to X\cdot\pi(\omega)$ is one-to-one. Hence, if

$$\vartheta(X\cdot\pi(\omega)|\omega) = P(X), \tag{6}$$

then δ is well-defined. As is evident, δ is a measurable, tail-disintegration of P, and, as is not difficult to verify, for every ω, $\delta(\cdot|\omega)$ is purely finitely additive. If σ is any tail-disintegration of P, and $X \in C$, then $\sigma(X|\cdot)$ is almost certainly constant, as is essentially well-known, and, in any event, not difficult to verify. This constant can be no other than P(X). Use the fact that C has a countable, dense subset and conclude that $\sigma(\cdot|\omega) = \delta(\cdot|\omega)$ for almost all ω. Hence, σ is purely finitely additive, which proves the theorem.

Let P' be the usual extension of P to C', the set of bounded, Borel functions. Of course, any tail-disintegration of P' is, in particular, a tail-disintegration of P, and à fortiori, purely finitely additive. However, I suspect not only that there are no tail disintegrations of P', but also that P' is orthogonal to every tail-disintegrable prevision definable on C'.

It is of interest to find conditions on a P and a π to assure that P is π-disintegrable.

Call σ a <u>weak</u> π-disintegration of P if, for all X in the domain of P, $\sigma(X|\omega) \geq 0$ for all ω implies $P(X) \geq 0$. Somewhat surprixingly, weak disintegrations are actually disintegrations provided only that the domain of P is appropriately closed. In particular, such is the case if every random number X is in the domain of P. [3, Theorem 1]. The remainder of this talk is concerned only to present this result in somewhat greater generality. Let $\mathcal{U}$ be a field of subsets of Ω. Call X $\mathcal{U}$-<u>measurable</u> if, for every pair of real number y and z, the event $y<X<z$ is in $\mathcal{U}$. A π-strategy σ is $\mathcal{U}$-<u>measurable</u> if, for all X in the domain of P, $\sigma(X|\cdot)$ is $\mathcal{U}$-measurable. Call P $\mathcal{U}$-<u>measurable</u> if it is defined for all bounded, $\mathcal{U}$-measurable X.

<u>Theorem 3</u>. Every $\mathcal{U}$-measurable, weak, π-disintegration σ of a $\mathcal{U}$-measurable P is a π-disintegration of P.

The proof of Theorem 3 is an obvious modification of the proof of its special case [3, Theorem 1] but is recorded here in almost complete detail for ease of verification.

Proof of Theorem 3. Plainly, for all X in the domain of P, and for all real numbers u,

(7) $$\sigma(X|\omega) \leq u \text{ for all } \omega \text{ implies } P(X) \leq u.$$

Call an element of U π-measurable if it is a union of elements of π. Let S be π-measurable, and verify that (7) can be strengthed to

(8) $$\sigma(X|\omega) \leq u \text{ for all } \omega \in S \text{ implies } P(XS) \leq uP(S).$$

Now let $\varepsilon > 0$. For each integer j, let S_j be the set of ω such that

(9) $$(j-1)\varepsilon \leq \sigma(X|\omega) < j\varepsilon,$$

and calculate, thus.

(10) $$\begin{aligned} \int_{S_j} \sigma(X|\omega)\, dP(\omega) &\geq (j-1)\varepsilon P(S_j) \\ &= j\varepsilon P(S_j) - \varepsilon P(S_j) \\ &\geq P(XS_j) - \varepsilon P(S_j) \end{aligned}$$

where the last inequality holds by (8) and the second inequality of (9). Sum over the finite number of j for which S_j is nonempty, and conclude

(11) $$\int \sigma(X|\omega)\, dP(\omega) \geq P(X).$$

If X is replaced by $-X$, the reverse inequality is seen to hold, which proves the theorem.

REFERENCES

[1] De Finetti, Bruno, 1972. Probability, Induction and Statistics. Wiley, New York.

[2] De Finetti, Bruno, 1974. Theory of Probability. Wiley, New York.

[3] Dubins, Lester E., 1975. "Finitely additive conditional Probabilities, Conglomerability and Disintegrations", The Annals of Probability, Vol. 3, No. 1, pp. 89-99.

[4] Krauss, P. H., 1968. "Representation of conditional probability measures on Boolean algebras", Acta. Math. Sci. Hungar. Tomus 19, (3-4) pp. 229-241.

[5] Ulam, Stanislaw, 1930. "Zur Masstheorie in der allgemeinen Mengenlehre", Fundamenta Mathematicae, 16, pp. 140-150.

ON LIAPUNOV VECTOR MEASURES

by

Greg. Knowles*

Institut für Angewandte Mathematik
Universität Bonn
53 Bonn
West Germany

1. INTRODUCTION.

The study of the set of values of a vector measure dates perhaps back to Sierpinski ([12]), who showed that the set of values of a non-atomic (real-valued) measure is an interval. It then remained some years until Liapunov proved that the range of any non-atomic finite dimensional vector measure is compact and convex ([7]), and also gave an example to show that this need not be true for infinite-dimensional measures ([8]). The importance of this result in other areas of Mathematics is well known, in particular, the Neyman-Pearson Lemma ([9]), and the bang-bang principle for the linear time-optimal control problem ([2]).

In this note some of the newer results in this area for infinite-dimensional vector measures are surveyed, and their relevance to infinite dimensional control problems, specifically the control of systems governed by linear partial differential equations is indicated.

2. LIAPUNOV MEASURES.

Suppose T is an abstract set, $\mathcal{S}$ a σ-algebra of subsets of T, and X a quasi-complete locally convex topological vector space (l.c.t.v.s.), with dual X'. A vector measure m on $\mathcal{S}$ is a countably additive map m: $\mathcal{S} \to X$. For a set $E \in \mathcal{S}$, set $\mathcal{S}_E = \{F : F \subseteq E, F \in \mathcal{S}\}$. By the range of the measure m on E, we mean the set $m(\mathcal{S}_E) = \{m(F) : F \in \mathcal{S}_E\}$, and for simplicity denote $m(\mathcal{S}_T) = m(\mathcal{S})$. In this note we shall be interested in the properties of those measures for which $m(\mathcal{S}_E)$ is a weakly compact, convex subset of X, for every $E \in \mathcal{S}$. We call such measures <u>Liapunov</u>.

* Research supported by the Sonderforschungsbereich 72.

It is quite easy to show that if X is infinite-dimensional, the non-atomicity of m is insufficient for it to be Liapunov. Consider the following

EXAMPLE 1. ([13]).

Suppose $T = [0,1]$, $\mathcal{S}$ is the Borel σ-algebra on T, denoted by $\mathcal{B}(0,1)$, and $X = L^1(0,1)$ with Lebesgue measure. Then the vector measure $m: \mathcal{S} \to L^1(0,1)$ defined by

$$m(E) = \chi_E , \quad E \in \mathcal{S},$$

is non-atomic, and clearly its range is neither convex, nor weakly compact. It will be useful for later to observe that the integration mapping defined by this measure is (essentially) the identity, that is if f is a bounded measurable function on T, then $\int_T f\, dm = f$ a.e.

From this example it is clear that for general non-atomic vector measures the best that can be expected are results of the form

THEOREM 1 ([3]).

If $m: \mathcal{S} \to X$ is non-atomic and either X is metrizable or $\mathcal{S}$ is m-essentially countably generated, then the weak closure of $m(\mathcal{S})$ is weakly compact and convex in X.

A set $E \in \mathcal{S}$ is m-null or m-negligible if $m(\mathcal{S}_E) = \{0\}$. Two sets are called m-equivalent if their symmetric difference is m-null, and the class of sets in $\mathcal{S}$ m-equivalent to $E \in \mathcal{S}$ is denoted by $[E]_m$. Let $\mathcal{S}(m) = \{[E]_m : E \in \mathcal{S}\}$. Similarly a bounded measurable function is called m-null if it is zero m-a.e., and the set of equivalence classes of bounded measurable functions modulo equality m-a.e. is denoted by $L^\infty(m)$.

As any bounded measurable function f can be uniformly approximated by finite valued functions, and the space X is assumed quasi-complete, the $\int_T f\, dm$ can be easily defined, consistent with linearity, continuity, and the requirement that the integral of a characteristic function is the measure of the underlying set. In this way, for each set $E \in \mathcal{S}$, m induces a linear mapping $m_E: L^\infty(m) \to X$ given by $m_E(f) = \int_E f\, dm$, $f \in L^\infty(m)$. The study of this map plays a fundamental role in the following theory. For a set $V \subset R$, let $L_V(m) = \{f \in L^\infty(m) : f(t) \in V, t \in T\}$.

In this note we will only consider closed vector measures. These are defined in terms of the following topology on $\mathcal{S}(m)$. A net $\{[E_\alpha]_m\}_{\alpha \in A}$ is called $\tau(m)$-convergent to $[E]_m$ ($\tau(m)$-Cauchy) if for every neighbourhood U of the origin in X, there exists an $\alpha_U \in A$ such that for all $\alpha \geq \alpha_U$, $m(\mathcal{S}_{E_\alpha \Delta E}) \subset U$ (such that for all $\alpha, \beta \geq \alpha_U$, $m(\mathcal{S}_{E_\alpha \Delta E_\beta}) \subset U$). The measure m is closed if $\mathcal{S}(m)$ is $\tau(m)$-complete, that is every

$\tau(m)$-Cauchy net in $\mathcal{S}(m)$ is $\tau(m)$-convergent to an element of $\mathcal{S}(m)$.

Working inside this class of measures causes little or no inconvenience in practice, however, as it can be shown that if X is metrizable then every measure $m: \mathcal{S} \to X$ is closed, or if m is the indefinite integral of Pettis integrable function, it is also closed (see [5] Section IV.7). The main property of closed measures of interest here is the following: if $m: \mathcal{S} \to X$ is a closed vector measure, then $\overline{co}\, m(\mathcal{S}) = \{\int f\, dm : 0 \le f \le 1, f \text{ is measurable}\}$ is weakly compact, and so a closed vector measure m is Liapunov if and only if

$$m(\mathcal{S}_E) = m_E(\{f : f^2 = f\}) = \overline{co}\, m(\mathcal{S}_E) = m_E(\{f : 0 \le f \le 1\})$$

for every set $E \in \mathcal{S}$. Clearly then, if for any non-null set $E \in \mathcal{S}$, the integration mapping m_E is 1-1, m cannot be Liapunov. The next Theorem says, in effect, that the converse statement is also true. The proof is given in [6], and [5] Theorem V.1.1.

THEOREM 2.

Suppose $m: \mathcal{S} \to X$ is a closed vector measure. Then the following are equivalent.

(1) for every set $E \in \mathcal{S}$ which is not m-null, there exists a bounded measurable function f, not m-null on E, such that $\int_E f\, dm = 0$.

(2) for every set $E \in \mathcal{S}$ which is not m-null, the integration mapping $m_E: L^\infty(m_E) \to X$ is not injective

(3) m is Liapunov.

Proof. Suppose (i) holds and $x \in \overline{co}\, m(\mathcal{S})$. Set $H_o = \{g \in L^\infty_{[0,1]}(m) : m(g) = x\}$, this set is non-empty as m is closed. We can construct a Hausdorff locally convex topology on $L^\infty(m)$, which we call $\sigma(m)$, such that $L^\infty_{[0,1]}(m)$ is $\sigma(m)$-compact, and the integration mapping $m: L^\infty(m) \to X$ is continuous between the $\sigma(m)$ topology on $L^\infty(m)$ and the weak topology on X. Consequently the set H_o is $\sigma(m)$-compact and so has an extreme point h, say. If we can show h is a characteristic function, the proof is finished. Suppose the contrary. Then there exists an $\varepsilon > 0$ and a non m-null set $E \in \mathcal{S}$ such that $\varepsilon \le h(t) < 1-\varepsilon$ for $t \in E$. By (i) we can find a non m-null function f on E, which can be chosen with $|f| < \varepsilon$ and $f = 0$ outside E, such that $\int_E f\, dm = 0$. However, since $h \pm f \in H$, this contradicts the extremality of h, and the result follows.

As an immediate Corollary we have the Theorem of Liapunov. For, if m is a non-atomic finite dimensional vector measure, and E is not m-null, the vector space $L^\infty(m_E)$ is infinite dimensional and so the integration mapping $m_E: L^\infty(m_E) \to R^n$ cannot be 1-1.

On the other hand, to see how this Theorem works in infinite dimensions consider the

EXAMPLE 2.

Let $T = [0,1]$, $\mathcal{S}$ be the Borel σ-algebra on T, and $X = \ell_2$, and $\{r_n\}$ be the sequence of Rademacher functions. Then the mapping $m: \mathcal{S} \to \ell_2$ defined by

$$m(E) = \left(\int_E r_1(x)dx, \int_E r_2(x)dx, \dots \right), \quad E \in \mathcal{S},$$

is a vector measure. It is known ([11]) that for any set E of positive (Lebesgue) measure, the Rademacher functions are not complete in $L^1(E)$ (that is, their linear span is not dense in $L^1(E)$). Hence by the Hahn--Banach Theorem there must exist a non-zero function $f \in L^\infty(E)$ such that $\int_E fr_n\, dn = 0$ for every $n = 1,2,\dots$. Then by (1) of Theorem 2, m is Liapunov.

The next theorem settles the existence of Liapunov measures in arbitrary locally convex spaces.

THEOREM 3 ([5] Section V.5).

If $m: \mathcal{S} \to X$ is any vector measure there exists a set T_1, a σ-algebra $\mathcal{S}_1$ of subsets of T_1, and a Liapunov vector measure $m_1: \mathcal{S}_1 \to X$, such that $m_1(\mathcal{S}_1) = \overline{co}\, m_1(\mathcal{S}_1) = \overline{co}\, m(\mathcal{S})$. If m is closed and non-atomic, and $\mathcal{S}$ is m-essentially countably generated, then we can choose $T_1 = T$ and $\mathcal{S}_1 = \mathcal{S}$.

Proof. We sketch the details for the metrizable case only, as the general case requires too detailed a knowledge of the theory of closed vector measures to be given here.

Define $T_1 = T\times[0,1]$, $\mathcal{S}_1 = \mathcal{S}\otimes\mathcal{B}$ and the vector measure m_1 on $\mathcal{S}_1$ by $m_1(E) = \int_T \ell(E^t)dm(t)$, where $\mathcal{B}$ is the Borel σ-algebra on $[0,1]$, ℓ is Lebesgue measure and $E^t = \{y : (t,y) \in E\}$, $t \in T$. Clearly $m_1(\mathcal{S}_1) \subset \{\int_T f\, dm : f \in L^\infty_{[0,1]}(m)\} = \overline{co}\, m(\mathcal{S})$ (as m is closed), and conversely for any $f \in L^\infty_{[0,1]}(m)$, $\int_T f\, dm = m_1(E)$, where $E = \{(t,y) \in T_1 : 0 \le y \le f(t)\}$. The measure m_1 is closed (since X is metrizable, in the non-metrizable case this is non-trivial), and if $E \in \mathcal{S}_1$ is non m_1-null, it is not difficult to construct a bounded $\mathcal{S}_1$-measurable function f, not m_1-null on E, with

$$\int_0^1 f(t,y)\chi_E(t,y)dy = 0 \quad , \quad t \in T .$$

Consequently

$$\int_E f\, dm_1 = \int_{T_1} f\chi_E dm_1 = \int_T \left(\int_0^1 f(t,y)\chi_E(t,y)dy\right)dm(t) = 0 ,$$

and m_1 is Liapunov by Theorem 2.

The final result follows by an isomorphism Theorem in [5] (Theorem II.6.1).

A vector measure $m: \mathcal{S} \to X$ is called injective if the integration mapping $m: L^{\infty}(m) \to X$ is 1-1. Example 1 is such a measure, and the next Theorem shows it is in a sense the typical type of a "non-Liapunov" measure, as it proves that any closed measure can be decomposed into a direct sum of a Liapunov measure, and a family of injective measures. This decomposition is anlogous to the decomposition of a finite dimensional measure into its non-atomic and atomic parts.

THEOREM 4 ([4]).

Suppose $m: \mathcal{S} \to X$ is a closed measure. Then there exists an m-essentially unique set E in $\mathcal{S}$, and a family of pairwise m-essentially disjoint sets in $\mathcal{S}$ such that m_E is Liapunov, $E \cap F = \emptyset$, m_F is injective for every $F \in \mathcal{F}$, and the union of $\mathcal{F}$ in $\mathcal{S}(m)$ is $[T-E]_m$.

Proof. As m is closed, $\mathcal{S}(m)$ is a complete Boolean Algebra under the usual operations ([4] Lemma), and the proof follows by exhaustion applied to the family of sets $[F]_m \in \mathcal{S}(m)$ where $m|_F$ is injective, and Theorem 2.

3. CONTROL PROBLEMS.

As one illustration of the use of Theorem 1 in Control Theory we consider an open problem in the theory of time optimal control. Other examples of Liapunov measures arising in this way, and sufficient conditions for determining when a measure is Liapunov are given in [5] Section V.7.

Consider the 1-dimensional linear parabolic equation

(1) $$\frac{\partial y}{\partial t} = \frac{\partial}{\partial x}(p(x)\frac{\partial y}{\partial x}) + q(x)y \qquad 0 < x < L , 0 < t \le t_0$$

with the initial-boundary conditions

(2) $$y(x,0) = 0 \qquad 0 \le x \le L$$

(3) $$A_0 y(0,t) + B_0 \frac{\partial y}{\partial x}(0,t) = f(t) \qquad A_0^2 + B_0^2 \neq 0$$

(4) $$A_1 y(L,t) + B_1 \frac{\partial y}{\partial x}(L,t) = 0 \qquad A_1^2 + B_1^2 \neq 0$$

where p is a twice continuously differentiable function on [0,L], q is continuous on [0,L], and f is interpreted as the control function and restricted to be measurable and take values in [0,1].

If f were a smooth function the problem (1)-(4) would have a unique classical solution which could be written

$$(5) \qquad y(x,t) = \sum_{n=1}^{\infty} c_n v_n(x) \int_0^t f(t)\, e^{-\lambda_n(t-\tau)}\, d\tau ,$$

where $\{c_n\}$ are constants and $\{\lambda_n, v_n\}$ are the eigenvalues and normalized eigenvectors, respectively, of the Sturm-Liouville problem

$$(p(x)y'(x))' + q(x)y(x) = \lambda y(x) \qquad 0 < x < L$$

$$A_0 y(0) + B_0 y'(0) = A_1 y(L) + B_1 y'(L) = 0$$

When f is only a bounded, measurable function we regard (5) as the generalized solution of (1)-(4).

Let t > 0 be fixed. For any bounded, measurable function f, $y(.,t)$ in (5) belongs to $L^2(0,L)$, and the mapping $m: \mathcal{B}(0,t) \to L^2(0,L)$ defined by

$$m(E)(x) = \sum_{n=1}^{\infty} c_n v_n(x) \int_0^t \chi_E(t)\, e^{-\lambda_n(t-\tau)}\, d\tau ,$$

$$0 < x < L,\ E \in \mathcal{B}(0,t)$$

is a closed vector measure, with $m(f) = y(.,t)$ in (5).

We consider the bang-bang principle for this problem. Namely, if (1)-(4) is regarded as a problem of linear heating on the rod $[0,L]$, this principle can be stated, is every distribution of temperature on the rod that is produced by a (measurable) function f in (3) taking values between 0 and 1, produced by some function taking values only 0 or 1, that is by a characteristic function? For obvious reasons the latter type of control functions are called bang-bang. By our remarks above this problem can be restated, are the sets $\{m(f) : 0 \leq f \leq 1$, f measurable$\}$, $\{m(f) : f^2 = f$, f measurable$\}$ equal? It is clear that if m is Liapunov this is the case. Using the fact that the $\{v_n\}$ are an orthonormal system in $L^2(0,L)$ and the Hahn-Banach Theorem as in Example 2, m will be Liapunov if and only if the sequence of functions $\{e^{-\lambda_n \tau}\}$ is incomplete in $L^1(E)$, for any set E of positive measure. Since $\Sigma_{n=1}^{\infty} 1/|\lambda_n| < \infty$, this is certainly the case if E is an interval, by the Theorem of Müntz ([10] p.54). The general case appears still to be an open problem.

Some partial results on this problem were obtained in [1] by other methods.

BIBLIOGRAPHY

1. EGOROV, J.V., Some problems in the theory of optimal control, USSR COMP. MATH. 3 (1963), 1209-1232.
2. HERMES, H. and LASALLE, J.P., Functional analysis and time optimal control, ACADEMIC PRESS. New York 1969.
3. KLUVANEK, I., The range of a vector measure, MATH. SYSTEMS THEORY 7 (1973), 44-54.
4. KLUVANEK, I. and KNOWLES, G., Liapunov decomposition of a vector measure, MATH. ANN. 210 (1974), 123-127.
5. KLUVANEK, I. and KNOWLES, G., Vector measures and Control Systems, NORTH HOLLAND.
6. KNOWLES, G., Liapunov vector measures, MATH. SYSTEMS THEORY 13 (1975), 294-303.
7. LIAPUNOV, A., Sur les fonctions-vecteurs complètement additives, (Russian: French Summary). IZV. AKAD. NAUK. SSSR Ser. MAT. 4 (1940), 465-478.
8. LIAPUNOV, A., Sur les fonctions-vecteurs complètement additives, (Russian) IZV. AKAD. NAUK. SSSR Ser. MAT. 10 (1946), 277-279.
9. NEYMAN, J. and PEARSON, E.S., On the problem of the most efficient tests of statistical hypotheses, PHILOS. TRANS. ROY. SOC. LONDON Ser. A 231 (1933), 289-337.
10. SCHWARTZ, L., Etude des sommes d'exponentielles, Deuxième édition. HERMANN. Paris 1959.
11. SHIREY, J., Restricting a Schauder basis to a set of positive measure, TRANS. AMER. MATH. SOC. 184 (1973), 61-71.
12. SIERPINSKI, W., Sur les fonctions d'ensemble additives et continues, FUND. MATH. 3 (1922), 240-246.
13. UHL, J.J., The range of a vector-valued measure, PROC. AMER. MATH. SOC. 23 (1969), 158-163.

MEASURABILITY AND PETTIS INTEGRATION IN HILBERT SPACES

by

P. Masani*

University of Pittsburgh, Pittsburgh, Pa. 15260

1. Introduction

A continuous function $x(\cdot)$ on a non-second countable locally compact abelian group Λ to a non-separable Hilbert space $\mathcal{H}$, i.e. a continuous variety in $\mathcal{H}$, need not have a separable range. Hence its integral with respect to a non-negative measure μ over Λ will have to be of the Pettis type. But for all such varieties, $|x(\cdot)|_{\mathcal{H}}$ is continuous and therefore Borel measurable, and for many we find that $\int_\Lambda |x(\lambda)|_{\mathcal{H}}\mu(d\lambda) < \infty$. This is the case, for instance, when $x(\cdot)$ is a stationary variety (for which $|x(\cdot)|_{\mathcal{H}}$ = const.) and μ is a probability measure. The analysis of such varieties requires extensive manip-

---

* Work supported by the National Science Foundation U.S.A. under Grants GP43072 and MPS74-07302 A01.

ulation of Pettis integrals of this sort, cf. e.g. [12], and demands a preliminary theory for them, preferably one that is topology-free. The object of this paper is to provide such a theory.

More precisely, our principal purpose is to develop a theory of Pettis integrable functions f over a non-negative countably additive measure algebra $(\Lambda,\mathfrak{A},\mu)$ to the Hilbert space $\mathfrak{H}$, for which $|f(\cdot)|_{\mathfrak{H}}$ is $\mathfrak{A}$-measurable and $\int_\Lambda |x(\lambda)|_{\mathfrak{H}}\mu(d\lambda) < \infty$, but for which f need not be separably ranged, and therefore need not be Bochner measurable (§§5,6). Such functions seem to have been left out in the literature; for instance, recent basic works on the subject by Vakhania [18], Uhl [17] and Chatterji [2] deal exclusively with Pettis integrable f for which range f is separable and $\int_\Lambda |f(\lambda)|_{\mathfrak{H}}\mu(d\lambda) = \infty$.

Since in the Lebesgue treatment the notion of measurability precedes that of integration, a second and more primary objective of this paper is to study the measurability of $\mathfrak{H}$-valued functions over a measurable algebra $(\Lambda,\mathfrak{A})$. The measurability concept will depend of course on the σ-algebra $\mathfrak{B}$ chosen over $\mathfrak{H}$. These questions occupy §§2-4.

For non-separable Banach spaces $\mathfrak{X}$ we find that there are four relevant σ-algebras, viz. $\mathfrak{B}_{\mathfrak{N}_w}$, $\mathfrak{B}_{\mathfrak{N}}$, $\mathfrak{B}_{\tau_w}$, $\mathfrak{B}_{\tau}$, generated respectively by the base $\mathfrak{N}_w$ of weak neighborhoods (finite intersections of open "slabs"), the base $\mathfrak{N}$ of metric neighborhoods (i.e. the so-called open "balls"), the weak topology τ_w, and the metric topology τ. All four are equal when $\mathfrak{X}$ is separable, but in general

$$\mathfrak{B}_{\mathfrak{N}_w} \subseteq \mathfrak{B}_{\tau_w} \subseteq \mathfrak{B}_{\tau} \quad \& \quad \mathfrak{B}_{\mathfrak{N}} \subseteq \mathfrak{B}_{\tau_w} \subseteq \mathfrak{B}_{\tau}.$$

For the corresponding classes $\mathfrak{M}(\mathfrak{A},\mathfrak{B})$ of $\mathfrak{A},\mathfrak{B}$ measurable functions f in $\mathfrak{X}^\Lambda$, we find first that only $\mathfrak{M}(\mathfrak{A}, \mathfrak{B}_{\mathfrak{N}_w})$, which is easily seen to be the class of $\mathfrak{A}$ scalarly measurable f, is invariably a vector space. Secondly, we find (enlarging on Nedoma [13]) that when card $\mathfrak{X} > \underline{c}$, the other three spaces $\mathfrak{M}(\mathfrak{A},\mathfrak{B})$ can be non-linear. For non-separable $\mathfrak{X}$ of cardinality $\underline{c}$, the question of their linearity is still undecided (§2).

When $\mathfrak{X} = \mathfrak{H}$, a Hilbert space, the situation just described improves in three ways. First, because of the perfect roundness of the balls of $\mathfrak{H}$, all "slabs" in $\mathfrak{N}_W$ are countable unions of "lenses", i.e. intersection of balls in $\mathfrak{H}$, and consequently $\mathfrak{N}_W \subseteq \mathcal{B}_{\mathfrak{N}}$. This yields a single chain

$$\mathcal{B}_{\mathfrak{N}_W} \subseteq \mathcal{B}_{\mathfrak{N}} \subseteq \mathcal{B}_{\tau_W} \subseteq \mathcal{B}_{\tau},$$

and resulting simplifications. Secondly, each maximal orthonormal subset Λ of $\mathfrak{H}$ and the σ-algebra $\mathfrak{A}$ generated by its singletons, provide us with a measurable space $(\Lambda, \mathfrak{A})$ which is very useful for the study of the measurability spaces $\mathfrak{M}(\mathfrak{A}, \mathcal{B})$. We can show, for instance, that for a non-separable $\mathfrak{H}$ of cardinality $\underline{c}$, $\mathfrak{M}(\mathfrak{A}, \mathcal{B}_{\mathfrak{N}})$ can be nonlinear. Thirdly, the inner product of $\mathfrak{H}$ allows us to define a <u>correlation relation</u> ("corr") between certain pairs of functions $f, g \in \mathfrak{H}^{\Lambda}$ by:

$$f \text{ corr } g \quad \text{iff.} \quad (f(\cdot), g(\cdot))_{\mathfrak{H}} \text{ is } \mathfrak{A}\text{-measurable},$$

and to consider the <u>vector graph</u> $(\mathfrak{H}^{\Lambda}, \text{corr})$. These considerations in turn lead to the concept of <u>conditionally linear subspace</u> of $\mathfrak{H}^{\Lambda}$, and to the observation that the space $\mathfrak{M}(\mathfrak{A}, \mathcal{B}_{\mathfrak{N}})$ is conditionally linear (§§3,4).

For a Hilbert space $\mathfrak{H}$, functions f in $\mathfrak{M}(\mathfrak{A}, \mathcal{B}_{\mathfrak{N}})$ are $\mathfrak{A}$, scalarly measurable, and for them $|x(\cdot)|_{\mathfrak{H}}$ is $\mathfrak{A}$ measurable. This makes $\mathfrak{A}$, $\mathcal{B}_{\mathfrak{N}}$ measurability especially attractive for defining the integrability class $\mathcal{L}_{1,\mu}$ of the sort desired, and more generally the classes $\mathcal{L}_{p,\mu}$, $p \in [1, \infty]$. We show that each $\mathcal{L}_{p,\mu}$ is a <u>conditional Banach space</u> in a rather natural sense, and contains Bochner's $L_{p,\mu}$ as a genuine Banach subspace. Moreover, the Closed Graph Thm. in conjunction with that on the reflexivity of $\mathfrak{H}$ entails that every $f \in \mathcal{L}_{1,\mu}$ is Pettis integrable (§5), and immediately yields the Pettis integral

$E_\mu(f,A) = \int_A f(\lambda)\mu(d\lambda)$. We easily find that $E_\mu(\cdot,A)$ is a conditionally linear contraction on $\mathfrak{L}_{1,\mu}$ to $\mathfrak{X}$, and that $E_\mu(f,\cdot)$ is a $\mathfrak{X}$-valued countably additive measure of bounded variation on $\mathfrak{A}$ (§6).

In the applications we have in mind, the function $f(\cdot)$ on Λ to $\mathfrak{X}$ is the partial integral of a scalar-valued kernel $\phi(\lambda,\omega)$ with respect to a $\mathfrak{X}$-valued orthogonally scattered measure ξ over another space Ω, cf. [10], and it is important to be able to change the order of iterated integrations with respect to μ and ξ. In §7 we establish two such Fubini-type interchange theorems. In the first, Λ and Ω are abstract measure spaces and $\phi(\cdot\cdot)$ is measurable with respect to the product σ-algebra. This theorem may be regarded as a vectorial extension of the classical Fubini theorem. In the second theorem Λ and Ω are locally compact Hausdorff spaces, and $\phi(\cdot\cdot)$ on $\Lambda \times \Omega$ is Borel measurable but not necessarily measurable with respect to the product σ-algebra. This second theorem may be viewed as a vectorial extension of the Bourbaki version of the Fubini theorem, cf. [16, Ch.I, §9].

In the organization of the paper we have been guided by Halmos's dictums on measurability and measure [6, p.78], and the path issuing therefrom. No measure is introduced until the investigation of the four σ-algebras $\mathfrak{B}$ over $\mathfrak{X}$ and of the corresponding measurability spaces $\mathfrak{M}(\mathfrak{A},\mathfrak{B})$ is completed. Then a measure μ is brought in, but only to study the corresponding conditional Banach spaces $\mathfrak{L}_{p,\mu}$ (§5). Integration is defined last, as a linear operation on $\mathfrak{L}_{1,\mu}$ to $\mathfrak{X}$, with maximal use of operator theory (§6). Since several questions answered in this paper would not have even arisen had we veered from this path, the paper owes much to the remarks on p.78 of Halmos's book.*

---

* The path pursued is geared to the Lebesgue approach. There is an alternative approach, as or even more profound, associated with the names of Cauchy, Riemann, Frechet, G. Birkhoff and Phillips, in which measure and integration come first.

The paper also bears the impress of very useful conversations on vectorial issues with Professors I. Kluvanek and K. Lau in the early stages of the work*, and on irksome measurability problems with Professors S.D. Chatterji, J.P.R. Christensen and J.R. Choksi in the final stages. The latter conversations, which took place in the congenial surroundings of Lausanne and Oberwolfach, were made possible by invitations from the Ecole Polytechnique Federale de Lausanne and the Forschungsinstitut, Oberwolfach, for which I am most grateful to Professors S.D. Chatterji, D. Kolzow and A. Ionescu-Tulcea.

This paper contains a complete exposition of the theoretical development, short of proofs, as well as a list of unsettled questions. For want of time and space, the proofs have been deferred to an expanded version of the paper to appear elsewhere.

2. Borel Algebras and Measurability in Banach Spaces

The σ-algebras generated by a topology τ for a Banach space $\mathfrak{X}$ and by a neighborhood-base of τ will be different in general, as will be the σ-algebras generated by different topologies of $\mathfrak{X}$. Our interests are in the metric and weak topologies τ and τ_w of $\mathfrak{X}$ and in their neighborhood bases $\mathfrak{N}$ and $\mathfrak{N}_w$. To lay down these concepts clearly, and maintain the necessary distinctions, we shall adhere to the following notation.

2.1 Notation

(a) $\mathfrak{X}$, Y are Banach spaces over the field $\mathbb{F}$#,

---

* Professor Lau has since succeeded in obtaining Banach space extensions of some theorems in §§3,4.

\# In this paper $\mathbb{F}$ will refer to either the real number field $\mathbb{R}$ or the complex number field $\mathbb{C}$, and $\mathbb{N}$ to the set of all integers. $\mathbb{N}_+$, $\mathbb{R}_+$, and $\mathbb{N}_{o+}$, $\mathbb{R}_{o+}$ will denote the subsets of positive elements, and subsets of non-negative elements of $\mathbb{N}$ and $\mathbb{R}$.

$CL(\mathfrak{X},Y)$ is the Banach space of continuous linear operators on $\mathfrak{X}$ to Y;

(b) $\mathfrak{X}'$ is the dual of $\mathfrak{X}$, i.e. $\mathfrak{X}' \underset{d}{=} CL(\mathfrak{X},\mathbb{F})$;

(c) $\forall x_o \in \mathfrak{X}$ & $\forall r \in \mathbb{R}_{o+}$, $N(x_o,r) \underset{d}{=} \{x: x \in \mathfrak{X}$ & $|x-x_o| < r\}$,

$$\overline{N}(x_o,r) \underset{d}{=} \{x: x \in \mathfrak{X} \ \& \ |x-x_o| \leq r\};$$

(d) $\forall x_o \in \mathfrak{X}$, $\forall \Psi \in \mathfrak{X}'$, $\forall F \subseteq \mathfrak{X}'$ & $\forall r \in \mathbb{R}_{o+}$,

$$W(x_o,\Psi,r) \underset{d}{=} \Psi^{-1}[N\{\Psi(x_o), r\}],$$

$$\overline{W}(x_o,\Psi,r) \underset{d}{=} \Psi^{-1}[\overline{N}\{\Psi(x_o), r\}],$$

$$W(x_o,F,r) \underset{d}{=} \bigcap_{\Psi\in F} W(x_o,\Psi, r);$$

(e)
$$\mathfrak{N} \underset{d}{=} \{N(x_o,r): x_o \in \mathfrak{X} \ \& \ r \in \mathbb{R}_+\},$$

$$\mathfrak{N}_w^o \underset{d}{=} \{W(x_o,\Psi_o,r): x_o \in \mathfrak{X}, \ \Psi_o \in \mathfrak{X}', \ |\Psi_o| = 1 \ \& \ r \in \mathbb{R}_+\},$$

$$\mathfrak{N}_w \underset{d}{=} \{W(x_o,F,r): x_o \in \mathfrak{X}, \ F \subseteq \mathfrak{X}\sim\{0\}, \ F \text{ is finite} \ \& \ r \in \mathbb{R}_+\};$$

(f) $\tau \underset{d}{=} top(\mathfrak{N})$, $\tau_w \underset{d}{=} top(\mathfrak{N}_w)$, where $top(\mathfrak{F})$ is the topology generated by a family $\mathfrak{F}$ of subsets of $\mathfrak{X}$.

2.2 <u>Remarks</u>. (a) τ and τ_w are the metric and weak topologies for $\mathfrak{X}$, and $\mathfrak{N}$ and $\mathfrak{N}_w$ are neighborhood bases for τ and τ_w. Also, $\mathfrak{N}_w^o$ is a subbase for τ_w. Hence

$$\mathfrak{N}_w^o \subseteq \mathfrak{N}_w \subseteq \tau_w \subseteq \tau \quad \& \quad top(\mathfrak{N}_w^o) = top(\mathfrak{N}_w) = \tau_w.$$

(b) Let $x \in \mathfrak{X}$, $\Psi \in \mathfrak{X}'$ & $r \in \mathbb{R}_+$. Then $\overline{N}(x,r)$ is the τ-closure of $N(x,r)$:

$$\overline{N}(x,r) = cls.N(x,r).$$

Also
$$\overline{W}(x,\Psi,r) = \bigcap_{n=1}^{\infty} W(x,\Psi, r + 1/n).$$

The following lemma, a simple consequence of Mazur's Thm. [7, p.36], is needed:

2.3 <u>Lma</u>. $\forall x \in \mathfrak{X}$ & $\forall r \in \mathbb{R}_+$, $N(x,r)$ is in $\mathfrak{F}_\sigma$ of the weak topology τ_w.

We turn next to the four σ-algebras of interest:

(2.4) $\mathfrak{B}_{\mathfrak{N}_w} \underset{d}{=} \sigma\text{-alg}(\mathfrak{N}_w)$, $\mathfrak{B}_{\mathfrak{N}} \underset{d}{=} \sigma\text{-alg}(\mathfrak{N})$, $\mathfrak{B}_{\tau_w} \underset{d}{=} \sigma\text{-alg}(\tau_w)$, $\mathfrak{B}_{\tau} \underset{d}{=} \sigma\text{-alg}(\tau)$.

The following result is fundamental. The proof of part (a) rests on Lma.2.3; the proof of (b) hinges on the result [7, p.34, 2.8.5] that for separable $\mathfrak{X}$, $\mathfrak{X}'$ has a countable determining set, and thus ultimately depends on the Hahn-Banach Theorem.

2.5 Thm. (a) For any $\mathfrak{X}$, $\mathfrak{B}_{\mathfrak{N}_w}$ & $\mathfrak{B}_{\mathfrak{N}} \subseteq \mathfrak{B}_{\tau_w} \subseteq \mathfrak{B}_{\tau}$.

(b) For separable $\mathfrak{X}$, $\mathfrak{B}_{\mathfrak{N}_w} = \mathfrak{B}_{\mathfrak{N}} = \mathfrak{B}_{\tau_w} = \mathfrak{B}_{\tau}$.

The equality $\mathfrak{B}_{\mathfrak{N}_w} = \mathfrak{B}_{\mathfrak{N}}$ in Thm.2.5(b) is the core of the Pettis Measurability Theorem, as will be clear from 2.13 below. As for the converse of 2.5(b), we do not yet know if for an arbitrary Banach space $\mathfrak{X}$, the equality of two or more of $\mathfrak{B}_{\mathfrak{N}_w}$, $\mathfrak{B}_{\mathfrak{N}}$, $\mathfrak{B}_{\tau_w}$, $\mathfrak{B}_{\tau}$ entails the separability of $\mathfrak{X}$. Our knowledge for Hilbert spaces is much more definitive, cf. Thm.3.13.

We turn next to the study of the measurability of functions on a set Λ to $\mathfrak{X}$. We shall adhere to the following:

2.6 Notation. For $\emptyset \neq \mathfrak{A} \subseteq 2^{\Lambda}$ & $\emptyset \neq \mathfrak{B} \subseteq 2^{\mathfrak{X}}$,

$$\mathfrak{M}(\mathfrak{A},\mathfrak{B}) \underset{d}{=} \{f: f \in \mathfrak{X}^{\Lambda} \ \& \ \forall B \in \mathfrak{B},\ f^{-1}(B) \in \mathfrak{A}\},$$

i.e. $\mathfrak{M}(\mathfrak{A},\mathfrak{B})$ is the set of all $\mathfrak{A},\mathfrak{B}$ measurable functions on Λ to $\mathfrak{X}$.

From Thm.2.5, we at once infer:

2.7 Cor. (a) For any $\mathfrak{X}$,

$$\mathfrak{M}(\mathfrak{A}, \mathfrak{B}_{\tau}) \subseteq \mathfrak{M}(\mathfrak{A}, \mathfrak{B}_{\tau_w}) \subseteq \mathfrak{M}(\mathfrak{A}, \mathfrak{B}_{\mathfrak{N}}) \ \& \ \mathfrak{M}(\mathfrak{A}, \mathfrak{B}_{\mathfrak{N}_w}).$$

(b) For separable $\mathfrak{X}$, all four spaces are equal.

We leave it to the reader to verify the following very useful triviality:

2.8 Triv. Let $\mathfrak{A}$ be a σ-algebra over Λ. Then[#]

(a) $$\mathfrak{M}(\mathfrak{A}, \mathfrak{B}_{\mathfrak{N}_W}) = \{f: f \in \mathfrak{X}^\Lambda \ \& \ \forall \Phi \in \mathfrak{X}', \Phi \circ f \in \mathfrak{M}(\mathfrak{A}, \mathrm{Bl}(\mathbb{F}))\},$$

i.e. $\mathfrak{M}(\mathfrak{A}, \mathfrak{B}_{\mathfrak{N}_W})$ is the class of $\mathfrak{A}$, scalarly measurable* measurable functions on Λ to $\mathfrak{X}$;

(b) $$\mathfrak{M}(\mathfrak{A}, \mathfrak{B}_{\mathfrak{N}}) = \{f: f \in \mathfrak{X}^\Lambda \ \& \ \forall x_o \in \mathfrak{X}, |f(\cdot)-x_o|_{\mathfrak{X}} \in \mathfrak{M}(\mathfrak{A}, \mathrm{Bl}(\mathbb{R}))\},$$

i.e. $\mathfrak{M}(\mathfrak{A}, \mathfrak{B}_{\mathfrak{N}})$ is the class of functions on Λ to $\mathfrak{X}$, all displacements of which have $\mathfrak{A}$-measurable absolute values.

We turn next to the topological and algebraic structure of the classes $\mathfrak{M}(\mathfrak{A},\mathfrak{B})$. Our study of the topological structure rests on the concept of a τ-stratifiable σ-algebra and a lemma governing it:

2.9 Def. Let τ be a topology for a space $\mathfrak{X}$, and $\mathfrak{B}$ be a σ-algebra over $\mathfrak{X}$. We say that $\mathfrak{B}$ is τ-stratifiable, iff. $\exists\, \tau_o \subseteq \tau$ such that $\mathfrak{B} = \sigma\text{-alg}(\tau_o)$, and

$$\forall V \in \tau_o, \ \exists \{V_r\}_1^\infty \subseteq \tau_o \ \ni \ \bigcup_{r=1}^\infty \mathrm{cls}.V_r = V = \bigcup_{r=1}^\infty V_r.$$

2.10 Lma. Let (i) Λ be any set, and $\mathfrak{X}$ a Hausdorff space with topology τ, (ii) $\mathfrak{A}$ be any σ-algebra over Λ, and $\mathfrak{B}$ be a τ-stratifiable σ-algebra over $\mathfrak{X}$. Then $\mathfrak{M}(\mathfrak{A},\mathfrak{B})$ is τ-sequentially closed, i.e.

$$(f_n)_1^\infty \text{ is in } \mathfrak{M}(\mathfrak{A},\mathfrak{B}) \ \& \ f(\cdot) = \tau\lim_{n\to\infty} f_n(\cdot) \in \mathfrak{X}^\Lambda \ \Rightarrow \ f \in \mathfrak{M}(\mathfrak{A},\mathfrak{B}).$$

Reverting to the case of a Banach space $\mathfrak{X}$ with metric topology τ, it is easy to check that the σ-algebras $\mathfrak{B}_\tau$, $\mathfrak{B}_{\mathfrak{N}}$, $\mathfrak{B}_{\mathfrak{N}_W}$ are τ-stratifiable. This fact along with Lma.2.10 immediately yields the following theorem:

2.11 Thm. For any σ-algebra $\mathfrak{A}$ over Λ, the spaces $\mathfrak{M}(\mathfrak{A}, \mathfrak{B}_\tau)$, $\mathfrak{M}(\mathfrak{A}, \mathfrak{B}_{\mathfrak{N}})$, $\mathfrak{M}(\mathfrak{A}, \mathfrak{B}_{\mathfrak{N}_W})$ are τ-sequentially closed.

$\mathrm{Bl}(\mathbb{F}) \underset{d}{=} \sigma\text{-alg}\{\mathrm{top}(\mathbb{F})\}$ is the family of Borel subsets of the field $\mathbb{F}$.

* Confusingly termed "weakly measurable" in much of the literature.

An easy corollary of this is the following result on multiplication by scalar-valued measurable functions. Its proof rests on the fact that every ψ in $\mathfrak{M}(\mathfrak{A}, \mathrm{Bl}(\mathbb{F}))$ is the τ-limit of a sequence of $\mathfrak{A}$-simple functions on Λ to $\mathbb{F}$.

2.12 Cor. Let (i) $\mathfrak{A}$ be any σ-algebra over Λ, (ii) $\mathfrak{B}$ be a τ-stratifiable σ-algebra over $\mathfrak{X}$. Then

$$f \in \mathfrak{M}(\mathfrak{A},\mathfrak{B}) \ \&\ \psi \in \mathfrak{M}(\mathfrak{A}, \mathrm{Bl}(\mathbb{F})) \quad \Rightarrow \quad \psi(\cdot)f(\cdot) \in \mathfrak{M}(\mathfrak{A},\mathfrak{B}).$$

In particular, this implication holds for $\mathfrak{B} = \mathfrak{B}_\tau, \mathfrak{B}_{\mathfrak{N}}, \mathfrak{B}_{\mathfrak{N}_w}$.

A combination 2.7(b), 2.8 and 2.11 yields the following useful omnibus theorem on Bochner measurability given by Khalili [8].

2.13 Thm. (on Bochner measurability). For any σ-algebra $\mathfrak{A}$ over Λ, the following conditions on $f \in \mathfrak{X}^\Lambda$ are equivalent:

(α) $\exists$ a sequence $(s_n)_1^\infty$ of $\mathfrak{A}$-simple functions in $\mathfrak{X}^\Lambda$ such that $f(\cdot) = \tau\lim_{n\to\infty} s_n(\cdot)$

(β) f is $\mathfrak{A}$, Bochner measurable, i.e.

$$f \in \mathfrak{M}(\mathfrak{A}, \mathfrak{B}_\tau) \quad \& \quad \mathfrak{R}_f \underset{d}{=} \text{range } f \ \text{ is separable}$$

(γ) $\forall x \in \mathfrak{X}$, $|f(\cdot)-x|_{\mathfrak{X}} \in \mathfrak{M}(\mathfrak{A}, \mathrm{Bl}(\mathbb{R}))$ & $\mathfrak{R}_f$ is separable

(δ) f is $\mathfrak{A}$, scalarly measurable, cf. 2.8(a), & $\mathfrak{R}_f$ is separable

(ϵ) $\exists$ a sequence $(s_n)_1^\infty$ of $\mathfrak{A}$, σ-simple functions in $\mathfrak{X}^\Lambda \ni \forall n \geq 1$,

$$|s_n(\cdot)| \leq |f(\cdot)| \quad \& \quad |s_n(\cdot) - f(\cdot)| < 1/n \quad \text{on } \Lambda.$$

The (δ) $\Rightarrow$ (α) part of Thm.2.13, due originally to Pettis [14, p.278], is sometimes referred to as Pettis's Measurability Theorem. Our appeal to Cor.2.7(b) in proving it, shows that the core of this theorem lies in the equality $\mathfrak{B}_{\mathfrak{N}_w} = \mathfrak{B}_{\mathfrak{N}}$ for separable $\mathfrak{X}$, which is asserted in Thm.2.5(b).

To turn to the algebraic structure of the spaces $\mathfrak{M}(\mathfrak{A},\mathfrak{B})$, we first note the triviality that

(2.14) $\forall$ σ-algebras $\mathfrak{A}$, $\mathfrak{M}(\mathfrak{A}, \mathfrak{B}_{\mathfrak{N}_w})$ is a vector space over $\mathbb{F}$.

An immediate consequence of this, in view of 2.7(b), is that

$$(2.15)\begin{cases} \forall \text{ separable } \mathfrak{X} \ \& \ \forall\ \sigma\text{-algebras } \mathfrak{A},\ \mathfrak{M}(\mathfrak{A},\mathfrak{B}) \text{ is the same vector} \\ \text{space over } \mathbb{F} \text{ for } \mathfrak{B} = \mathfrak{B}_{\mathfrak{N}_W},\ \mathfrak{B}_{\mathfrak{N}},\ \mathfrak{B}_{\tau_W},\ \mathfrak{B}_{\tau}. \end{cases}$$

Every separable Banach space $\mathfrak{X}$ has cardinality $\underline{c} \underset{d}{=} 2^{\aleph_0}$. But there are non-separable $\mathfrak{X}$ also having cardinality $\underline{c}$. Nedoma [13] has shown that for non-separable $\mathfrak{X}$ of cardinality exceeding $\underline{c}$, there exist σ-algebras $\mathfrak{A}$ for which $\mathfrak{M}(\mathfrak{A}, \mathfrak{B}_\tau)$ is not a vector space. He does not say what happens when $\mathfrak{B}_\tau$ is replaced by $\mathfrak{B}_{\tau_W}$ or $\mathfrak{B}_{\mathfrak{N}}$. Actually, the lemma he uses can be adapted to prove the following extended version of his theorem:

2.16 <u>Thm.</u> (Nedoma) Let card $\mathfrak{X} > \underline{c} \underset{d}{=} 2^{\aleph_0}$. Then $\exists$ a σ-algebra $\mathfrak{A}$ over a set Λ, such that $\forall$ σ-algebras $\mathfrak{B}$ satisfying $\mathfrak{B}_{\mathfrak{N}} \subseteq \mathfrak{B} \subseteq 2^{\mathfrak{X}}$, $\mathfrak{M}(\mathfrak{A},\mathfrak{B})$ is not a vector space. In particular with this $\mathfrak{A}$, the spaces $\mathfrak{M}(\mathfrak{A}, 2^{\mathfrak{X}})$, $\mathfrak{M}(\mathfrak{A}, \mathfrak{B}_\tau)$, $\mathfrak{M}(\mathfrak{A}, \mathfrak{B}_{\tau_W})$, $\mathfrak{M}(\mathfrak{A}, \mathfrak{B}_{\mathfrak{N}})$ are non-linear, the first three being non-linear subspaces of the vector space $\mathfrak{M}(\mathfrak{A}, \mathfrak{B}_{\mathfrak{N}_W})$. More specifically, we let $\Lambda = \mathfrak{X}\mathrm{x}\mathfrak{X}$, $\mathfrak{A} \underset{d}{=} \sigma\text{-alg}(2^{\mathfrak{X}}\mathrm{x}2^{\mathfrak{X}})$, $p_i(x_1;x_2) = x_i$, $i = 1,2$; then $p_1(\cdot), p_2(\cdot) \in \mathfrak{M}(\mathfrak{A},\mathfrak{B})$ but $p_1(\cdot) + p_2(\cdot) \notin \mathfrak{M}(\mathfrak{A},\mathfrak{B})$.

This theorem leaves unsettled the question of the linearity of $\mathfrak{M}(\mathfrak{A},\mathfrak{B})$, where $\mathfrak{B}_{\mathfrak{N}} \subseteq \mathfrak{B} \subseteq 2^{\mathfrak{X}}$, for a non-separable Banach space $\mathfrak{X}$ of cardinality $\underline{c}$. In 4.3, we shall settle this question for $\mathfrak{B} = \mathfrak{B}_{\mathfrak{N}}$ and $\mathfrak{X}$ = a Hilbert space.

3. <u>Borel Algebras over a Hilbert Space</u>

In this section we shall show how two results in §2 governing the four σ-algebras over a Banach space $\mathfrak{X}$ strengthen when $\mathfrak{X} = \mathfrak{H}$.

We shall denote by $\mathfrak{H}$ a Hilbert space over the field $\mathbb{F}$. In view of the existence of the semi-linear isometry $u \to (\cdot,u)_{\mathfrak{H}}$ on $\mathfrak{H}$ onto $\mathfrak{H}'$, it is reasonable to write $W(x_o,u,r)$ in place of $W(x_o, (\cdot,u)_{\mathfrak{H}}, r)$ for the subbasic weak neighborhoods of $\mathfrak{H}$, cf. 2.1(d). Thus $\forall x_o, u \in \mathfrak{H}$ & $\forall r \in \mathbb{R}_{o+}$,

$$(3.1)\begin{cases} W(x_o,u,r) \underset{d}{=} \{x: x\in\mathfrak{H} \ \& \ |(x-x_o,u)_{\mathfrak{H}}| < r\} \\ \mathfrak{N}_w^o = \{W(x_o,u,r): x_o,u \in \mathfrak{H} \ \& \ |u| = 1 \ \& \ r \in \mathbb{R}_{o+}\}. \end{cases}$$

Our first objective is to exploit the roundness of the balls of $\mathfrak{H}$ in order to strengthen Thm.2.5(a). We claim that every subbasic weak neighborhood $W \underset{d}{=} W(x_o,u,r)$ is in $\mathfrak{B}_{\mathfrak{N}}$. This crucial feature of Hilbert spaces* is easy to visualize when $\mathbb{F}$ is $\mathbb{R}$. For cf. Fig., every such neighborhood W of an $\mathfrak{H}$ over $\mathbb{R}$ is an open "slab" bounded by a pair of parallel affine hyperplanes, and is therefore the union of a sequence of expanding spherical "lenses" L_n, i.e. of intersections of overlapping open balls of radii n which are tangential to the bounding hyperplanes of W at a fixed pair of opposite points. The formal statement of this feature, provable by an easy rigorization of the graphic evidence, is as follows:

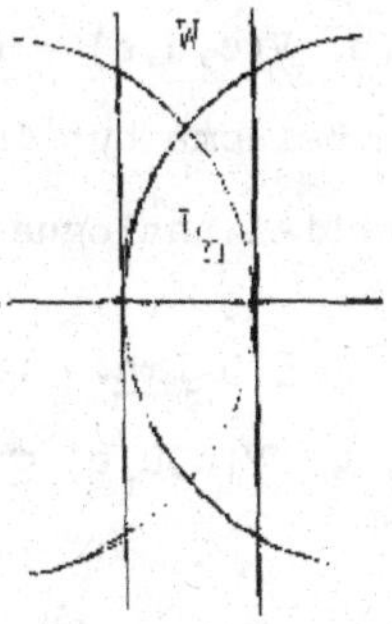

Fig.

3.2 <u>Lma</u>. Let $\mathfrak{H}$ be any Hilbert space over $\mathbb{R}$, and $x,u \in \mathfrak{H}$, $|u| = 1$ and $r \in \mathbb{R}_{o+}$. Then

$$W(x_o,u,r) = \bigcup_{n=1}^{\infty}[N(x_o-nu, n+r) \cap N(x_o+nu, n+r)] \in \mathfrak{B}_{\mathfrak{N}}.$$

The corresponding result when $\mathbb{F} = \mathbb{C}$ is more complicated. In addition to our subbasic weak neighborhoods W, which are Φ^{-1} images of open disks in $\mathbb{C}$, where $\Phi \in \mathfrak{H}'$, we have to consider the <u>Φ^{-1} images of open squares</u> in $\mathbb{C}$, and to express these <u>new</u> weak neighborhoods as countable unions of spherical "lenses", i.e. intersections of <u>four</u> balls in the (complex) $\mathfrak{H}$, along the lines of Lma.3.2. Let

---

* and presumably of other Banach spaces having "smooth" balls

$\forall c \in \mathbb{C}$ & $\forall \epsilon > 0$,

(3.3) $$S(c,\epsilon) = \{z: z \in \mathbb{C} \ \& \ -\epsilon < \mathrm{real}(z-c), \mathrm{imag}(z-c) < \epsilon\}$$

be the open square in $\mathbb{C}$ with centre c and edge length 2ϵ, and let

(3.4) $\forall c\in\mathbb{C}$, $\forall u\in\mathfrak{X}$ & $\forall \epsilon > 0$, $V(c,u,\epsilon) = \{x: x\in\mathfrak{X} \ \& \ (x,u) \in S(c,\epsilon)\}$

i.e. $V(c,u,\epsilon) = \Phi^{-1}\{S(c,\epsilon)\}$, where $\Phi(x) \underset{d}{=} (x,u)_{\mathfrak{X}}$. Then a somewhat cumbersome but intrinsically elementary argument yields the following partial analogue of 3.2 for the new weak neighborhoods V:

3.5 <u>Lma.</u> Let $c = a+ib \in \mathbb{C}$, $\epsilon > 0$ & $u \in \mathfrak{X}$ with $|u| = 1$. Then

$$V(c,u,\epsilon) \subseteq \bigcup_{n>|c|} [N\{-(n-a)u, n+\epsilon\} \cap N\{(n+a)u, n+\epsilon\} \cap N\{-(n-b)iu, n+\epsilon\} \cap N\{(n+b)iu, n+\epsilon\}] \subseteq V(c,u,\epsilon) \cup W(0,u, |c|+\epsilon).$$

Now by Lindelof's Thm. every open disk in $\mathbb{C}$ with centre 0 is a countable union of open squares:

(3.6) $$\forall r \in \mathbb{R}_+,\quad N(0,r) = \bigcup_{k=1}^{\infty} S(c_k,\epsilon_k),\quad c_k \in \mathbb{C} \ \& \ \epsilon_k > 0;$$

whence obviously

$$\forall u \in \mathfrak{X},\quad W(0,u,r) = \bigcup_{k=1}^{\infty} V(c_k,u,\epsilon_k).$$

Now apply Lma.3.5 to each $V(c_k,u,\epsilon_k)$ and observe that the resulting terms on the extreme right, viz. $V(c_k,u,\epsilon_k) \cup W(0,u, |c_k|+\epsilon)$ are all included in $W(0,u,r)$. On taking the union over $k \in \mathbb{N}_+$, we therefore get the equality for $W(0,u,r)$ given in the following lemma, which is the counterpart for $\mathbb{F} = \mathbb{C}$ of Lma.3.2:

3.7 <u>Lma.</u> Let $r \in \mathbb{R}_+$ and, cf. (3.6), $N(0,r) = \bigcup_{k=1}^{\infty} S(c_k,\epsilon_u)$, where $0 \in \mathbb{C}$, $c_k = a_k + ib_u \in \mathbb{C}$ and $\epsilon_k > 0$. Then $\forall u \in \mathfrak{X}$ with $|u| = 1$, we have

$$W(0,u,r) = \bigcup_{k=1} \bigcup_{n>|c_k|} [N\{-(n-a_k)u, n+\epsilon_k\} \cap N\{(n+a_k)u, n+\epsilon_k\} \cap N\{-(n-b_k)iu, n+\epsilon_k\} \cap N[(n+b_k)iu, n+\epsilon_k\}].$$

It follows from Lma.3.7 that $W(0,u,r) \in \mathfrak{B}_{\mathfrak{N}}$. Since $\mathfrak{B}_{\mathfrak{N}}$ is obviously translation invariant, it follows that $\forall x_o, u \in \mathfrak{H}$ with $|u| = 1$ and $\forall r \in \mathbb{R}_+$, $W(x_o,u,r) \in \mathfrak{B}_{\mathfrak{N}}$. This result for $\mathbb{F} = \mathbb{C}$, along with that in 3.2 for $\mathbb{F} = \mathbb{R}$, shows that invariably $\mathfrak{N}_w^o \subseteq \mathfrak{B}_{\mathfrak{N}}$. Since each neighborhood in $\mathfrak{N}_w$ is a finite intersection of neighborhoods in $\mathfrak{N}_w^o$, we see that $\mathfrak{N}_w \subseteq \mathfrak{B}_{\mathfrak{N}}$. This inclusion together with those in Thm.2.5 yields the following improvement of the latter:

3.8 Thm. (a) For any $\mathfrak{H}$, $\quad \mathfrak{B}_{\mathfrak{N}_w} \subseteq \mathfrak{B}_{\mathfrak{N}} \subseteq \mathfrak{B}_{\tau_w} \subseteq \mathfrak{B}_{\tau}$.

(b) For separable $\mathfrak{H}$, $\quad \mathfrak{B}_{\mathfrak{N}_w} = \mathfrak{B}_{\mathfrak{N}} = \mathfrak{B}_{\tau_w} = \mathfrak{B}_{\tau}$.

Our second objective in this section is to exploit the presence of orthonormal (o.n.) bases for $\mathfrak{H}$ to obtain a converse for Thm.3.8(b), and thereby settle for $\mathfrak{X} = \mathfrak{H}$ the questions asked apropos of Thm.2.5(b). For this we have to find for non-separable $\mathfrak{H}$, sets in $\mathfrak{B}_{\mathfrak{N}} \setminus \mathfrak{B}_{\mathfrak{N}_w}$, $\mathfrak{B}_{\tau_w} \setminus \mathfrak{B}_{\mathfrak{N}}$, $\mathfrak{B}_{\mathfrak{N}} \setminus \mathfrak{B}_{\mathfrak{N}_w}$. Some intermediate results are required. The entire argument hinges on the set-theoretic relationship between a given o.n. subset Λ of $\mathfrak{H}$ and the members of $\mathfrak{B}_{\mathfrak{N}_w}$, $\mathfrak{B}_{\mathfrak{N}}$ and τ_w, stated in the following lemma, the proof of which is routine:

3.9 Geometric Lma. Let (i) Λ be any o.n. subset of $\mathfrak{H}$, (ii) $x_o, y_o \in \mathfrak{H}$ & $r > 0$, (iii)

$$\forall x \in \mathfrak{H}, \quad \sigma(x) \underset{d}{=} \{\lambda: \lambda \in \Lambda \ \& \ (x,\lambda) \neq 0\}.$$

Then

(a) $\quad 0 \notin W(x_o,y_o,r) \Rightarrow \Lambda \cap W(x_o,y_o,r) \subseteq \sigma(y_o)$,

$\quad 0 \in W(x_o,y_o,r) \Rightarrow \Lambda \setminus W(x_o,y_o,r) \subseteq \sigma(y_o)$;

(b) $\quad 1+|x_o|^2-r^2 \geq 0 \Rightarrow \Lambda \cap N(x_o,r) \subseteq \sigma(x_o)$,

$\quad 1+|x_o|^2-r^2 < 0 \Rightarrow \Lambda \setminus N(x_o,r) \subseteq \sigma(x_o)$;

(c) $\quad \forall \Lambda_1, \Lambda_o \subseteq \Lambda, \quad \Lambda_1 \cap \bigcup_{\lambda \in \Lambda_o} W(\lambda,\lambda,1) = \Lambda_1 \cap \Lambda_o$.

The last lemma enables us to establish in a straightforward way the relationships that subsist between the subsets of a given maximal o.n. set Λ in $\mathfrak{H}$ and the members of the σ-algebras $\mathfrak{B}_{\mathfrak{N}_w}$, $\mathfrak{B}_{\mathfrak{N}}$. These are stated in the following very useful result:

3.10 Main Lma. Let (i) Λ be a maximal o.n. subset of $\mathfrak{H}$, (ii) $\mathfrak{A} \underset{d}{=} \sigma\text{-alg}\{\{\lambda\}: \lambda\in\Lambda\}$. Then

(a) $\forall A \subseteq \Lambda, \qquad A \in \mathfrak{A} \Leftrightarrow A$ or $\Lambda \setminus A$ is countable,

$\forall \Lambda_o \subseteq \Lambda \qquad 2^{\Lambda_o} \subseteq \mathfrak{A} \Leftrightarrow \Lambda_o$ is countable;

(b) $\forall \Lambda_o \subseteq \Lambda, \qquad 0 \notin B \in \mathfrak{B}_{\mathfrak{N}_w} \Rightarrow \Lambda_o \cap B \in \mathfrak{A}$,

$0 \in B \in \mathfrak{B}_{\mathfrak{N}_w} \Rightarrow \Lambda_o \setminus B \in \mathfrak{A}$;

(c) $\forall \Lambda_o \in \mathfrak{A}, \qquad B \in \mathfrak{B}_{\mathfrak{N}} \Rightarrow \Lambda_o \cap B \;\&\; \Lambda_o \setminus B \in \mathfrak{A}$.

3.11 Remark. In proving the last lemma we have to consider the functions f_{Λ_o} defined by $f_{\Lambda_o}(\lambda) \underset{d}{=} \chi_{\Lambda_o}(\lambda)\cdot\lambda$, where $\Lambda_o \subseteq \Lambda$ and $\lambda\in\Lambda$, and to observe that

(a) $$\Lambda_o \subseteq \Lambda \;\&\; 0 \in B \subseteq \mathfrak{H} \Rightarrow \Lambda \setminus f^{-1}_{\Lambda_o}(B) = \Lambda_o \setminus B$$

(b) $$\Lambda_o \in \mathfrak{A} \Rightarrow f_{\Lambda_o} \in \mathfrak{M}(\mathfrak{A}, \mathfrak{B}_{\mathfrak{N}}).$$

A closed subspace $\mathfrak{M}$ of $\mathfrak{H}$ is of course in $\mathfrak{B}_\tau$, and being also weakly closed, it is in $\mathfrak{B}_{\tau_w}$. The next theorem tells us exactly when $\mathfrak{M}$ will be in $\mathfrak{B}_{\mathfrak{N}_w}$ or $\mathfrak{B}_{\mathfrak{N}}$. Its proof rests on applying Lma.3.10 to a maximal o.n. subset Λ of $\mathfrak{H}$, chosen wisely in relation to $\mathfrak{M}$; dim $\mathfrak{M}$ and dim $\mathfrak{M}^\perp$ come in of course via the cardinality of subsets of Λ.

3.12 Thm. Let $\mathfrak{M}$ be a (closed linear) subspace of $\mathfrak{H}$. Then

(a) $$\mathfrak{M} \in \mathfrak{B}_{\mathfrak{N}} \Leftrightarrow \dim \mathfrak{M} \le \aleph_0 \text{ or } \dim \mathfrak{M}^\perp \le \aleph_0;$$

(b) $$\mathfrak{M} \in \mathfrak{B}_{\mathfrak{N}_w} \Leftrightarrow \dim \mathfrak{M}^\perp \le \aleph_0.$$

This theorem in conjunction with Lma.3.10 yields the following partial converse to Thm.3.8(b):

3.13 Thm. If $\mathfrak{B}_{\mathfrak{N}_W} = \mathfrak{B}_{\mathfrak{N}}$ or $\mathfrak{B}_{\mathfrak{N}} = \mathfrak{B}_{\tau_W}$, then $\mathfrak{H}$ is separable. Thus for any non-separable Hilbert space we have $\mathfrak{B}_{\mathfrak{N}_W} \subset \mathfrak{B}_{\mathfrak{N}} \subset \mathfrak{B}_{\tau_W} \subseteq \mathfrak{B}_{\tau}$.

This theorem fails for arbitrary Banach spaces. It also leaves open the question as to whether or not the equality $\mathfrak{B}_{\tau_W} = \mathfrak{B}_{\tau}$ entails the separability of $\mathfrak{H}$.

4. Measurability in Hilbert Spaces; the Vector Graph ($\mathfrak{H}^{\Lambda}$, Corr)

We shall now study the spaces $\mathfrak{M}(\mathfrak{A},\mathfrak{B})$ of $\mathfrak{A},\mathfrak{B}$ measurable functions on a set Λ to a Hilbert space $\mathfrak{H}$, where $\mathfrak{A}$ is a σ-algebra over Λ and $\mathfrak{B}$ is one of the four σ-algebras over $\mathfrak{H}$ considered in §§2,3, our objective being to strengthen the results of §2.

In view of Thm.3.8, the result corresponding to Cor.2.7 improves to:

4.1 Thm. (a) For any Hilbert space $\mathfrak{H}$,

$$\mathfrak{M}(\mathfrak{A}, \mathfrak{B}_{\tau}) \subseteq \mathfrak{M}(\mathfrak{A}, \mathfrak{B}_{\tau_W}) \subseteq \mathfrak{M}(\mathfrak{A}, \mathfrak{B}_{\mathfrak{N}}) \subseteq \mathfrak{M}(\mathfrak{A}, \mathfrak{B}_{\mathfrak{N}_W}).$$

(b) For separable $\mathfrak{H}$, all four spaces are equal.

We turn next to the improvement for $\mathfrak{X} = \mathfrak{H}$ of our version 2.16 of Nedoma's result. It is convenient first to state a completed form of the result 3.11(b):

4.2 Triv. Let (i) $\mathfrak{H}$ be a Hilbert space over $\mathbb{F}$, (ii) Λ & $\mathfrak{A}$ be as in 3.10(i),(ii), and (iii) $\forall \Lambda_o \subseteq \Lambda$, $f_{\Lambda_o}(\lambda) \underset{d}{=} \chi_{\Lambda_o}(\lambda)\cdot\lambda$, $\lambda \in \Lambda$. Then

$$f_{\Lambda_o} \in \mathfrak{M}(\mathfrak{A}, \mathfrak{B}_{\mathfrak{N}}) \iff \Lambda_o \in \mathfrak{A}.$$

4.3 Thm. Let (i) $\mathfrak{H}$ be any non-separable Hilbert space over $\mathbb{F}$, (ii) Λ be a maximal o.n. subset of $\mathfrak{H}$, (iii) $\mathfrak{A} = \sigma\text{-alg}\{\{\lambda\}: \lambda \in \Lambda\}$. Then $\mathfrak{M}(\mathfrak{A}, \mathfrak{B}_{\mathfrak{N}})$ is not a vector space. More specifically, with f_{Λ_o} as in 4.2(iii) and $g_{\Lambda_o} \underset{d}{=} 2f_{\Lambda_o} - f_{\Lambda}$, we have $\forall \Lambda_o \in 2^{\Lambda} \sim \mathfrak{A}$,

$$f_{\Lambda} \ \& \ g_{\Lambda_o} \in \mathfrak{M}(\mathfrak{A}, \mathfrak{B}_{\mathfrak{N}}) \quad \text{but} \quad f_{\Lambda} + g_{\Lambda_o} \notin \mathfrak{M}(\mathfrak{A}, \mathfrak{B}_{\mathfrak{N}}).$$

The last theorem suggests the following somewhat more general result, which lays down a sufficient condition on $\mathfrak{A}$ in order that $\mathfrak{M}(\mathfrak{A}, \mathfrak{B}_{\mathfrak{N}})$ be non-linear for any non-separable $\mathfrak{H}$. We leave its proof to the reader.

4.4 Thm. Let (i) $\mathfrak{H}$ be any non-separable Hilbert space over $\mathbb{F}$, (ii) $\mathfrak{A}$ be a σ-algebra over a set Λ such that $\forall\lambda \in \Lambda$, $\{\lambda\} \in \mathfrak{A} \subset 2^{\Lambda}$. Then $\mathfrak{M}(\mathfrak{A}, \mathfrak{B}_{\mathfrak{N}})$ is not a vector space.

Note that although Thms.4.3, 4.4 are improvements of Thm.2.16, they leave unanswered the question of the linearity of the spaces $\mathfrak{M}(\mathfrak{A}, \mathfrak{B}_{\tau})$ and $\mathfrak{M}(\mathfrak{A}, \mathfrak{B}_{\tau_W})$ for non-separable $\mathfrak{H}$ of cardinality $\underline{c}$.

We shall now exploit the existence of an inner product in $\mathfrak{H}$ to introduce a relation corr from the vector space $\mathfrak{H}^{\Lambda}$ to itself, i.e. corr $\subseteq \mathfrak{H}^{\Lambda} \times \mathfrak{H}^{\Lambda}$, and to consider $(\mathfrak{H}^{\Lambda}$, corr) as a graph (in the modern sense [1]). We shall then show that the space $\mathfrak{M}(\mathfrak{A}, \mathfrak{B}_{\mathfrak{N}})$, which in general is a non-linear subset of the vector space $\mathfrak{H}^{\Lambda}$, is always "conditionally linear" in the vector graph $(\mathfrak{H}^{\Lambda}$, corr) in a rather natural sense.

4.5 Def. Let (i) $\mathfrak{H}$ be any Hilbert space over $\mathbb{F}$, (ii) $\mathfrak{A}$ be a σ-algebra over a set Λ, (iii) $f,g \in \mathfrak{H}^{\Lambda}$ and $\emptyset \neq \mathfrak{F},\mathfrak{G} \subseteq \mathfrak{H}^{\Lambda}$. Then by definition

(a) f is correlated (more fully, $\mathfrak{A}$-correlated) to g, in symbols
f corr g, iff. $(f(\cdot), g(\cdot))_{\mathfrak{H}} \in \mathfrak{M}(\mathfrak{A}, \mathrm{Bl}(\mathbb{F}))$;

(b) $\mathfrak{F}$ corr $\mathfrak{G}$ $\underset{d}{\Leftrightarrow}$ $\forall f \in \mathfrak{F}$ & $\forall g \in \mathfrak{G}$, f corr g;

(c) $\mathfrak{F}$ is self-correlated $\Leftrightarrow$ $\mathfrak{F}$ corr $\mathfrak{F}$;

(d) $\mathfrak{F}^{c} \underset{d}{=} \mathfrak{F}^{\mathrm{corr}} \underset{d}{=} \{g: g \in \mathfrak{H}^{\Lambda}$ & $\{g\}$ corr $\mathfrak{F}\} \underset{d}{=}$ the corona of $\mathfrak{F}$;

(e) $\mathfrak{S} \underset{d}{=} \{f: f \in \mathfrak{H}^{\Lambda}$ & f corr f$\}$.

From the sesquilinearity and continuity of the inner product in

$\mathfrak{H}$, and the linearity and sequential closure of $\mathfrak{M}(\mathfrak{U}, \mathrm{Bl}(\mathbb{F}))$, we easily get the following result:

4.6 Thm. ($\mathfrak{H}^\Lambda$, corr) is a convergence vector graph over $\mathbb{F}$ under τ-sequential convergence, ordinary addition and scalar multiplication, and the relation corr. More fully,

(a) $\mathfrak{H}^\Lambda$ is a τ-sequentially closed vector space over $\mathbb{F}$;

(b) corr is a symmetric relation, with $\mathfrak{H}^\Lambda$ as domain and range;

(c) $\forall f_1, f_2, g \in \mathfrak{H}^\Lambda$ & $\forall c_1, c_2 \in \mathbb{F}$,

$$f_1 \text{ corr } g \ \& \ f_2 \text{ corr } g \Rightarrow c_1 f_1 + c_2 f_2 \text{ corr } g;$$

(d) $\forall f_n, g_n, f, g \in \mathfrak{H}^\Lambda$,

$$\forall n \geq 1,\ f_n \text{ corr } g_n,\ f = \tau\lim_{n\to\infty} f_n \ \&\ g = \tau\lim_{n\to\infty} g_n \Rightarrow f \text{ corr } g.$$

For vector graphs the following concept is very natural and in many instances very useful:

4.7 Def. (a) $\mathfrak{M}$ is called a conditionally linear manifold (briefly, cond. lin. mnfd.) of $\mathfrak{H}^\Lambda$, iff.

$$\emptyset \neq \mathfrak{M} \subseteq \mathfrak{H}^\Lambda; \qquad \forall c \in \mathbb{F},\ c\cdot\mathfrak{M} \subseteq \mathfrak{M}$$

$$f, g \in \mathfrak{M} \ \&\ f \text{ corr } g \Rightarrow f+g \in \mathfrak{M}.$$

(b) $\mathfrak{M}$ is called a conditionally linear subspace of $\mathfrak{H}^\Lambda$, iff. $\mathfrak{M}$ is a cond. lin. mnfd. of $\mathfrak{H}^\Lambda$, and $\mathfrak{M}$ is τ-sequentially closed.

The following result does not depend on the specific definition 4.5(a) of corr, but only on the attributes 4.6(b)-(d). It accordingly holds for all convergence vector graphs satisfying the postulates 4.6(b)-(d), cf. [9]. Its proof is straightforward.

4.8 Triv.

(a)# $\emptyset \neq \mathfrak{J}, \mathfrak{Q} \subseteq \mathfrak{H}^\Lambda$ & $\mathfrak{J}$ corr $\mathfrak{Q}$ $\Rightarrow$ $\mathfrak{S}(\mathfrak{J})$ corr $\mathfrak{S}(\mathfrak{Q})$.

$\mathfrak{S}(\mathfrak{J}) \underset{d}{=}$ the least closed linear subspace spanned by $\mathfrak{J}$.

(b) $\forall \mathfrak{M} \subseteq \mathfrak{H}^{\Lambda}$, $\mathfrak{M}$ is cond. lin. & $\mathfrak{M}$ corr $\mathfrak{M}$ $\Rightarrow$ $\mathfrak{M}$ is a lin. mnfd.

(c) $\mathfrak{S}$ is conditionally linear.

(d) $\emptyset \neq \mathfrak{J} \subseteq \mathfrak{H}^{\Lambda}$ $\Rightarrow$ $\mathfrak{J}^{c}$ is a closed linear subspace of $\mathfrak{H}^{\Lambda}$.

We now turn to results which depend on the specific definition of corr given in 4.5(a). The notation for the class of constant-valued functions:

$$(4.9) \qquad \underline{\mathfrak{H}} \underset{d}{=} \{f: f \in \mathfrak{H}^{\Lambda} \ \& \ f(\cdot) = y, \ y \in \mathfrak{H}\}$$

is rather convenient for the formulation of our results, which follow easily from 2.8(a) and 4.1(a).

4.10 Thm. (a) $\mathfrak{S} = \{f: f \in \mathfrak{H}^{\Lambda} \ \& \ |f(\cdot)|_{\mathfrak{H}} \in \mathfrak{M}(\mathfrak{A}, \mathrm{Bl}(\mathbb{R}))\}$.

(b) $\underline{\mathfrak{H}}^{c} = \mathfrak{M}(\mathfrak{A}, \mathfrak{B}_{\mathfrak{N}_{W}})$.

(c) $\mathfrak{S} \cap \underline{\mathfrak{H}}^{c} = \mathfrak{M}(\mathfrak{A}, \mathfrak{B}_{\mathfrak{N}})$.

Now by 4.8(c),(d), $\mathfrak{S}$ is conditionally linear, and $\underline{\mathfrak{H}}^{c}$ is linear. Hence obviously $\mathfrak{S} \cap \underline{\mathfrak{H}}^{c}$ is conditionally linear, i.e. by 4.10(a) we have

(4.11) $\forall$ σ-algebras $\mathfrak{A}$ over Λ, $\mathfrak{M}(\mathfrak{A}, \mathfrak{B}_{\mathfrak{N}})$ is conditionally linear.

It is not known if the non-linear spaces $\mathfrak{M}(\mathfrak{A}, \mathfrak{B}_{\tau_W})$, $\mathfrak{M}(\mathfrak{A}, \mathfrak{B}_{\tau})$ are also conditionally linear.

From Thm. 2.16 we know that if card $\mathfrak{X} > \underline{c}$, then $p_1(\cdot)$ & $p_2(\cdot) \in \mathfrak{M}(\mathfrak{A}, \mathfrak{B}_{\mathfrak{N}})$ but $p_1(\cdot) + p_2(\cdot) \notin \mathfrak{M}(\mathfrak{A}, \mathfrak{B}_{\mathfrak{N}})$. For $\mathfrak{X} = \mathfrak{H}$ it follows from (4.11) and 4.8(b) that $p_1(\cdot)$ is not correlated to $p_2(\cdot)$, i.e. $(p_1(\cdot), p_2(\cdot))_{\mathfrak{H}} \notin \mathfrak{M}(\mathfrak{A}, \mathrm{Bl}(\mathbb{F}))$. But

$$\forall x_1, x_2 \in \mathfrak{H}, \quad (p_1(x_1;x_2), p_2(x_1;x_2))_{\mathfrak{H}} = (x_1, x_2)_{\mathfrak{H}}.$$

We thus arrive at the following conclusion:

4.12 Cor. Let card $\mathfrak{H} > \underline{c} = 2^{\aleph_0}$. Then the inner product $(\cdot,-)_{\mathfrak{H}}$ on $\mathfrak{H} \times \mathfrak{H}$ to $\mathbb{F}$ is not $\sigma\text{-alg}(2^{\mathfrak{H}} \times 2^{\mathfrak{H}})$, $\mathrm{Bl}(\mathbb{F})$ measurable.

5. Integrability in $\mathfrak{H}$; the Spaces $\mathfrak{L}_{p,\mu}$

In this section we shall introduce a non-negative measure μ on $\mathfrak{A}$, and study the resulting integrability concepts, especially the classes $\mathfrak{L}_{p,\mu}$ of $\mathfrak{A},\mathfrak{B}_{\mathfrak{H}}$ measurable functions f on Λ to $\mathfrak{H}$ for which $|f(\cdot)|_{\mathfrak{H}}^p$ is integrable. We shall adopt the following notation:

5.1 Notation

(i) $\mathfrak{H}$ is a Hilbert space over $\mathbb{F}$

(ii) $\mathfrak{A}$ is a σ-algebra over a set Λ

(iii) μ is a c.a. measure on $\mathfrak{A}$ to $[0,\infty]$

(iv) $\mathfrak{A}_\mu = \{A: A \in \mathfrak{A} \ \& \ \mu(A) < \infty\}$

(v) $\mathfrak{A}_\mu^{loc} = \{C: C \subseteq \Lambda \ \& \ \forall A \in \mathfrak{A}_\mu, A \cap C \in \mathfrak{A}\}$.

Obviously $\mathfrak{A}_\mu$ is a δ-ring and $\mathfrak{A}_\mu^{loc}$ a σ-algebra, and

(5.2) $$\mathfrak{A}_\mu \subseteq \sigma\text{-ring}(\mathfrak{A}_\mu) \subseteq \sigma\text{-alg}(\mathfrak{A}_\mu) \subseteq \mathfrak{A} \subseteq \mathfrak{A}_\mu^{loc}.$$

Let us recall the definition of Pettis integrability for Banach spaces $\mathfrak{X}$, cf. [7, p.77]:

5.3 Def. Let $\mathfrak{X}$ be a Banach space over $\mathbb{F}$, $f \in \mathfrak{X}^\Lambda$, $C \in \mathfrak{A}_\mu^{loc}$ and $\mathcal{C} \subseteq \mathfrak{A}_\mu^{loc}$. We say that

(a) f is Pettis integrable on C with respect to μ, iff. $\exists x_C \in \mathfrak{X} \ni \forall \Phi \in \mathfrak{X}'$, $\Phi \circ f \in L_1(\Lambda,\mathfrak{A},\mu;\mathbb{F})$ & $\Phi(x_C) = \int_C \Phi\{f(\lambda)\}\mu(d\lambda)$,#

(b) f is Pettis integrable over $\mathcal{C}$ with respect to μ, iff. $\forall C \in \mathcal{C}$, f is Pettis integrable on C with respect to μ,

(c) $\mathcal{P}_{1,\mu} = \{f: f \in \mathfrak{X}^\Lambda$ & f is Pettis integrable over $\mathfrak{A}$ with respect to $\mu\}$.

\# Obviously, this x_C is unique.

For the reasons given in §1 we seek the class $\mathcal{L}_{1,\mu}$ of functions $f(\cdot)$ on Λ to $\mathfrak{H}$, which are Pettis integrable with respect to μ, and for which $|f(\cdot)|_{\mathfrak{H}}$ is $\mathfrak{A}$, Bl(IR) measurable and $\int_\Lambda |f(\lambda)|_{\mathfrak{H}}\, \mu(d\lambda) < \infty$. By Def.5.3 such an f is of course $\mathfrak{A}$, scalarly measurable, and hence by Triv.2.8(a) and Thm.4.10(a) it must belong to $\underline{\mathfrak{H}}^c$. Also by Thm.4.10 (b), f must belong to $\mathfrak{S}$. Thus $f \in \underline{\mathfrak{H}}^c \cap \mathfrak{S}$, i.e. by Thm.4.10(c), $f \in \mathfrak{M}(\mathfrak{A}, \mathfrak{B}_{\mathfrak{N}})$. It turns out that any such f for which $\int_\Lambda |f(\lambda)|_{\mathfrak{H}}\mu(d\lambda) < \infty$ is automatically in $\mathcal{P}_{1,\mu}$, cf. 5.19 below. We are thus led to accept $\mathfrak{A},\mathfrak{B}_{\mathfrak{N}}$ measurability as the pertinent measurability concept for defining the class $\mathcal{L}_{1,\mu}$, and more generally the classes $\mathcal{L}_{p,\mu}$ for $p \in [1,\infty]$. We shall state this definition for any Banach space $\mathfrak{X}$, even though its motivation comes from Hilbert spaces $\mathfrak{H}$, and it is only with $\mathfrak{H}$ that we shall be concerned:#

5.4 <u>Def.</u> Let $\mathfrak{X}$ be a Banach space over IF and $f \in \mathfrak{X}^\Lambda$. Then (a), for $f \in \mathfrak{M}(\mathfrak{A}, \mathfrak{B}_{\mathfrak{N}})$ we define

$$\forall p \in \mathbb{R}_+, \quad |f|_{p,\mu} \underset{d}{=} \{\int_\Lambda |f(\lambda)|^p_{\mathfrak{H}}\mu(d\lambda)\}^{1/p} \in [0,\infty],$$

$$|f|_{\infty,\mu} \underset{d}{=} \mu\text{-ess.lub } |f(\cdot)|_{\mathfrak{H}} \in [0,\infty],$$

(b) for $p \in [0,\infty]$,

$$L_p(\Lambda,\mathfrak{A},\mu;\mathfrak{X},\mathfrak{B}_{\mathfrak{N}}) \underset{d}{=} \{f: f \in \mathfrak{M}(\mathfrak{A}, \mathfrak{B}_{\mathfrak{N}}) \ \& \ |f|_{p,\mu} < \infty\}.$$

The abbreviations $\mathcal{L}_{p,\mu}$, $\mathcal{L}_p$ will be used for the LHS.

Some simple properties of functions in $\mathcal{L}_{p,\mu}$ in the case $\mathfrak{X} = \mathfrak{H}$ are listed in the following result:

5.5 <u>Triv.</u> Let $f \in \mathcal{L}_{p,\mu}$, where $p \in \mathbb{R}_+$. Then

---

G.B. Price [15] seems to have been the first to consider $\mathfrak{A},\mathfrak{B}_{\mathfrak{N}}$ measurability, as appropriate for the purposes of integration of $\mathfrak{X}$-valued functions.

(a) $f(\cdot)$ is $\mathfrak{A}$, scalarly measurable;

(b) supp $f(\cdot) \in \sigma\text{-ring}(\mathfrak{A}_\mu)$;

(c) $\forall\phi \in \mathfrak{M}(\mathfrak{A}_\mu^{loc}, Bl(\mathbb{F}))$, $\quad \phi(\cdot)f(\cdot) \in \mathfrak{M}(\mathfrak{A}, \mathcal{B}_\mathfrak{N})$;

(d) $\forall C \in \mathfrak{A}_\mu^{loc}$, $\quad \chi_C(\cdot)f(\cdot) \in \mathfrak{L}_{p,\mu}$ & $|f(\cdot)\chi_C(\cdot)|_{p,\mu} \leq |f|_{p,\mu}$.

By the usual classical argument we can show that

$$f,g \;\&\; f+g \in \mathfrak{M}(\mathfrak{A}, \mathcal{B}_\mathfrak{N})$$
$$(5.6) \qquad \Rightarrow \forall p \in [1,\infty], \; |f+g|_{p,\mu} \leq |f|_{p,\mu} + |g|_{p,\mu}.$$

This Minkowski inequality together with (4.11) shows that

(5.7) $\forall p \in [1,\infty]$, $\mathfrak{L}_{p,\mu}$ is a conditionally linear subspace of $\mathfrak{M}(\mathfrak{A}, \mathcal{B}_\mathfrak{N})$.

Moreover, with a slightly widened concept of "Cauchy sequence", $\mathfrak{L}_{p,\mu}$ is topologically complete. This emerges from the following analogue of the classical lemma of Weyl, provable by essentially classical argumentation:

5.8 <u>Lma.</u> Let $(f_n)_1^\infty$ be a "Cauchy sequence" in the non-linear space $\mathfrak{L}_{p,\mu}$, where $p \in [1,\infty)$, in the sense that

$$\forall m,n \geq 1, \quad f_m \;\&\; f_m - f_n \in \mathfrak{L}_{p,\mu} \quad \& \quad \lim_{m,n\to\infty} |f_m - f_n|_{p,\mu} = 0.$$

Then $\exists$ a subsequence $(f_{n_k})_{k=1}^\infty$ of $(f_n)_1^\infty$, and $\exists\, f \in \mathfrak{M}(\mathfrak{A}, \mathcal{B}_\mathfrak{N})$

$$f(\lambda) = \tau\lim_{k\to\infty} f_{n_k}(\lambda), \text{ a.e. } \mu \quad \& \quad f - f_n \in \mathfrak{M}(\mathfrak{A}, \mathcal{B}_\mathfrak{N}).$$

Now let $p \in [1,\infty)$ and f_n, f, f_{n_k} be as in Lma. 5.8. Then 5.8 and a simple application of Fatou's Lma. yield

$$\forall k \geq 1, \quad f - f_k \in \mathfrak{M}(\mathfrak{A}, \mathcal{B}_\mathfrak{N}) \quad \& \quad |f - f_{n_k}|_{p,\mu} < \infty.$$

From this we easily conclude that f & $f - f_k \in \mathfrak{L}_{p,\mu}$, and furthermore that $|f - f_k|_{p,\mu} \to 0$, as $k \to \infty$. Thus, our "Cauchy sequence" has a limit in $\mathfrak{L}_{p,\mu}$. For $p = \infty$, the same conclusion is reached by an even simpler argument from the $p = \infty$ version of Lma. 5.8 which we leave to the reader to formulate. We can thus establish the following result:

5.9 Thm. (Riesz-Fischer). For $p \in [1,\infty]$, $\mathfrak{L}_{p,\mu}$ is a complete, conditionally linear subspace of $\mathfrak{M}(\mathfrak{A}, \mathfrak{B}_{\mathfrak{N}})$.

In order to get a genuinely normed (non-linear) space, we have to supplant functions f in $\aleph^{\Lambda}$ by their <u>μ-equivalence classes $\dot{f}$</u>:

(5.10) $\forall f \in \aleph^{\Lambda}, \quad \dot{f} \underset{d}{=} \{g: g \in \aleph^{\Lambda} \ \& \ \text{supp}\,(g-f) \text{ is } \mu\text{-negligible}\}$.#

Note that μ is not assumed to be complete; consequently we may have $f \in \mathfrak{M}(\mathfrak{A}, \mathfrak{B}_{\mathfrak{N}})$, $g \in \dot{f}$, but $g \notin \mathfrak{M}(\mathfrak{A}, \mathfrak{B}_{\mathfrak{N}})$. As with $\mathbb{F}$-valued functions, we define $\dot{f}+\dot{g}$, $c\dot{f}$ and $|\dot{f}|_{\mu,p}$; furthermore we say that

(5.11) $$\dot{f} \text{ corr } \dot{g} \ \underset{d}{\Leftrightarrow} \ \exists f_1 \in \dot{f} \ \& \ \exists g_1 \in \dot{g} \ \ni \ f_1 \text{ corr } g_1.$$

In place of $\mathfrak{L}_{p,\mu}$ we now consider:

(5.12) $$\dot{\mathfrak{L}}_{p,\mu} \underset{d}{=} \{\dot{f}: f \in \mathfrak{L}_{p,\mu}\}.$$

Using the term Cauchy sequence in $\dot{\mathfrak{L}}_{p,\mu}$ in the sense of Lma.5.8, we can restate the last theorem without terminological abuse as follows:

5.13 Thm. (Riesz-Fischer). For $p \in [1,\infty]$, $\dot{\mathfrak{L}}_{p,\mu}$ is a conditional Banach space over $\mathbb{F}$ under the norm $|\ |_{p,\mu}$. More fully, $\forall \dot{f},\dot{g} \in \dot{\mathfrak{L}}_{p,\mu}$ & $\forall c \in \mathbb{F}$

$$|\dot{f}|_{p,\mu} = 0 \ \Rightarrow \ \dot{f} = \dot{0}$$

$$c\dot{f} \in \dot{\mathfrak{L}}_{p,\mu} \ \& \ |c\dot{f}|_{p,\mu} = |c|\,|\dot{f}|_{p,\mu}$$

$$\dot{f} \text{ corr } \dot{g} \ \Rightarrow \ \dot{f}+\dot{g} \in \dot{\mathfrak{L}}_{p,\mu} \ \& \ |\dot{f}+\dot{g}|_{p,\mu} \le |\dot{f}|_{p,\mu} + |\dot{g}|_{p,\mu}$$

$$(\dot{f}_n)_1^{\infty} \text{ is Cauchy in } \dot{\mathfrak{L}}_{p,\mu} \ \Rightarrow \ \exists \dot{f} \in \dot{\mathfrak{L}}_{p,\mu} \ \ni \ \lim_{n\to\infty} |\dot{f}_n - \dot{f}|_{p,\mu} = 0.$$

The corresponding <u>Bochner classes</u> are defined by

(5.14) $$\begin{cases} L_{p,\mu} = \{f: f \in \aleph^{\Lambda} \ \& \ f \text{ is } \mathfrak{A}, \text{ Bochner measurable } \& \ |f|_{p,\mu} < \infty\} \\ \dot{L}_{p,\mu} = \{\dot{f}: f \in L_{p,\mu}\}. \end{cases}$$

---

i.e. $S \underset{d}{=} \text{supp}\,(f-g) \in \mathfrak{A}$ & $\mu(S) = 0$.

Note that if $g \in \dot{f} \in \dot{L}_{p,\mu}$, then Range g may not be separable, but it is μ-essentially separable, i.e. $g(\Lambda \setminus N)$ is separable where N is μ-negligible. The facts (cf. [7, pp.88-89]) that for $p \in [1,\infty]$, $L_{p,\mu}$ is a Banach space over $\mathbb{F}$, and (cf. Thm.2.13(β), Cor.2.7(a)) that any $\mathfrak{A}$, Bochner measurable f is in $\mathfrak{M}(\mathfrak{A}, \mathfrak{B}_{\mathfrak{N}})$ at once yield the following:

5.15 Cor. (a) For $p \in [1,\infty]$, $\dot{L}_{p,\mu}$ is a closed linear subspace of the conditional Banach space $\dot{\mathfrak{L}}_{p,\mu}$.

(b) For separable $\mathfrak{H}$, $\dot{L}_{p,\mu} = \dot{\mathfrak{L}}_{p,\mu}$.

Even for non-separable $\mathfrak{H}$ it may transpire that $L_{p,\mu} = \mathfrak{L}_{p,\mu}$. The following result shows that this accident cannot occur if the measure space $(\Lambda,\mathfrak{A},\mu)$ is "sufficiently non-atomic". Its proof is left to the reader.

5.16 Thm. Let (i) $\mathfrak{H}$ be any non-separable Hilbert space, (ii) Λ be uncountable, (iii) $\mathfrak{A}$ contain all singletons, (iv) μ be not purely-atomic and (v) $\exists \phi \in L_1(\Lambda,\mathfrak{A},\mu;\mathbb{R}_{0+})$ such that

$$\text{card}(\text{supp}\ \phi) \leq \dim \mathfrak{H} \quad \& \quad \mu_c(\text{supp}\ \phi) > 0,$$

where μ_c is the continuous part of μ. Then $L_{p,\mu} \subset \mathfrak{L}_{p,\mu}$.

To understand the relationship between the classes $\mathfrak{L}_{1,\mu}$ and $\mathcal{P}_{1,\mu}$ for a Hilbert space it is revealing to consider an arbitrary Banach space $\mathfrak{X}$ and any $p \in [1,\infty]$, and to ask for the relationship between $\mathfrak{L}_{p,\mu}$ and the class of "scalarly L_p" functions on Λ to $\mathfrak{X}$:

$$(5.17) \qquad \mathfrak{L}^{(s)}_{p,\mu} \underset{d}{=} \{f: f \in \mathfrak{M}(\mathfrak{A}, \mathfrak{B}_{\mathfrak{N}_w}) \ \& \ \forall \Phi \in \mathfrak{X}', \ \Phi \circ f \in L_p(\Lambda,\mathfrak{A},\mu;\mathbb{F})\}.$$

First, observe that it follows from the Closed Graph Thm., cf. [7, p.78], that

$$(5.18) \qquad \forall \text{ reflexive } \mathfrak{X}, \ \& \text{ so for } \mathfrak{X} = \mathfrak{H}, \quad \mathcal{P}_{1,\mu} = \mathfrak{L}^{(s)}_{1,\mu}.$$

For $\mathfrak{X} = \mathfrak{H}$, it also follows, since $\mathfrak{M}(\mathfrak{A}, \mathfrak{B}_{\mathfrak{N}}) \subseteq \mathfrak{M}(\mathfrak{A}, \mathfrak{B}_{\mathfrak{N}_w})$, cf. 4.1(a), that $\mathfrak{L}_{p,\mu} \subseteq \mathfrak{L}^{(s)}_{p,\mu}$. Recalling (5.7) and the obvious fact that $\mathfrak{L}^{(s)}_{p,\mu}$ is a linear manifold in the vector space $\mathfrak{M}(\mathfrak{A}, \mathfrak{B}_{\mathfrak{N}_w})$, we may conclude as follows.

5.19 Thm. For a Hilbert space $\mathfrak{H}$, we have (a) $\forall p \in [1,\infty]$, $\mathcal{L}_{p,\mu}$ is a conditionally linear submanifold of the vector space $\mathcal{L}_{p,\mu}^{(s)}$; (b) in particular $\mathcal{L}_{1,\mu}$ is a conditionally linear submanifold of $\mathcal{P}_{1,\mu}$.

The inclusions in 5.19 are proper for all infinite dimensional $\mathfrak{H}$ and all but rather trivial μ, as the next result shows.

5.20 Triv. Let the c.a. measure μ on $\mathfrak{A}$ to $[0,\infty]$ be such that $\exists$ a sequence $(A_n)_1^\infty$ of disjoint sets in $\mathfrak{A}$ of positive μ measure. Then for every infinite dimensional $\mathfrak{H}$ and every $p \in [1,\infty)$, $\mathcal{L}_{p,\mu} \subset \mathcal{L}_{p,\mu}^{(s)}$.

Vakhania [18] has shown that we can have $\mathcal{L}_{2,\mu} \subset \bigcap_{p=1}^{\infty} \mathcal{L}_{p,\mu}^{(s)}$. Actually $\mathcal{L}_{2,\mu}$ is replaceable by any $\mathcal{L}_{r,\mu}$. However, the case $r = 2$ is important in the theory of covariance operators of probability measures over $\mathfrak{X}$, and the strict inclusion shows that for this theory our $\mathcal{L}_{2,\mu}$ class is too restrictive, cf. Vakhania [19].

Another interesting consequence of (5.18) and 5.5(d) is that for $\mathfrak{H}$, and in fact all reflexive $\mathfrak{X}$, Pettis integrability over $\mathfrak{A}$ with respect to μ is equivalent to Pettis integrability over $\mathfrak{A}_\mu^{loc}$. We have

5.21 Thm. $f \in \mathcal{P}_{1,\mu}$, iff. f is Pettis integrable over $\mathfrak{A}_\mu^{loc}$ with respect to μ, provided that $\mathfrak{X}$ is reflective.

6. Integration on $\mathcal{L}_{1,\mu}$ to $\mathfrak{H}$

Let us recall the definition of the Pettis integral of a function f in the class $\mathcal{P}_{1,\mu}$ for a Hilbert space $\mathfrak{H}$, cf. Thm.5.21, (5.18) and Def.5.3.

6.1 Def. Let $f \in \mathcal{P}_{1,\mu}$. Then $\forall C \in \mathfrak{A}_\mu^{loc}$, the unique vector x_C mentioned in 5.3(a) is called the Pettis integral of f on C with respect to μ, and denoted by $\int_C f(\lambda)\mu(d\lambda)$ or $E_\mu(f,C)$. We shall write $E_\mu(f)$ for $E_\mu(f,\Lambda)$.

The fundamental property of $E_\mu(\cdot\cdot)$ is thus:

$$(6.2)\begin{cases} \forall f \in \mathcal{P}_{1,\mu} \ \& \ \forall C \in \mathfrak{A}_\mu^{loc}, & E_\mu(f,C) \in \mathcal{H}, \\ \& \ \forall y \in \mathcal{H}, & (E_\mu(f,C), y)_{\mathcal{H}} = \int_C (f(\lambda),y)_{\mathcal{H}}\mu(d\lambda). \end{cases}$$

Obviously,

$$(6.3) \quad \forall C \in \mathfrak{A}_\mu^{loc}, \quad E_\mu(\cdot,C) \text{ is a linear operator on } \mathcal{P}_{1,\mu} \text{ to } \mathcal{H}.$$

If $f \in \mathcal{P}_{1,\mu}$, $C \in \mathfrak{A}_\mu^{loc}$ and $y \in \mathcal{H}$, then obviously

$$\forall g \in \dot{f}, \quad (E_\mu(f,C), y)_{\mathcal{H}} = \int_C (g(\lambda),y)_{\mathcal{H}}\mu(d\lambda).$$

This suggests the definitions:

$$(6.1')\begin{cases} \dot{\mathcal{P}}_{1,\mu} \underset{d}{=} \{\dot{f}: f \in \mathcal{P}_{1,\mu}\} \\ \forall \dot{f} \in \dot{\mathcal{P}}_{1,\mu} \ \& \ \forall C \in \mathfrak{A}_\mu^{loc}, \quad E_\mu(\dot{f},C) \underset{d}{=} E_\mu(f,C), \end{cases}$$

in the second of which we have for simplicity omitted dotting the E. Corresponding to the results (6.2) and (6.3) we now have:

$$(6.2')\begin{cases} \forall \dot{f} \in \dot{\mathcal{P}}_{1,\mu} \ \& \ \forall C \in \mathfrak{A}_\mu^{loc}, & E_\mu(\dot{f},C) \in \mathcal{H}, \\ \& \ \forall y \in \mathcal{H}, & (E_\mu(\dot{f},C), y)_{\mathcal{H}} = \int_C (f(\lambda),y)_{\mathcal{H}}\mu(d\lambda). \end{cases}$$

$$(6.3') \quad \forall C \in \mathfrak{A}_\mu^{loc}, \quad E_\mu(\cdot,C) \text{ is a linear operator on } \dot{\mathcal{P}}_{1,\mu} \text{ to } \mathcal{H}.$$

We shall, however, follow the usual practice of dealing with functions f rather than their μ-equivalence classes $\dot{f}$ wherever possible.

The happy circumstance that for Hilbert spaces, $\mathcal{L}_{1,\mu} \subseteq \mathcal{P}_{1,\mu}$ (5.19(b)) obviates the need for defining integration on $\mathcal{L}_{1,\mu}$ de novo. We just restrict $E_\mu(\cdot,B)$ from $\mathcal{P}_{1,\mu}$ to $\mathcal{L}_{1,\mu}$, and use the same symbol for this restriction. $\mathcal{L}_{1,\mu}$ is not of course a linear subspace cf. (5.7). Nevertheless, with a slight and very natural widening of the term "linear operator" we can assert the following result:

6.4 Triv. $\forall C \in \mathfrak{A}_\mu^{loc}$, the operator $E_\mu(\cdot, C)$ is a "linear" contraction of (Banach-) norm 1 on the conditional Banach space $\dot{\mathcal{L}}_{1,\mu}$ to the Hilbert space $\mathfrak{H}$, in the following sense:

$$\dot{f} \in \dot{\mathcal{L}}_{1,\mu} \ \& \ a \in \mathbb{F} \quad \Rightarrow \quad a\dot{f} \in \dot{\mathcal{L}}_{1,\mu} \ \& \ E_\mu(a\dot{f}, C) = aE_\mu(\dot{f}, C)$$

$$\dot{f}, \dot{g} \ \& \ \dot{f}+\dot{g} \in \dot{\mathcal{L}}_{1,\mu} \quad \Rightarrow \quad E_\mu(\dot{f}+\dot{g}, C) = E_\mu(\dot{f}, C) + E_\mu(\dot{g}, C)$$

$$|E_\mu(\cdot, C)| \ \overset{=}{\mathrm{d}} \ \sup \{\frac{|E_\mu(f,C)|_{\mathfrak{H}}}{|f|_{1,\mu}} : 0 \neq f \in \mathcal{L}_{1,\mu}\} = 1.$$

It is obvious that for a Banach space $\mathfrak{X}$ and $f \in \mathcal{P}_{1,\mu}$, $E_\mu(f, \cdot)$ is an $\mathfrak{X}$-valued c.a. measure on $\mathfrak{A}$, and hence, cf. [4, p.50,#17] bounded on $\mathfrak{A}$. For $\mathfrak{X} = \mathfrak{H}$ the domain of this measure is $\mathfrak{A}_\mu^{loc}$, cf. Thm.5.21 and (6.2). More information concerning it is given in the next result, in which the notation $|\xi|(\cdot)$ refers to the <u>(total) variation measure</u> of the measure $\xi(\cdot)$:

6.5 <u>Thm.</u> (a) Let $f \in \mathcal{P}_{1,\mu}$. Then $E_\mu(f, \cdot)$ is a bounded c.a. measure on $\mathfrak{A}_\mu^{loc}$ to $\mathfrak{H}$, and its variation measure satisfies

$$\forall C \in \mathfrak{A}_\mu^{loc}, \quad |E(f,\cdot)|(C) \leq \int_C |f(\lambda)|_{\mathfrak{H}} \mu(d\lambda) \in [0, \infty].$$

(b) Let $f \in \mathcal{L}_{1,\mu}$. Then $E_\mu(f, \cdot)$ is a c.a. measure on $\mathfrak{A}_\mu^{loc}$ to $\mathfrak{H}$ which is of bounded variation on $\mathfrak{A}_\mu^{loc}$:

$$\forall C \in \mathfrak{A}_\mu^{loc}, \quad |E(f,\cdot)|(C) \leq \int_C |f(\lambda)|_{\mathfrak{H}} \mu(d\lambda) \leq |f|_{1,\mu} < \infty.$$

(c) For $f \in L_{1,\mu}$ we have equality in the first step of the last relation.

For $f \in \mathcal{L}_{1,\mu} \setminus L_{1,\mu}$ the first inequality in 6.5(b) can be strict. This is shown in the next result, which deals with a simple case of the general situation described in Thm.5.16.

6.6 <u>Thm.</u> Let (i) $\Lambda = \mathbb{R}$, $\mathfrak{A} = \mathrm{Bl}(\mathbb{R})$, μ be an atomless probability measure on $\mathfrak{A}$, (ii) $\dim \mathfrak{H} = c$, (iii) $(u(\lambda): \lambda \in \Lambda)$ be an o.n. basis for $\mathfrak{H}$. Then $u(\cdot) \in \mathcal{L}_{1,\mu}$, and $\forall C \in \mathfrak{A}_\mu^{loc}$ ($= \mathfrak{A}$ by (i))

$$|E_\mu(u,\cdot)|(C) = 0 \quad \text{but} \quad \int_C |u(\lambda)|_{\mathfrak{H}} \mu(d\lambda) = \mu(C).$$

Le $f \in \mathfrak{L}_{1,\mu}$. Then by 6.5(b), the $\mathfrak{H}$-valued c.a. measure $E_\mu(f,\cdot)$ is of bounded variation on $\mathfrak{U}_\mu^{loc}$. Since $\mathfrak{H}$ has the Radon-Nikodym property, there exists a μ-essentially unique $\tilde{f} \in L_{1,\mu}$ such that

$$(6.7) \qquad \forall C \in \mathfrak{U}_\mu^{loc}, \quad E_\mu(\tilde{f},C) = E_\mu(f,C).$$

We shall call this $\tilde{f}$ <u>the Bochner associate of f</u>. For instance for the function $u(\cdot)$ in 6.6, we see that $\tilde{u}(\cdot) = 0$. This illustrates that in general f and $\tilde{f}$ are far from being in the same μ-equivalence class. The correspondence: $f \to \tilde{f}$ on the conditional Banach space $\mathfrak{L}_{1,\mu}$ to the Banach subspace $L_{1,\mu}$ vaguely resembles a conditional expectation and has some similar properties. We shall not explore these in this paper.

Since every $f \in \mathfrak{L}_{1,\mu}$ is thus replaceable by a function $\tilde{f} \in L_{1,\mu}$ without affecting its indefinite integral, cf. (6.7), it may be asked why one should even consider $f \in \mathfrak{L}_{1,\mu} \setminus L_{1,\mu}$. The answer is that such functions f and their Pettis integrals are encountered in applications, and theorems are needed to deal with them, e.g. interchange theorems of the Fubini type. This happens, for instance, in the study of stationary varieties in a non-separable Hilbert space parametrized over a non-second countable l.c.a. group, cf. [12]. The knowledge that there is an $\tilde{f} \in L_{1,\mu}$ corresponding to f does not help, since we do not as yet have methods to obtain $\tilde{f}$ from f and to infer the properties of $\tilde{f}$ from those of f. To sum up, <u>we cannot dispense with the direct study of functions in $\mathfrak{L}_{1,\mu} \setminus L_{1,\mu}$ until more is known about the correspondence $f \to \tilde{f}$ on $\mathfrak{L}_{1,\mu}$ to $L_{1,\mu}$</u>. This remark should be kept in mind in reading §7.

7. Interchange Theorem for the Partial Integral of an Orthogonally Scattered Measure

For the reader's convenience we shall reproduce here the definition of a Hilbert space-valued c.a.o.s. measure $\xi(\cdot)$, and list the properties of integration with respect to $\xi(\cdot)$ which we will require. We adhere to the Notation 5.1 for the symbols $\mathfrak{H}$, Λ, $\mathfrak{A}$, μ, etc.

7.1 Def. (a)# We shall call $\xi(\cdot)$ a $\mathfrak{H}$-valued, countably-additive, orthogonally scattered (c.a.o.s.) measure on the δ-ring $\mathfrak{A}_\mu$ with control measure $\text{Rstr.}_{\mathfrak{A}_\mu}\mu$, iff. $\xi(\cdot)$ is a function on $\mathfrak{A}_\mu$ to $\mathfrak{H}$ such that

$$\forall A,B \in \mathfrak{A}_\mu, \qquad (\xi(A),\xi(B))_{\mathfrak{H}} = \mu(A\cap B).$$

(b) $\mathfrak{S}_\xi \underset{d}{=} \mathfrak{S}\{\xi(A): A \in \mathfrak{A}_\mu\}$ is called the subspace of ξ.

(c) $L_{2,\mu}(\mathbb{F}) \underset{d}{=} L_2(\Lambda,\mathfrak{A},\mu;\mathbb{F})$.

The theory of such measures ξ and of integration of $\mathbb{F}$-valued functions with respect to them is expounded in [10] and also [11]. It is shown there that $\xi(\cdot)$ is indeed c.a. on $\mathfrak{A}_\mu$ to $\mathfrak{H}$. The following theorem lists the basic properties of integration which we require, cf. [10, 5.7-5.12]:

7.2 Thm. Let (i) $\xi(\cdot)$ be a $\mathfrak{H}$-valued c.a.o.s. measure on $\mathfrak{A}_\mu$ with control measure $\text{Rstr.}_{\mathfrak{A}_\mu}\mu$, (ii) $\forall y \in \mathfrak{H}$, $m_y(\cdot) \underset{d}{=} (y, \xi(\cdot))_{\mathfrak{H}}$ on $\mathfrak{A}_\mu$. Then

(a) the correspondence $\Sigma_\xi: \phi \to \int_\Lambda \phi(\lambda)\xi(d\lambda)$ is a unitary operator on $L_{2,\mu}(\mathbb{F})$ onto $\mathfrak{S}_\xi \subseteq \mathfrak{H}$;

(b) $\forall y \in \mathfrak{H}$, $\psi_y(\cdot) \underset{d}{=} dm_y/d\mu \in L_{2,\mu}(\mathbb{F})$, &

$$P_{\mathfrak{S}_\xi}(y) = \int_\Lambda \psi_y(\lambda)\xi(d\lambda), \qquad |P_{\mathfrak{S}_\xi}(y)| = |\psi_y(\cdot)|_{2,\mu},$$

where $P_{\mathfrak{S}_\xi}$ is the orthogonal projection on $\mathfrak{H}$ onto $\mathfrak{S}_\xi$;

C.a.o.s. measures are definable on any pre-ring. The definition given here is a specialized for the purposes at hand.

(c) $\forall \phi \in L_{2,\mu}(\mathbb{F})$ & $\forall y \in \mathcal{S}_\xi$,

$$(\int_\Lambda \phi(\lambda)\xi(d\lambda),\ y)_{\mathfrak{H}} = \int_\Lambda \phi(\lambda)\overline{\psi_y(\lambda)}\mu(d\lambda).$$

If the integrand $\phi(\cdot)$ depends on a parameter ω in Ω, then its integral with respect to $\xi(\cdot)$ will be a function on Ω to $\mathfrak{H}$. To prove its measurability (Thm.7.4) we have to appeal to a triviality on the measurability of a function and its restriction, the proof of which is obvious:

7.3 <u>Triv.</u> Let (i) $\mathcal{B}$ be a σ-algebra over a set Ω, (ii) $\Omega_o \in \mathcal{B}$ and $\mathcal{B}_o \underset{d}{=} \mathcal{B} \cap 2^{\Omega_o}$, (iii) $F_o(\cdot) \in \mathfrak{H}^{\Omega_o}$, (iv) $F(\cdot) \in \mathfrak{H}^{\Omega}$ be defined by $F(\cdot) \underset{d}{=} F_o(\cdot)$ on Ω_o & $F(\cdot) \underset{d}{=} 0$ on $\Omega \setminus \Omega_o$. Then

$$F_o(\cdot) \in \mathfrak{M}(\mathcal{B}_o, \mathcal{B}_{\mathfrak{H}}) \Rightarrow F(\cdot) \in \mathfrak{M}(\mathcal{B}, \mathcal{B}_{\mathfrak{H}}).$$

Part (a) of the next theorem is obvious. As for part (b), we appeal to 7.3 and 2.8(b) to reduce the question to the measurability of $|F(\cdot)-x|_{\mathfrak{H}}$, $x \in \mathfrak{H}$, and then show this by using Thm.7.2(a) and Tonelli's Thm.

7.4 <u>Thm</u>. (Measurability of $\int_\Lambda f(\lambda,\cdot)\xi(d\lambda)$). Let

(i) $\mathfrak{A}$, $\mathcal{B}$ be σ-algebras over Λ, Ω, and $\mathcal{C} \underset{d}{=} \sigma\text{-alg}(\mathfrak{A}\times\mathcal{B})$

(ii) μ, ν be σ-finite c.a. measures on $\mathfrak{A}$, $\mathcal{B}$ to $[0,\infty]$

(iii) $\xi(\cdot)$ be a $\mathfrak{H}$-valued c.a.o.s. measure on $\mathfrak{A}_\mu$ with control measure $\text{Rstr.}_{\mathfrak{A}_\mu}\mu$

(iv) $f(\cdot\cdot) \in \mathfrak{M}(\mathcal{C}, Bl(\mathbb{F}))$

(v) $\exists$ a carrier Ω_o of ν such that $\forall \omega \in \Omega_o$, $f(\cdot,\omega) \in L_{2,\mu}(\mathbb{F})$.

Then

(a) $\forall \omega \in \Omega_o$, $F(\omega) \underset{d}{=} \int_\Lambda f(\lambda,\omega)\xi(d\lambda)$ exists & $\in \mathfrak{H}$;

(b) letting $F(\cdot) \underset{d}{=} 0$ on $\Omega \setminus \Omega_o$, we have $F(\cdot) \in \mathfrak{M}(\mathcal{B}, \mathcal{B}_{\mathfrak{H}})$.

Now by Thm.4.1(a) and Triv.2.8(a), any $\mathcal{B}$, $\mathcal{B}_{\mathfrak{H}}$ measurable function is $\mathcal{B}$, scalarly measurable. Thus, Thm.7.4 shows that under very reasonable conditions the partial integral $F(\cdot) \underset{d}{=} \int_\Lambda f(\lambda,\cdot)\xi(d\lambda)$ is $\mathfrak{B}$, scalarly measurable on Ω. With an extra condition on $f(\cdot\cdot)$ we can make $F(\cdot)$ Pettis integrable on Ω with respect to the measure ν on $\mathfrak{B}$. The question of change of order of integration then arises, viz.

$$\int_\Omega\{\int_\Lambda f(\lambda,\omega)\xi(d\lambda)\}\nu(d\omega) = \int_\Lambda\{\int_\Omega f(\lambda,\omega)\nu(d\omega)\}\xi(d\lambda).$$

Our objective is to find conditions sufficient to ensure this. The proof of our theorem (7.7 below) is rather long and beset by technicalities pertaining to measurability. To help the reader we have segregated some of these technicalities into two classical lemmas. The first lemma asserts the measurability of the tensor product of measurable functions; the second narrates a consequence of Tonelli's theorem.

7.5 Lma. Let Λ, Ω, $\mathfrak{A}$, $\mathfrak{B}$, $\mathcal{C}$, $f(\cdot\cdot)$ be as in Thm.7.4, and

$$\forall(\lambda,\omega,\omega') \in \Lambda\times\Omega\times\Omega, \quad g(\lambda,\omega,\omega') \underset{d}{=} |f(\lambda,\omega)\cdot f(\lambda,\omega')|.$$

Then

(a) $\qquad \forall\lambda \in \Lambda, \quad g(\lambda,\cdot\cdot) \in \mathfrak{M}\{\sigma\text{-alg}(\mathfrak{B}\times\mathfrak{B}),\ \mathrm{Bl}(\mathbb{R}_{0+})\}$

(b) $\qquad g(\cdot\cdot\cdot) \in \mathfrak{M}\{\sigma\text{-alg}(\mathfrak{A}\times\mathfrak{B}\times\mathfrak{B}),\ \mathrm{Bl}(\mathbb{R}_{0+})\}$.

7.6 Lma. Let Λ, Ω, $\mathfrak{A}$, $\mathfrak{B}$, $\mathcal{C}$, μ, ν, $f(\cdot\cdot)$ be as in Thm.7.4, and let $\forall\lambda \in \Lambda$, $f(\lambda,\cdot) \in L_{1,\nu}(\mathbb{F}) \underset{d}{=} L_1(\Omega,\mathfrak{B},\nu;\mathbb{F})$. Then

$$G(\cdot) \underset{d}{=} \int_\Omega f(\cdot,\omega)\nu(d\omega) \in \mathfrak{M}(\mathfrak{A},\ \mathrm{Bl}(\mathbb{F})).$$

7.7 Interchange Thm. Let (i)-(v) be as in Thm.7.4, and (vi)

$$\int_\Omega|f(\cdot,\omega)|_{2,\mu}\nu(d\omega) < \infty.$$

Then

(a) $\int_\Lambda f(\lambda,\cdot)\xi(d\lambda) \in \mathfrak{L}_{1,\nu} \underset{d}{=} L_1(\Omega,\mathfrak{B},\nu;\mathfrak{H},\mathfrak{B}_{\mathfrak{M}})$, and is therefore Pettis integrable with respect to ν, cf. 5.19(b);

(b) $\exists$ a carrier Λ_0 of μ such that $\forall\lambda \in \Lambda_0$,

$$f(\lambda,\cdot) \in L_1(\Omega,\mathfrak{B},\nu;\mathbb{F}) \ \ \&\ \ \int_\Lambda|f(\lambda,\cdot)|^2_{1,\nu}\mu(d\lambda) < \infty;$$

(c) $\int_\Omega f(\cdot,\omega)\nu(d\omega) \in L_2(\Lambda,\mathfrak{A},\mu;\mathbb{F})$;

(d) $\int_\Omega\{\int_\Lambda f(\Lambda,\omega)\xi(d\lambda)\}\nu(d\omega) = \int_\Lambda\{\int_\Omega f(\lambda,\omega)\nu(d\omega)\}\xi(d\lambda)$.

Thm.7.7 subsumes an improved version of the interchange theorem 5.20 of [10] involving Bochner integrals.* On the other hand, Thm.7.7 is not applicable in situations in which $f(\cdot\cdot)$ violates the premiss (iv). For instance, if Λ is a l.c.a. group with card $\Lambda > 2^{\aleph_0} = \underline{c}$, Ω is the character group of Λ, and $f(\lambda,\omega) \underset{d}{=} \omega(\lambda)$, $\lambda\in\Lambda$ and $\omega\in\Omega$, then with $\mathfrak{A} \underset{d}{=} Bl(\Lambda)$ and $\mathfrak{B} \underset{d}{=} Bl(\Omega)$ we find that

$$C \underset{d}{=} \sigma\text{-alg}(\mathfrak{A}\times\mathfrak{B}) \subset Bl(\Lambda\times\Omega)$$

and

$$f \in \mathfrak{M}(Bl(\Lambda\times\Omega), Bl(\mathbb{C})) \setminus \mathfrak{M}(C, Bl(\mathbb{C})), ^{\#}$$

in violation of 7.4(iv), i.e. 7.7(iv). Fortunately, in such topological situations, if the measures μ,ν on $\mathfrak{A},\mathfrak{B}$ are Radon, then the product measure $\mu\times\nu$ on C can be extended to a c.a. measure π on $Bl(\Lambda\times\Omega)$, and the classical Tonelli, Fubini theorems, on which the proof of Thm.7.7 hinges, admit versions valid for this π and for functions $f(\cdot\cdot)$ in $\mathfrak{M}(Bl(\Lambda\times\Omega), Bl(\mathbb{F}))$. This has been shown by the Bourbaki School, cf. L. Schwartz, [16, pp.63-73].**

---

* The improvement being the elimination of the premiss [10, 5.20(v)]. In the enunciation it is claimed that the partial integral with respect to ξ is Bochner integrable, but in the proof it is not shown that its range is separable. This lacuna is most easily removed by hypothesizing the separability of $\mathfrak{S}_\xi$ in [10, 5.20].

\# Professor J.P.R. Christensen has given a simple demonstration of this.

** I am very grateful to Professor J.R. Choksi for alerting me to this work and to its bearing on the questions at hand, and for checking the adaptations made of it in 7.11 - 7.13.

We proceed to demonstrate that these variants of the Tonelli, Fubini theorems yield an interchange theorem for $f(\cdot\cdot)$ in $\mathfrak{M}(Bl(\Lambda x \Omega), Bl(\mathbb{F}))$, which is an exact analogue of Thm.7.7, with analogous proof. Crucial to this demonstration is the following classical fact, and the following concept due to Bourbaki:

7.8 <u>Triv.</u> Sections of Borel measurable functions are Borel measurable. More precisely, let (i) Λ, Ω be topological spaces, (ii) $\mathfrak{F}$ be a σ-algebra over a space $\mathfrak{X}$, (iii) $f(\cdot\cdot) \in \mathfrak{M}(Bl(\Lambda x \Omega), \mathfrak{F})$, where the Borel algebra is generated by the weak product topology for $\Lambda x \Omega$. Then $\forall \lambda \in \Lambda$ & $\forall \omega \in \Omega$

$$f(\cdot, \omega) \in \mathfrak{M}(Bl(\Lambda), \mathfrak{F}), \quad f(\lambda, \cdot) \in \mathfrak{M}(Bl(\Omega), \mathfrak{F}).$$

7.9 <u>Def.</u> Let Λ be a topological space. Then a c.a. measure μ on $Bl(\Lambda)$ to $[0,\infty]$ is called <u>moderate</u>, iff. $\Lambda = \bigcup_1^\infty \Lambda_n$, where Λ_n are open and such that $\mu(\Lambda_n) < \infty$.

Throughout the sequel we shall assume that

(7.10)
- (i) Λ, Ω are locally compact Hausdorff spaces and $\mathfrak{A} \underset{d}{=} Bl(\Lambda)$, $\mathfrak{B} \underset{d}{=} Bl(\Omega)$
- (ii) μ, ν are moderate, c.a. measures on $\mathfrak{A}$, $\mathfrak{B}$ to $[0,\infty]$ which are "Radon", i.e. finite on compact sets and inner regular on $\mathfrak{A}$, $\mathfrak{B}$.

We then have the following three theorems, as adaptations to the set-up (7.10) of more general results due to the Bourbaki school, cf. [16, p.73, Thm.].

7.11 <u>Thm.</u> (Extension of product measure) Under the assumption of (7.10), $\exists$ a unique moderate c.a. measure π on $Bl(\Lambda x \Omega)$ to $[0,\infty]$ which is Radon and such that $\mu x \nu \subseteq \pi$.

This unique π is called the <u>tensor product</u> of μ, ν and denoted by

$\mu\otimes\nu$ to distinguish it from its restriction $\mu\times\nu$.

7.12 Tonelli's Thm. (for $\mu\otimes\nu$) Assuming (7.10), let $f(\cdot\cdot) \in \mathfrak{M}(Bl(\Lambda\times\Omega), Bl(\mathbb{R}_{0+}))$. Then

(a) $\forall\lambda\in\Lambda$ & $\forall\omega\in\Omega$, $f(\cdot,\omega) \in \mathfrak{M}(\mathfrak{A}, Bl(\mathbb{R}_{0+}))$, $f(\lambda,\cdot) \in \mathfrak{M}(\mathfrak{B}, Bl(\mathbb{R}_{0+}))$;

(b) $\int_\Omega f(\cdot,\omega)\nu(d\omega) \in \mathfrak{M}(\mathfrak{A}, Bl[0,\infty])$ & $\int_\Lambda f(\lambda,\cdot)\mu(d\lambda) \in \mathfrak{M}(\mathfrak{B}, Bl[0,\infty])$;

(c) the following integrals have the same value in $[0,\infty]$:

$$\int_\Lambda\{\int_\Omega f(\lambda,\omega)\nu(d\omega)\}\mu(d\lambda), \quad \int_\Omega\{\int_\Lambda f(\lambda,\omega)\mu(d\lambda)\}\nu(d\omega)$$
$$\int_{\Lambda\times\Omega} f(\lambda,\omega)(\mu\otimes\nu)\{d(\lambda,\omega)\}.$$

7.13 Fubini's Thm. (for $\mu\otimes\nu$) Assuming (7.10), let $f(\cdot\cdot) \in L_1(\Lambda\times\Omega, Bl(\Lambda\times\Omega), \mu\otimes\nu; \mathbb{F})$. Then

(a) for ν almost all $\omega\in\Omega$, $f(\cdot,\omega) \in L_1(\Lambda,\mathfrak{A},\mu;\mathbb{F})$
for μ almost all $\lambda\in\Lambda$, $f(\lambda,\cdot) \in L_1(\Omega,\mathfrak{B},\nu;\mathbb{F})$;

(b)# $\int_\Omega f(\cdot,\omega)\nu(d\omega) \in L_1(\Lambda,\mathfrak{A},\mu;\mathbb{F})$ & $\int_\Lambda f(\lambda,\cdot)\mu(d\lambda) \in L_1(\Omega,\mathfrak{B},\nu;\mathbb{F})$;

(c) the three integrals in 7.12(c) now have the same value in $\mathbb{F}$.

We leave it to the reader to deduce these three results from [16, p.73, Thm.].

We prove next the following result corresponding to Thm.7.4 in the same way as the latter, except for an appeal to the version 7.12 of Tonelli's Thm. rather than to the classical version.

7.14 Thm. (Measurability of $\int_\Lambda f(\lambda,\cdot)\xi(d\lambda)$).
Let (i),(ii) be as in (7.10)

(iii) $\xi(\cdot)$ be a $\mathfrak{H}$-valued c.a.o.s. measure on $\mathfrak{A}_\mu$ with control measure $\mathrm{Rstr.}_{\mathfrak{A}_\mu}\mu$

Here it is understood that we define the first integral to be 0 on the μ-negligible set of λ for which $f(\lambda,\cdot) \notin L_1(\Omega,\mathfrak{B},\nu;\mathbb{F})$, and similarly with the second.

(iv) $f(\cdot\cdot) \in \mathfrak{M}(Bl(\Lambda x\Omega), Bl(\mathbb{F}))$

(v)# $\exists$ a carrier Ω_o of ν such that $\forall\omega \in \Omega_o$, $f(\cdot,\omega) \in L_{2,\mu}(\mathbb{F})$.

Then

(a) $\forall\omega \in \Omega_o$, $F(\omega) = \int_\Lambda f(\lambda,\cdot)\xi(d\lambda)$ exists & $\in \mathfrak{H}$;

(b) letting $F(\cdot) = 0$ on $\Omega \sim \Omega_o$, we have $F(\cdot) \in \mathfrak{M}(\mathcal{B}, \mathcal{B}_{\mathfrak{H}})$.

For the interchange theorem under the set-up (7.10) we also need lemmas corresponding to 7.5(b) and 7.6. Their proofs are again classical, and therefore omitted.

7.15 Lma. Under the set-up (7.10), let

(i) $$f(\cdot\cdot) \in \mathfrak{M}(Bl(\Lambda x\Omega), Bl(\mathbb{F}))$$

(ii) $$\forall(\lambda,\omega,\omega') \in \Lambda x\Omega x\Omega, \quad g(\lambda,\omega,\omega') \underset{d}{=} |f(\lambda,\omega)\cdot f(\lambda,\omega')|.$$

Then $$g(\cdot\cdot\cdot) \in \mathfrak{M}\{Bl(\Lambda x\Omega x\Omega), Bl(\mathbb{R}_{0+})\}.$$

7.16 Lma. Under the set-up (7.10), let

(i) $$f(\cdot\cdot) \in \mathfrak{M}(Bl(\Lambda x\Omega), Bl(\mathbb{F}))$$

(ii) $$\forall\lambda \in \Lambda, \quad f(\lambda,\cdot) \in L_{1,\nu}(\mathbb{F}) \underset{d}{=} L_1(\Omega,\mathcal{B},\nu;\mathbb{F}).$$

Then $$G(\cdot) \underset{d}{=} \int_\Omega f(\cdot,\omega)\nu(d\omega) \in \mathfrak{M}(\mathfrak{A}, Bl(\mathbb{F})).$$

We can now prove the following variant of the Interchange Thm. 7.7 in the same way as the latter except for appeal to our new lemmas in place of the old.

7.17 Interchange Thm. Let (i)-(v) be as in Thm.7.14, and (vi)

$$\int_\Omega |f(\cdot,\omega)|_{2,\mu}\nu(d\omega) < \infty.$$

Then the conclusions 7.7(a)-(d) are valid.

By (iv) and Triv.7.8, $\forall\omega \in \Omega$, $f(\cdot,\omega) \in \mathfrak{M}(\mathfrak{A}, Bl(\mathbb{F}))$.

Appendix: Open Questions and Problems

1. Characterize all Banach spaces $\mathfrak{X}$ for which $\mathfrak{B}_{\mathfrak{X}_w} \subseteq \mathfrak{B}_{\mathfrak{X}}$. (We know, cf. 3.8(a) & 2.5(b), that all Hilbert spaces and all separable Banach spaces are of this type.)
2. (a) Does there exist a non-separable Hilbert space for which $\mathfrak{B}_{\tau_w} = \mathfrak{B}_{\tau}$? (b) Does there exist a Hilbert space for which $\mathfrak{B}_{\tau_w} \neq \mathfrak{B}_{\tau}$? Cf. 3.13.
3. Does there exist a non-separable Banach space $\mathfrak{X}$ such that for all σ-algebras $\mathfrak{A}$, $\mathfrak{M}(\mathfrak{A}, \mathfrak{B}_{\mathfrak{X}})$ is a vector space? (We know from 4.3 that $\mathfrak{X}$ cannot be a Hilbert space. The answer seems to be: no.)
4. Does there exist a σ-algebra $\mathfrak{A}$ and a Hilbert space $\mathfrak{H}$ of cardinality $\underline{c}$ for which (a) $\mathfrak{M}(\mathfrak{A}, \mathfrak{B}_{\tau})$ is not a vector space? (b) $\mathfrak{M}(\mathfrak{A}, \mathfrak{B}_{\tau_w})$ is not a vector space? (Cf. 2.16; we know from (2.15) that $\mathfrak{H}$ will have to be non-separable.)
5. Does there exist a non-separable Hilbert space such that for all σ-algebras $\mathfrak{A}$, (a) $\mathfrak{M}(\mathfrak{A}, \mathfrak{B}_{\tau})$ is conditionally linear? (b) $\mathfrak{M}(\mathfrak{A}, \mathfrak{B}_{\tau_w})$ is conditionally linear? Cf. (4.11).
6. Does there exist a non-separable Hilbert space $\mathfrak{H}$ for which the inner product $(\cdot\ \cdot)_{\mathfrak{H}}$ on $\mathfrak{H}\times\mathfrak{H}$ to $\mathbb{F}$ is $\sigma\text{-alg}(2^{\mathfrak{H}} \times 2^{\mathfrak{H}})$, $Bl(\mathbb{F})$ measurable? (By 4.12 such an $\mathfrak{H}$ must have cardinality $\underline{c}$.)

N. B. Colleagues versed in foundations have suggested that some of these questions might be undecidable.

REFERENCES

1. C. Berge, The theory of graphs, Methuen, London, 1962.
2. S.D. Chatterji, Sur l'intégrabilité de Pettis, Math. Zeit. 136 (1974), 53-58.
3. J.P.R. Christensen, Topology and Borel structure, North-Holland, Amsterdam, 1974.

4. N. Dinculeanu, Vector measures, Pergamon Press, Oxford, 1967.

5. N. Dunford and J.T. Schwartz, Linear operators I, Interscience, New York, 1958.

6. P.R. Halmos, Measure theory, van Nostrand, New York, 1950.

7. E. Hille and R.S. Phillips, Functional analysis and semi-groups, Amer. Math. Soc. Colloq. Publ., vol. 31, Amer. Math. Soc., Providence, R.I., 1957.

8. S. Khalili, Measurability of Banach space valued functions and the Bochner integral, (to appear).

9. P. Masani, Graph-theoretic aspects of generalized harmonic analysis, (Abstract) Notices, Amer. Math. Soc. 14 (1967), 407-408.

10. P. Masani, Orthogonally scattered measures, Adv. in Math. 2 (1968), 61-117. (Originally, Technical Report #738, Mathematics Research Center, University of Wisconsin, 1967).

11. P. Masani, Quasi-isometric measures and their applications, Bull. Amer. Math. Soc. 76 (1970), 427-528.

12. P. Masani, Generalizations of P. Levy's inversion theorem, (to appear).

13. J. Nedoma, Note on generalized random variables, Trans. of the First Prague Conference in Information Theory, Statistical Decision Functions, Random Processes, (1956), 139-141.

14. B.J. Pettis, On integration in vector spaces, Trans. Amer. 44 (1938), 277-304.

15. G.B. Price, The theory of integration, Trans. Amer. Math. Soc. 47 (1940), 1-50.

16. L. Schwartz, Radon measures on arbitrary topological spaces and cylindrical measures, Oxford Univ. Press, London, 1973.

17. J.J. Uhl, A characterization of strongly measurable Pettis integrable functions, Proc. Amer. Math. Soc. 34 (1972), 425-427.

18. N.N. Vakhania, On a certain condition on existence of Pettis integral, Studia Math. 29 (1968), 243-249 (Russian).

19. N.N. Vakhania, Covariance operators of probability measures in Banach spaces, Bull. Georgian Acad. Sci. USSR, 51 (1968), No. 1 (Russian).

VECTOR VALUED INNER MEASURES

M. K. Nayak and T. P. Srinivasan
The University of Kansas
Lawrence, Kansas

Scalar valued inner measures are being used more widely now than in the past. The non-negative case was considered by one of us a long time ago [4] and was later used by us in [1] and [2] to extend a pre-measure on a lattice of sets to a countably additive measure on a σ-field. In this paper we attempt a vector valued analogue. Just as sub additivity is basic to a non-negative outer measure, 'supra additivity' is a basic feature of a non-negative inner measure. This property involves the order relation among the reals and has no counter part in the vector valued case. We present below a possible formulation of a vector valued inner measure and use it to extend a vector valued pre-measure on a lattice of sets to a vector valued measure on a σ-field. We give an outline of the proof. For further details and related results we refer to [3].

We take the range vector space to be a Banach Space $\mathcal{E}$. Recall that a family $\mathcal{C}$ of subsets of an arbitrary set X is a <u>hereditary ring</u> if $\mathcal{C}$ contains the union of every pair of its members and it contains all the subsets of each its members. An $\mathcal{E}$-valued function μ on a family $\mathcal{C}$ of sets is continuous from above at a member A if for each decreasing sequence $\{A_n\}_n$ of members of $\mathcal{C}$ whose intersection is A we have $\lim_n \mu(A_n) = \mu(A)$. For an $\mathcal{E}$-valued function μ on a hereditary ring $\mathcal{C}$ of sets, a set S (not necessarily in $\mathcal{C}$) is μ-<u>additive</u> or simply <u>additive</u> if it splits each member of $\mathcal{C}$ additively, that is if for each $E \in \mathcal{C}$, $\mu(E) = \mu(E \cap S) + \mu(E \setminus S)$.

<u>DEFINITION</u> μ is an inner measure on a set X if the domain of μ is a hereditary ring $\mathcal{C}$ of sets in X, μ takes values in a Banach space $\mathcal{E}$, $\mu(\Phi) = 0$ and μ is continuous from above at each member.

<u>THEOREM</u> 1. Let μ be an inner measure with domain $\mathcal{C}$. The family M of μ-additive sets is a σ-field and the restriction of μ to the δ-ring $M \cap \mathcal{C}$ is a countably additive measure.

PROOF The fact that M is a field and the restriction $\mu|M \cap \mathcal{C}$ is finitely additive, is standard [Lemma 1, [1]]. We will show that for each decreasing sequence $\{S_n\}_n$ of members of M, the intersection $S = \bigcap_n S_n$ is again a member. It then follows that M is a σ-field, $M \cap \mathcal{C}$ is a δ-ring and since μ is continuous, it is countably additive on $M \cap \mathcal{C}$.

To establish that $S \in M$ we need to show that for each member $E \in \mathcal{C}$,

$$\mu(E) = \mu(E \cap S) + \mu(E \setminus S)$$

The corresponding equality is true of S_n in place of S for each n and so,

$$\mu(E) = \mu(E \cap S_n) + \mu(E \setminus S_n)$$

and

$$\mu(E \setminus S) = \mu((E \setminus S) \cap S_n) + \mu(E \setminus S_n)$$

for each n. Since $\bigcap_n (E \cap S_n) = E \cap S$ and $\bigcap_n ((E \setminus S) \cap S_n) = \phi$ and μ is continuous from above, we have $\lim_n \mu(E \cap S_n) = \mu(E \cap S)$ and $\lim_n \mu((E \setminus S) \cap S_n) = 0$. It follows that

$$\mu(E) - \mu(E \cap S) - \mu(E \setminus S) = (\mu(E \cap S_n - \mu(E \cap S)) - \mu((E \setminus S \cap S_n)$$

$$= 0 \quad \text{(by passing to the limit).}$$

The difficulty in the application of the preceding theorem lies in the requirement of continuity of μ at each member of its domain $\mathcal{C}$. The following proposition shows that the continuity of μ at each member of an approximating sub family and additional mild condition imply its continuity on all of $\mathcal{C}$.

An $\mathcal{E}$-valued function μ on a family $\mathcal{C}$ of sets has inner approximation in a sub family $\mathcal{a}$ if for each member $E \in \mathcal{C}$ and $e > 0$ there exists a member $A_0 \subset E$ in $\mathcal{a}$ so that for all $A \in \mathcal{a}$ where $A_0 \subset A \subset E$ we have $\| \mu(E) - \mu(A) \| < e$. μ is said to be modular on a subfamily $\mathcal{a}$ which is a <u>lattice</u> if

$$\mu(A \cup B + \mu(A \cap B) = \mu(A) + \mu(B)$$

for every pair of members A,B in $\mathcal{a}$.

PROPOSITION 2. Suppose that an $\mathcal{E}$-valued function μ on a family of sets has inner approximation in a subfamily $\mathcal{A}$ which is a lattice closed under countable intersections. Suppose that $\mu|\mathcal{A}$ is continuous from above at each member and is further modular. Then μ is continuous from above at each member of $\mathcal{C}$.

PROOF Let E be an arbitrary member of $\mathcal{C}$ and let $\{E_n\}_n$ be an arbitrary decreasing sequence of members of $\mathcal{C}$ with $\bigcap_n E_n = E$. We need to show that $\mu(E) = \lim_n \mu(E_n)$. Let $e > 0$. By inner approximation we can choose $A_0 \in \mathcal{A}$ and then A_n in $\mathcal{A}$ containing A_0 for each n so that

$$\|\mu(E) - \mu(A)\| < e \text{ for all } A \in \mathcal{A} \text{ with } A_0 \subset A \subset E$$

and

$$\|\mu(E_n) - \mu(A)\| < e/2^{n+1} \text{ for all } A \in \mathcal{A} \text{ with } A_n \subset A \subset E_n$$

Replacing A in the last inequality by A_k and by $A_k \cup (A_{k+1} \cap \ldots \cap A_n)$ in succession and using the modularity of μ on $\mathcal{A}$ we see that for each n and each $k < n$,

$$\|\mu(E_n) - \mu(A_1 \cap A_2 \cap \ldots \cap A_n)\| < \frac{e}{2} + \frac{e}{2^2} + \ldots + \frac{e}{2^n}$$

We illustrate the argument taking $n = 3$.

$$\|\mu(E_3) - \mu(A_1 \cap A_2 \cap A_3)\|$$

$$\leq \|\mu(E_3) - \mu(A_2 \cap A_3)\| + \|\mu(A_1 \cup (A_2 \cap A_3)) - \mu(A_1)\|$$

$$\leq \|\mu(E_3) - \mu(A_3)\| + \|\mu(A_2 \cup A_3) - \mu(A_2)\| + \|\mu(A_1 \cup (A_2 \cap A_3)) - \mu(A_1)\|$$

$$\leq \frac{e}{2} + \frac{e}{2^2} + \frac{e}{2^3}.$$

Now

$$\|\mu(E) - \mu(E_n)\| \leq \|\mu(E) - \mu(\bigcap_{n=1}^{\infty} A_n)\| + \|\mu(\bigcap_{n=1}^{\infty} A_n) - \mu(A_1 \cap A_2 \cap \ldots A_n)\| + \|\mu(A_1 \cap A_2 \cap \ldots A_n) - \mu(E_n)\|$$

The first two summands are small by the inequalities we stated above while the third summand is small for all large n by the continuity of $\mu|\mathcal{A}$. It follows that $\mu(E) = \lim_n \mu(E_n)$.

The preceding proposition applies to the lattice of compact sets in a topological space but not to the lattice of finite unions of intervals in R^n since the latter is not closed under countable intersection. The following is an extension of the preceding proposition to lattices not necessarily closed under countable intersection.

An &-valued function μ on a family $\mathcal{C}$ of sets has <u>outer approximation</u> in a subfamily $\mathcal{A}$ if for each member $E \in \mathcal{C}$ and $e > 0$ there exists a member $A_0 \supset E$ in $\mathcal{A}$ so that for all $A \in \mathcal{A}$ with $E \subset A \subset A_0$, we have $\|\mu(E) - \mu(A)\| < e$.

<u>PROPOSITION</u> 3. Suppose that an &-valued function μ on a family $\mathcal{C}$ of sets has inner approximation in a subfamily $\mathcal{A}^1 = \mathcal{A}_\delta$, the family of all countable intersections of members of a lattice $\mathcal{A}$. Suppose that $\mu|\mathcal{A}$ is continuous from above at each member and is modular. Suppose further that $\mu|\mathcal{A}_\delta$ has outer approximation in $\mathcal{A}$. Then μ is continuous from above at each member of $\mathcal{C}$.

<u>PROOF</u> In view of the preceding proposition it suffices to show that $\mu|\mathcal{A}_\delta$ is continuous from above at each member. In turn it suffices to show that for each decreasing sequence $\{A_n\}_n$ of members of $\mathcal{A}$, $\mu(\bigcap_n A_n) = \lim_n \mu(A_n)$. Let $A^1 = \bigcap_{m=1}^\infty A_n$ and let $e > 0$. Since $\mu|\mathcal{A}_\delta$ has outer approximation in $\mathcal{A}$, there exists $A_0 \in \mathcal{A}$ so that $\|\mu(A) - \mu(A_0)\| < e$ for all A in $\mathcal{A}$ with $A^1 \subset A \subset A_0$. Since $\{A_n \cup A_0\}_n$ is a decreasing sequence of members of $\mathcal{A}$ with $\bigcap_n (A_n \cup A_0) = A_0 \in \mathcal{A}$, we have $\lim_n \mu(A_n \cup A_0) = \mu(A_0)$. Consequently for all large n and for each p,

$$\|\mu(A_n \cup A_0) - \mu(A_{n+p} \cup A_0)\| < e.$$

By using the modularity of $\mu|\mathcal{A}$,

$$\|\mu(A_n) - \mu(A_{n+p})\|$$

$$= \| \mu(A_n \cup A_0) + \mu(A_n \cap A_0) - \mu(A_0) - \mu(A_{n+p} \cup A_0)$$

$$- \mu(A_{n+p} \cap A_0) + \mu(A_0) \|$$

$$\leq \| \mu(A_n \cup A_0) - \mu(A_{n+p} \cup A_0) \| + \| \mu(A_n \cap A_0) - \mu(A_0) \|$$

$$+ \| \mu(A_{n+p} \cap A_0) - \mu(A_0) \|$$

$< e + e + e$ for all large n and each p.

It follows that $\lim_n \mu(A_n)$ exists. The value of this limit can be seen to be equal to the limit of the net $\{\mu(A): A \in \mathcal{A}, A \supset A^1\}$ which in turn is $\mu(A^1)$ because of the outer approximation of $\mu|\mathcal{A}_\delta$ in the subfamily $\mathcal{A}$.

The way is now clear for the proof of our final theorem. It generalizes the known theorems on extension of vector valued measures. Besides, our proof is direct (not using the scalar theorem) and is completely elementary (not using any Banach space theory, not even the idea of a linear functional).

An $\mathcal{E}$-valued function μ on a family $\mathcal{A}$ of sets is <u>inner tight</u> if for each pair of members A_1, A_2 in $\mathcal{A}$ with $A_1 \supset A_2$ and for each $e > 0$ there exists a member A_3 in $\mathcal{A}$, $A_3 \subset A_1 \setminus A_2$ so that

$$\| \mu(A_1) - \mu(A_2) - \mu(A_3) \| < e$$

μ is <u>strongly bounded</u> on $\mathcal{A}$ iff for each sequence $\{A_n\}_n$ of mutually disjoint members of $\mathcal{A}$, $\lim_n \mu(A_n) = 0$. If μ is inner tight then strong boundedness implies that for each increasing sequence $\{A_n\}_n$ of members of $\mathcal{A}$, $\lim_n \mu(A_n)$ exists in $\mathcal{E}$, and this is the form in which we use strong boundedness in what follows.

<u>THEOREM</u> 4. An $\mathcal{E}$-valued function μ on a lattice $\mathcal{A}$ of sets extends to an $\mathcal{E}$-valued countably additive measure $\bar{\mu}$ on a δ-ring $\bar{\mathcal{A}} \supset \mathcal{A}$ with inner approximations in the family $\mathcal{A}_\delta$ provided μ is

i) finitely additive
ii) continuous at ϕ
iii) inner tight
iv) for each increasing sequence $\{A_n\}_n$ of members of $\mathcal{A}$ whose union is contained in a member of $\mathcal{A}$, $\lim_n \mu(A_n)$ exists in &.

If in place of iv), μ satisfies the stronger hypothesis of strong boundedness on $\mathcal{A}$, then $\bar{\mathcal{A}}$ can be taken to be a σ-field instead of a δ-ring.

PROOF An outline of the proof is as follows.

STEP 1 The assumption iv) on μ implies that for each A_1, A_2 in where $A_1 \supset A_2$, the net $\{\mu(A) : A \in \mathcal{A},\ A \subset A_1 \setminus A_2\}$ ordered by inclusion, has a limit in &. The value of the above limit is $\mu(A_1) - \mu(A_2)$, by inner tightness and additivity of μ.

STEP 2 Using step 1, μ can be shown to be modular, and continuous from above at each member of $\mathcal{A}$.

STEP 3 Using assumption iv) above which is a compactness type of hypothesis on the range of μ, μ can be extended through outer approximation to the members of $\mathcal{A}_\delta$. Denote the resulting function on $\mathcal{A}_\delta$ by μ^1.

STEP 4 For each increasing sequence $\{A_n^1\}_n$ of members of $\mathcal{A}_\delta$ whose union is contained in a member of $\mathcal{A}$, the sequence $\{\mu^1(A^1 n)\}_n$ can be shown to have a limit in &. In case μ satisfies the hypothesis of strong boundedness, $\lim_n \mu^1(A^1 n)$ exists for every increasing sequence $\{A^1{}_n\}_n$ of members of $\mathcal{A}_\delta$

STEP 5 Using the conclusion in Step 4, μ^1 can be extended through inner approximation to the hereditary ring $\mathcal{C}$ of all subsets of the members of $\mathcal{A}$ (on alternatively to the family P(X) of all subsets of X if μ satisfies the hypothesis of strong boundedness). Denote the resulting extension of μ^1 to $\mathcal{C}$ (or to P(X)) by μ_*

STEP 6 Using proposition 3, μ_* can be seen to be an &-valued inner measure.

STEP 7 Using the inner tightness of μ, each member of $\mathcal{A}$ can be shown to be μ_*-additive

STEP 8 By theorem 1, the restriction of μ to the family $\bar{\mathcal{A}}$ of μ_* - additive sets in its domain yields the desired coutably additive extension of μ.

A brief sketch of the proofs of the different steps is given below.

STEP 2 We need to establish the continuity of μ at each member. Let $\bar{A} \in \mathcal{A}$ and let $\{A_n\}_n$ be an arbitrary sequence of members of whose intersection is $\bar{A}$. By Step 1, choose $\bar{A}_n$ in $\mathcal{A}$ for each n so that $\bar{A}_n \subset A_n \setminus \bar{A}$ and $\| \mu(A_n) - \mu(\bar{A}) - \mu(A) \| < e/2^{n+1}$

for all $A \in \mathcal{A}$ with $\bar{A}_n \subset A \subset A_n \setminus \bar{A}$. Replacing A in the above inequality by the sets $\bar{A}_k$ and $\bar{A}_k \cup (\bar{A}_{k+1} \cap \bar{A}_{k+2} \cap .. \cap \bar{A}_n)$ in succession we see that for each n and for each k < n,

$$\| \mu(\bar{A}_k \cup (\bar{A}_{k+1} \cap \bar{A}_{k+2} \cap .. \cap \bar{A}_n) - \mu(\bar{A}_k) \| < e/2^k$$

By a repeated application of the modularity of μ we can then show that

$$\| \mu(A_n) - \mu(\bar{A}) - \mu(\bar{A}_1 \cap \bar{A}_2 \cap .. \cap \bar{A}_n) \| < \frac{e}{2} + \frac{e}{2^2} + \cdots + \frac{e}{2^n}$$

for each n. It follows that

$$\| \mu(A_n) - \mu(\bar{A}) \| \leq \| \mu(A_1 \cap \bar{A}_2 \cap .. \cap \bar{A}_n) + e$$

The continuity of μ at ϕ implies that $\lim_n \mu(\bar{A}_1 \cap \bar{A}_2 \cap .. \cap \bar{A}_n = 0$

Consequently $\lim_n \mu(A_n) = \mu(\bar{A})$.

STEP 3 We first show that for each decreasing sequence $\{A_n\}_n$ in $\mathcal{A}$ (With $\cap_n A_n$ not necessarily in $\mathcal{A}$), $\lim_n \mu(A_n)$ exists in $\mathcal{E}$. We know from Step 1 that $\lim \{\mu(A): A \in \mathcal{A}, A \subset A_1 \setminus A_n\} = \mu(A_1) - \mu(A_n)$ for each n. We may then choose A_{1n} in $\mathcal{A}$, $A_{1n} \subset A_1 \setminus A_n$ so that $\| \mu(A_1) - \mu(A_n) - \mu(A) \| < e/_3$ for all A in $\mathcal{A}$ where $A_{1n} \subset A \subset A_1 \setminus A_n$. Since $\{A_{1n}\}_n$ is increasing and $\cup_n A_{1n} \subset A_1$, $\lim_n \mu(A_{\ n})$ exists in $\mathcal{E}$ by assumption iv) in the theorem. So $\| \mu(A_{1n}) - \mu(A_{1,n+p}) \| < e/_3$ for all large n and all p. Now

$$\| \mu(A_n) - \mu(A_{n+p}) \| \leq \| \mu(A_1) - \mu(A_n) - \mu(A_{1n}) \| +$$

$$\| \mu(A_1) - \mu A_{n+p}) - \mu(A_{1\ n+p}) \| + \| \mu(A_{1n}) - \mu(A_{1\ n+p}) \|$$

$$\leq e/_3 + e/_3 + e/_3$$

Consequently $\lim_n \mu(A_n)$ exists. It is easy to see that this limit is the same as the limit of the net $\{\mu(A): A \in \mathcal{A}, A \supset \bigcap_{n=1}^{\infty} A_n\}$. It then follows that the definition $\mu^1(\bigcap_{n=1}^{\infty} A_n) = \lim_n \mu(A_n)$ extends μ unambiguously to the members of $\mathcal{A}_\delta$. It can also be seen that μ^1 is continuous from above at each member of $\mathcal{A}_\delta$.

STEP 4 Let $\{A_n^{\ 1}\}_n$ be any increasing sequence of members of $\mathcal{A}_\delta$ with $\cup_n A_n^{\ 1} \subset A_0$ where A_0 is some member of $\mathcal{A}$, the last restriction being unnecessary if μ is strongly bounded. We need to show that $\lim_n \mu^1(A_n^{\ 1})$ exists in $\mathcal{E}$. Let $e > 0$. Choose $A_n \in \mathcal{A}$ for each n, $A_n \in \mathcal{A}$ so that

$$\| \mu^1(A_n^{\ 1}) - \mu(A) \| < {}^e/_{2^{n+1}} \quad \text{for all } A \in \quad \text{where}$$

$A_0 \supset A_n \supset A \supset A_n^{\ 1}$. Choosing for A the sets A_k and

$A_k \cap (A_{k+1} \cup A_{k+2} \cup .. \cup A_n)$ is succession we see that for each n and each $k < n$,

$$\| \mu(A_k \cap (A_{k+1} \cup A_{k+2} \cup \ldots \cup A_n)) - \mu(A_k) \| < e/_{2^k}$$

Then by a repeated application of the modularity of μ we can deduce that

$$\| \mu^1(A_n{}^1) - \mu(A_1 \cup A_2 \cup \ldots \cup A_n) \| < \frac{e}{2} + \frac{e}{2^2} + .. + \frac{e}{2^n}$$

By hypothesis iv) in the theorem (or the stronger hypothesis of strong boundedness as the case may be), the sequence $\{\mu(A_1 \cup A_2 \ldots \cup A_n)\}_n$ has a limit in $\mathcal{E}$. It then follows that the sequence $\{\mu^1(A_n{}^1)\}_n$ has the same limit in $\mathcal{E}$.

STEPS 5,6. Let E be an arbitrary subset of some member of $\mathcal{A}$ (or an arbitrary subset of X if μ is strongly bounded). Define $\mu_*(E)$ as

$$\mu_*(E) = \lim \{\mu^1(A^1): A^1 \in \mathcal{A}_\delta, A^1 \subset E\}.$$

The above limit exists by the conclusion in Step 4. It is easily shown that μ^1 is modular and continuous from above at each member of $\mathcal{A}_\delta$. Then by proposition 2, μ_* is continuous from above at each member of its domain and is consequently an $\mathcal{E}$-valued inner measure.

STEP 7 Let $S \in \mathcal{A}$ and let E be an arbitrary member of the domain of μ_*. We need to show that

$$\mu_*(E) = \mu_*(E \cap S) + \mu_*(E \setminus S).$$

If $E \in \mathcal{A}$ then the above relation follows from the inner tightness of μ. The same relation then extends to each member $E \in \mathcal{A}_\delta$, by the continuity of μ_*, and then to an arbitrary member E in the domain of μ_* by inner approximation by members of $\mathcal{A}_\delta$.

STEP 8 completes the proof of the theorem.

REFERENCES

1. J. L. Kelley, M. K. Nayak and T. P. Srinivasan, Pre-measures on lattices of sets. II. Sympos. on Vector Measures, Salt Lake City, Utah, 1972.

2. M. K. Nayak and T. P. Srinivasan, Scalar and Vector Valued Pre-measures, Proc. Amer. Math. Soc. 47(1975).

3. M. K. Nayak, Vector Valued Pre-measure on Lattices of Sets, Thesis, Panjab University, Chandigarh, India(1974).

4. T. P. Srinivasan, On Extensions of Measures, J. Ind. Math. Soc. 19(1955), 31-60.

Department of Mathematics, University of Kansas, Lawrence, Kansas 66045.

Totally Summable Functions with Values in Locally Convex Spaces

G. Erik F. Thomas

The fundamental theorems of integration theory such as: the dominated convergence theorem, Fubini's theorem, Lebesgue's theorem on differentiation of the indefinite integral, are known to be valid for Bochner's integral but not for Pettis' integral. This article shows that these theorems hold for totally summable functions, a class of Pettis integrable functions generalising the Bochner class, and comprising most cases encountered in analysis (for instance in group representations, semigroups of operators, spectral theory).

Introduction. The integral introduced by Bochner [1] probably owes its success to the fact that the important and most useful theorems of the ordinary Lebesgue integration theory remain valid for the Bochner integral. This is the case in particular for: the dominated convergence theorem and its consequences (completeness of L^1), Fubini's theorem, Lebesgue's theorem on differentiation of the indefinite integral.

On the other hand, the Pettis integral [7], particularly useful when one considers functions with values in locally convex spaces, has not been so widely used, which may be due to the fact that none of the theorems mentioned above have analogues for this integral[1].

In spectral theory the need arises to integrate functions with values in locally convex spaces which in general are neither separable nor metrisable, and to have at one's disposal, as far as possible, the theorems mentioned above.

The object of this article is to describe a class of functions with values in general locally convex spaces, for which the above theorems do hold, and which at the same time is large enough for all applications. Roughly described these are the functions which are the product of a bounded measurable vector valued function and a scalar integrable function. We call these functions totally summable functions in analogy with the class of totally summable sequences introduced by A. Pietsch [9] p.29.

An example of totally summable functions is the following: Consider a function with values in the space of bounded linear operators in Hilbert space, equipped with the strong operator topology. If this function is measurable with respect to this topology, in a sense to be made precise later, the composition with the operator norm, $\|f(\cdot)\|$, is measurable, and if this is an integrable function, the function f will be totally summable. In order for f to be Bochner integrable f

(1) We discuss some negative results, such as the space of Pettis integrable functions with values in an infinite dimensional Banach space being incomplete, in an addendum to this article.

would have to be measurable with respect to the operator norm topology, which in many important examples is not the case (e.g. functions associated with strongly continuous group representations).

In general we shall make use of a norm to measure boundedness, 'the size of a function', but a topology weaker than the one defined by the norm to define measurability.

Actually it is not obvious what the definition of measurability ought to be in the case of functions with values in an arbitrary locally convex space E. A continuous function f: [0,1] → E ought to be measurable, but one cannot in general approximate such functions by a sequence of simple (finite valued) functions. On the other hand a limit of a sequence of continuous or simple functions does not necessarily have the kind of regularity property that one expects of 'measurable' functions. Scalar (weak) measurability or even Borel measurability does not in general give sufficient coherence to the function to furnish easily applicable integrability criteria. If the linear space is a Suslin space no such difficulties arise and integration over arbitrary abstract measure spaces is possible (cf. Thomas [12]). But in the present article we limit the situation to that of measures defined on the Borel sets of a topological space, not necessarily locally compact; measurability can then be defined in terms of continuity in the manner suggested by Bourbaki. The use of general topological spaces (as opposed to locally compact spaces) not only seems the most natural but makes it possible, as is desirable in certain contexts, to transport the measure from the parameter space onto the vector space. Since the available accounts of integration in general topological spaces (cf. [3] [10]) are quite involved, containing much more than is needed here, we briefly recapitulate the essential results needed here in §1, with as starting point the standard abstract measure theory.

1. Measure Theory and Integration of Scalar Functions.

Notations. We use the following abbreviations: S a topological Hausdorff space(2); B(S) the set of Borel subsets of S (i.e. the σ algebra generated by the open sets); $\mathcal{K}(S)$ the set of compact subsets of S; μ a Radon measure on S, that is: a countable additive set function μ: B(S) → [0,+ ∞] satisfying the conditions:

RM1 $\mu(A) = \sup_{\substack{K \subset A \\ K \in \mathcal{K}(S)}} \mu(K) \qquad \forall A \in B(S).$

RM2 S is the union of open sets of finite measure.

---

(2) If the need arises the rectriction "Hausdorff" can be removed at the price of some technical complications (cf. L. Schwartz [10])

In this article we shall always assume RM2': S is a countable union of open sets of finite measure[3].

We denote by B^{μ} the Lebesgue completion of B(S) with respect to the measure μ, which is extended to B^{μ} in the usual way. Condition RM_1 then remains valid for all $A \in B^{\mu}$. The sets belonging to B^{μ} are called μ-measurable sets[4].

The integration theory for scalar functions to be used in the sequal is just the standard theory relative to the measure space (S, B^{μ}, μ). Accordingly a function $f: S \to C$ or $\overline{R}$ is said to be μ-measurable if $f^{-1}(B) \in B^{\mu}$ $\forall B \in B(C)$ (resp. $B(\overline{R})$). The integral of a μ-measurable function $f: S \to [0, +\infty]$ is defined as the supremum of integrals of simple Borel or μ-measurable functions $\leq f$. $\mathcal{L}^1(\mu)$ stands for $\mathcal{L}^1_C(S, B^{\mu}, \mu)$ i.e. the space of real or complex μ-measurable functions f such that $\int |f| d\mu < +\infty$.

Besides the results valid for arbitrary measure spaces (e.g. Lebesgue's dominated convergence theorem). There are some special consequences of the regularity assumption RM1, of which we mention the following:

1) The existance of the support, the complement of the largest open nul set (the union of the set of all open nul sets is an open nul set by RM1).

2) For any μ-measurable function $f: S \to [0, +\infty]$

(1-1) $$\int_A f d\mu = \sup_{\substack{K \subset A \\ K \in \mathcal{K}}} \int_K f d\mu \qquad \forall A \in B^{\mu} .$$

3) (Lusin's theorem) A function $f: S \to C$ is μ-measurable iff for every $K \in \mathcal{K}(S)$ and $\epsilon > 0$ there is a compact subset $K' \subset K$ with $\mu(K \setminus K') \leq \epsilon$ such that the restriction of f to K' is continuous[5].

---

(3) A Radon measure satisfying RM2' is called is called moderated. This implies $\mu(A) = \hat{\mu}(A) \underset{\text{def}}{=} \inf_{\substack{A \subset O \\ O \text{ open}}} \mu(O)$, for all $A \in B(S)$.
For a general Radon measure one has to distinguish between μ and $\hat{\mu}$.

(4) It follows from RM1 and RM2' that $A \in B^{\mu}$ iff $A \cap K \in B^{\mu}$ for all $K \in \mathcal{K}(S)$. For a general Radon measure B^{μ} would be defined as the set of all A such that $A \cap K$ belongs to the Lebesgue completion for all K, and μ would be extended to B^{μ} by setting $\mu(A) = \sup_{K \subset A} \mu(K)$, for $A \in B^{\mu}$.

(5) In the original form of the theorem due to Lusin, f was said to coincide on K with a continuous function defined on the whole space. This is true here if S is normal or a subspace of a normal space, but for the present form of the theorem no abundance of continuous function defined on S is necessary. See L. Schwartz [10].

Recall the following definition due to Bourbaki [2]: A class $\mathcal{C}$ of compact subsets of S is <u>μ-dense</u> if it satisfies conditions $\mathcal{C}_1$ $\mathcal{C}_2$ and $\mathcal{C}_3$:

$\mathcal{C}_1$: $K_1, K_2 \in \mathcal{C} \Rightarrow K_1 \cup K_2 \in \mathcal{C}$

$\mathcal{C}_2$: $K' \subset K,\ K \in \mathcal{C},\ K' \in \mathcal{K} \Rightarrow K' \in \mathcal{C}$

$\mathcal{C}_3$: $\forall K \in \mathcal{K}\ \forall \epsilon > 0\ \exists K' \in \mathcal{C},\ K' \subset K$: $\mu(K \setminus K') \leq \epsilon$.

The class $\mathcal{C}_f$ of all $K \in \mathcal{K}$ such that f/K is continuous in any case possesses the properties $\mathcal{C}_1$ and $\mathcal{C}_2$. Thus f is μ-measurable if and only if $\mathcal{C}_f$ is μ-dense.

Under the assumptions RM_1 RM_2' properties $\mathcal{C}_1$ $\mathcal{C}_2$ $\mathcal{C}_3$ are equivalent to $\mathcal{C}_1$ $\mathcal{C}_2$ $\mathcal{C}_4$:

$\mathcal{C}_4$: Every $A \in B^{\mu}$ has a partition

$$A = N + \sum_{n=1}^{\infty} K_n$$

with $K_n \in \mathcal{C}$ $\mu(N) = 0$.

Relation (1-1) remains valid if K is replaced by any μ-dense class $\mathcal{C}$. This leads to the following proposition which we use as a basis for generalization to vector valued functions:

<u>Proposition 1.</u> Let $f: S \to C$ be μ-measurable and let $\mathcal{C} \subset \mathcal{C}_f$ be any μ-dense class. Then the following conditions are equivalent:

1) $f \in \mathcal{L}^1(\mu)$

2) $\lim_{K \in \mathcal{C}} \int_K f\, d\mu$ exists(6).

3) $\sum_n \int_{K_n} f\, d\mu$ converges for any sequence of disjoint sets $K_n \in \mathcal{C}$.

If f satisfies these conditions $\int f d\mu = \lim_{K \in \mathcal{C}} \int_K f\, d\mu$.

<u>Proof</u> 1) implies 2) and the last assertion, by separating f into nonnegative components and applying (1.1) extended as indicated above. Also 1) implies 3) by the countable additivity of the integral. For the converses 2) ⇒ 1) and 3) ⇒ 1) we may assume f to be real because the real and imaginary parts separately satisfy 2) or 3). If $|L - \int_K f\, d\mu| \leq 1$ for $K \supset K_o$ then for any $K \in \mathcal{C}$ $\int_K f\, d\mu = \int_{K \cup K_o \setminus K} f\, d\mu$ whence

$$\left|\int_K f\, d\mu\right| \leq |L| + 1 + \int_{K_o} |f|\, d\mu \quad \text{and} \quad \sup_{K \in \mathcal{C}} \left|\int_K f\, d\mu\right| < +\infty .$$

(6) By condition $\mathcal{C}_1$, $\mathcal{C}$ is directed upwards under the inclusion, hence $K \to \int_K f\, d\mu$ is a net.

Let $A = \{s: f(s) \geq 0\}$ $\int f^+ = \sup_{\substack{K \subset A \\ K \in C}} \int_K f d\mu < +\infty$, similarly $\int f^- < +\infty$, and so 2) implies 1). Also 3) $\Rightarrow$ 1). With A as above let $A = N + \sum_n K_n$ be a decomposition as in C_4. Then

$$\int f^+ = \sum_{n=1}^{\infty} \int_{K_n} f^+ = \sum_{n=1}^{\infty} \int_{K_n} f\, d\mu < +\infty. \text{ Similarly } \int f^- < +\infty$$

One other very useful property of μ-dense classes is the following: if C_n $n \geq 1$ is a finite or countable family of μ-dense classes $C = \bigcap_{n \geq 1} C_n$ is also a μ-dense class (given $K \in \mathcal{K}$ one inductively constructs $K_n \in C_n$ with $K_n \subset K_{n-1}$ $K_o = K$ $\mu(K_{n-1} \setminus K_n) \leq \epsilon/2^n$ so $K' = \bigcap_{n \geq 1} K_n \in C$ and $\mu(K \setminus K') \leq \epsilon$).

(2) Summable Functions

Let E be a locally convex topological vector space over C or R. We assume E to be Hausdorff and quasi-complete: i.e. all closed bounded sets are complete.

Definition. We shall say that a function $f:S \to E$ is μ-measurable if C_f is μ-dense.

Clearly the sum of two μ-measurable functions is μ-measurable ($C_f \cap C_g \subset C_{f+g}$) and the product of a μ-measurable vector function with a μ-measurable scalar function is μ-measurable ($C_\rho \cap C_g \subset C_{\rho g}$).

If E is a Banach space μ-measurability is equivalent to strong measurability in the sense of Bochner (approximation a.e by simple functions).

Proposition 2. Let $f: S \to E$ be μ-measurable, and let $C \subset C_f$ be any μ-dense class. The following conditions are equivalent:

1) $\lim_{K \in C} \int_K f\, d\mu$ exists.

2) For any sequence of disjoint sets $K_n \in C$ $\sum_n \int_{K_n} f\, d\mu$ converges(7).

3) f is Pettis integrable (relative to (S, B^μ, μ)). If these conditions are satisfied $\int f\, d\mu = \lim_{K \in C} \int_K f\, d\mu$ and more generally $\int_A f\, d\mu = \lim_{\substack{K \subset A \\ K \in C}} \int_K f\, d\mu$.

The above proposition makes sense because for $K \in C_f$ $\int_K f\, d\mu$ exists a priori as a Pettis integral (bipolar argument) or as a limit of 'Riemann sums'.

For the purpose of this article we only need the equivalence of 2) and 3), and we shall give the proof of that only. Clearly 3) implies 2) by the countable additivity of the Pettis integral. Conversely 2) $\Rightarrow$ 3)). We have $C \subset C_f \subset C_{x' \circ f}$

(7) The series converges, or the series converges unconditionally. This comes to the same thing as any partial series which must have the same property.

for all $x' \in E'$ and $\sum_n < \int_{K_n} f\, d\mu, x' > = \sum_n \int_{K_n} < f, x' > d\mu$ exists. By proposition 1 $< f,x'> \in \mathcal{L}^1(\mu)$ for all $x' \in E'$. Let $A \in B^\mu$ and let $A = N + \sum_{n=1}^{\infty} K_n$ be a decomposition as in C_4. Then

$$< \sum_{n=1}^{\infty} \int_{K_n} f\, d\mu, x' > = \sum_{n=1}^{\infty} \int_{K_n} < f, x' > d\mu = \int_A < f, x' > d\mu,$$

which proves that f is Pettis integrable and that $\int_A f\, d\mu = \sum_{n=1}^{\infty} \int_{K_n} f\, d\mu$ (8).

Definition. We call μ-summable any μ-measurable function which is Pettis integrable.

As in the classical case of Banach spaces the measurability has, apart from the above integrability criteria, the advantage that the function f is defined up to a nul function by the integrals $\int_A f\, d\mu$.

Let β be a closed absolutely convex subset of E and let $|x|_\beta = \inf_{x \in \lambda\beta} \lambda$. Then $x \to |x|_\beta$ is a lower-semi-continuous, positively homogeneous symmetric function with values in $[0, +\infty]$; a continuous semi-norm if β is a neighbourhood of 0. Thus $s \to |f(s)|_\beta$ is μ-measurable if f is μ-measurable.

Proposition 3. Let f be μ-summable. Then

$$|\int_A f\, d\mu|_\beta \le \int_A |f(s)|_\beta d\mu \le +\infty .$$

This follows immediately from the Hahn-Banach theorem and the resulting identity $|x|_\beta = \sup_{x' \in \beta^o} |< x,x' >|$.

(Remark: In spite of the appearances the whole subject of vector integration can be treated without the Hahn-Banach theorem: for instance proposition 3 can be proved using the identity $|x|_\beta = \sup_{p \in \beta'} |x|_p$ where β' is the set of continuous semi-norms $p: x \to |x|_p$ such that $|x|_p \le |x|_\beta$ for all x. We cannot go into this further here).

(3) Totally Summable Functions

Let $B \subset E$ be a bounded, closed, absolutely convex subset of E.

Lemma a) $E_B = \{x: |x|_B < +\infty\} = \bigcup_{\lambda > 0} \lambda B$ is a linear subspace of E.

b) $x \to |x|_B$ is a norm on E_B; equipped with this norm E_B is a Banach space.

c) The inclusion operator $E_B \subsetneq E$ is continuous.

This is well known and easy to prove.

(8) Another treatment can be found in [11].

Definition. A μ-measurable function $f:S \to E$ is totally μ-summable if there exists $B \subset E$ as above such that $\int |f|_B \, d\mu < +\infty$.

Proposition 4. Every totally summable function is summable.

Proof. Let K_n be pairwise disjoint in C_f. Then

$$\sum_n \left| \int_{K_n} f \, d\mu \right|_B \leq \sum_n \int_{K_n} |f|_B \, d\mu \leq \int |f|_B \, d\mu < +\infty .$$

Thus $\sum_n \int_{K_n} f \, d\mu$ converges normally in E_B, a fortiori in E.

Remark. Since $\left| \int_A f \, d\mu \right|_B \leq \int_A |f|_B \, d\mu < +\infty$, $\int_A f \, d\mu \in E_B$, for all $A \in B^{\mu}$, and an argument similar to the above proof shows that $A \to \int_A f \, d\mu$ is countable additive in E_B.

We denote by $\mathcal{L}^1(\mu; E_B; E)$ the class of μ-measurable functions $f:S \to E$ such that $\int |f|_B \, d\mu < +\infty$. $L^1(\mu;E_B;E)$ denotes the corresponding set of equivalence classes of functions equal μ a.e. Clearly these are linear spaces.

If E is a Banach space a function f is totally summable if and only if f is Bochner integrable.

In general the definition implies $|f(s)|_B < +\infty$ μ a.e. which means $f(s) \in E_B$ μ a.e. The function $f: S \to E_B$ (defined a.e.) need not be μ-measurable however.

Proposition 5. A function $f: S \to E$ is totally μ-summable iff there exists $\rho \in L^1(\mu)$ and $g:S \to E$ bounded μ-measurable such that $f(s) = \rho(s)g(s)$ μ a.e.

Proof. If $\int |f|_B \, d\mu < +\infty$, let $\rho(s) = |f(s)|_B$ if $|f(s)|_B < +\infty$, $g(s) = \frac{1}{\rho(s)} f(s)$ if $\rho(s) > 0$. Then ρ and g (defined a.e.) have the desired property. Conversely if f has such a decomposition let B be the closed absolutely convex hull of g(S). Then $|g(s)|_B \leq 1$ and $\int |f(s)|_B \, d\mu(s) \leq \int |\rho(s)| d\mu(s) < +\infty$.

Example. See introduction.

If there is a set $B \subset E$ such that $\int |f|_B \, d\mu < +\infty$ and such that $f:S \to E_B$ defined a.e.) is μ-measurable we shall say that the function $f:S \to E$ is Bochner integrable. This extension of the notion of Bochner integrability (which reduces immediately to the Bochner integral property speaking) is only useful because in some cases it is not obvious what the set B may be (In [12] we have shown that any scalarly (weakly) integrable function with values in D' and other nuclear spaces is Bochner integrable).

Remark. If $\mu(S) < +\infty$ proposition 5 implies that every bounded μ-measurable function $f:S \to E$ is totally μ-summable. But even a continuous function $f:[0,1] \to E$ need not be Bochner integrable (with respect to Lebesgue measure). Examples: $E = L_s(\mathcal{H})$ the space of operators in a Banach space $\mathcal{H}$, with the strong operator topology $f(t) = P_t$ a strongly continuous but not uniformly continuous semi-group.

$E = R^I$ $f(s) = (f_i(s))_{i \in I}$ the family of all real continuous functions on [0,1]. In the last example f is not Bochner integrable with respect to any non atomic Radon measure on [0,1].

We now come to the main point of this article: the general theorems mentioned in the introduction.

Theorem 1. Let $f_n \in \mathcal{L}^1(\mu; E_B; E)$ be a sequence such that $f_n(s) \to f(s)$ in E_B μ a.e and such that there exists a function $g \in \mathcal{L}^1(\mu)$ with $|f_n(s)|_B \leq g(s)$ μ a.e. Then $f \in \mathcal{L}^1(\mu; E_B; E)$ and $\int |f - f_n|_B \, d\mu \to 0$; in particular $\int f_n d\mu \to \int f \, d\mu$ in E_B.

Proof. $F_n(s) = |f_n(s) - f(s)|_B = \lim_{m \to \infty} |f_n(s) - f_m(s)|_B$ a.e. is a measurable function of s; $F_n(s) \to 0$ μ a.e. Thus by Egoroff's theorem the set $\mathcal{C}$ of compact subsets of S such that each $f_{n/K}$ is continuous (E valued) and such that F_n/K tends to zero uniformly, is μ-dense. But for $K \in \mathcal{C}$ $f_n(s) \to f(s)$ in E uniformly on K, whence f/K is continuous and since $\mathcal{C} \subset \mathcal{C}_f$ f is μ-measurable. The remainder of the proof is obvious.

Corollary 1. Let $f_n \in \mathcal{L}^1(\mu; E_B; E)$ be a sequence such that $\sum_{n=1}^{\infty} \int |f_n|_B d\mu < +\infty$. Then $f(s) = \sum_{n=1}^{\infty} f_n(s)$ exists μ a.e., $f \in \mathcal{L}^1(\mu; E_B; E)$ and $\int f \, d\mu = \sum_{n=1}^{\infty} \int f_n \, d\mu$ (in fact: $\int |f - s_n|_B \, d\mu \to 0$ where $s_n = f_1 + f_2 + \dots + f_n$).

Corollary 2. $L^1(\mu; E_B; E)$ is a Banach space.

Proof: By corollary 1 every normally convergent series is convergent.

Remarks 1. If in theorem 1 we only assume $f_n(s) \to f(s)$ in E (rather than in E_B) the conclusion is false: f might not be μ-measurable. But if it is known that f is μ-measurable, (e.g. if E is metrisable) f is totally μ-summable and $\int f_n \, d\mu \to \int f \, d\mu$ in E.

2. The space $L^1(\mu; E_B)$ of E_B valued (classes of) Bochner integrable functions is in general a closed subspace of $L^1(\mu; E_B; E)$. If it is known that E_B is separable or reflexive the two spaces coincide.

More generally if $E \hookrightarrow F$ where F is a second locally convex Hausdorff space, $L^1(\mu; E_B; E)$ is a closed subspace of $L^1(\mu; E_B; F)$.

In most cases there seems to be an F so that $L^1(\mu; E_B; F)$ is maximal. Can it happen that there is no such F?

Next we consider the generalization of Fubini's theorem to totally summable functions. Given two Radon measures μ on T and ν on S there is a unique Radon measure m on $T \times S$ such that $m(A \times B) = \mu(A) \, \nu(B)$ for all $A \in B(T), B \in B(S)$. If we assume as always that μ and ν satisfy the condition RM_2' (i.e. are moderated) the same will be true for m.

Theorem 2. If $f:S \times T \to E$ is totally m-summable, $t \to f(t,s)$ is totally μ-summable for ν a.a $s \in S$, $s \to \int f(s,t)\, d\mu(t)$ is totally ν-summable and

$$\int f\, dm = \int d\nu(s) \int f(s,t)$$

Proof. (routine except for the measurability of the map $s \to \int f(s,t)d\mu(t)$): The function f being m-measurable there exists a partition $S \times T = N + \sum_{n=1}^{\infty} K_n$, with $K_n \in C_f$ and $m(N) = 0$. If we put $A_s = \{t:(s,t) \in A\}$, $T = N_s + \sum_{n=1}^{\infty} K_{n,s}$ and for ν a.a. s, $\mu(N_s) = 0$, that is for $s \notin N_1$ with $\nu(N_1) = 0$. Also $K_{n,s}$ is compact and $f_s/K_{n,s}$ is continuous (where $f_s(t) = f(s,t)$). Thus for $s \notin N_1$ f_s is μ-measurable.

By hypothesis there is B such that $\int |f|_B dm = \int d\nu(s) \int |f(s,t)|_B d\mu(t) < +\infty$, $\int |f(s,t)|_B d\mu(t) < +\infty$ ν a.e. Thus there is a ν-nul set $N_2 \supset N_1$ such that for $s \notin N_2$ f_s is totally μ-summable.

It remains to show that i) $s \to \int f(s,t)d\mu(t)$ is ν-measurable (on N_2^c), and ii) $\int d\nu(s) \int f(s,t)d\mu(t) = \int f\, dm$. It will be sufficient to prove these points for each function $f_n = 1_{K_n} f$ instead of f. For this done, we would have

$$\sum_{n=1}^{\infty} \left| \int f_n(s,t)d\mu(t)\right|_B \leq \sum_{n=1}^{\infty} \int |f_n(s,t)|_B d\mu(t) \leq \int |f(s,t)|_B d\mu(t)$$

as well as

$$f(s,t) = \sum_{n=1}^{\infty} f_n(s,t) \quad \text{for } \mu \text{ a.a } t$$

provided $s \notin N_2$, which implies

$$\int f(s,t)d\mu(t) = \sum_{n=1}^{\infty} \int f_n(s,t)d\mu(t)$$

by corollary 1 of theorem 1. The same corollary implies that $s \to \int f(s,t)d\mu(t)$ is ν-measurable (on N_2^c) and that

$$\int d\nu(s) \int f(s,t)d\mu(t) = \sum_{n=1}^{\infty} \int d\nu(s) \int f_n(s,t)d\mu(t)$$

which would be equal to

$$\sum_{n=1}^{\infty} \int f_n dm = \int f\, dm .$$

To prove points i) above we need the following two lemmas:

Lemma 1. Let $K \subset S \times T$ be compact. Then the map $s \to 1_K(s,-) \in L^1(T,\mu)$ is ν-measurable.

This easily follows from the fact that every function in $L^1(S \times T,m)$, in particular 1_K, has a series expansion $\Sigma\, \varphi_n(s)\, \psi_n(t)$ with $\sum_n \int |\varphi_n| d\nu \int |\psi_n| d\mu < +\infty$.

Thus according to this lemma, for every $S_1 \in \mathcal{K}(S)$ and $\epsilon > 0$ there is $S_o \in \mathcal{K}(S_1)$ with $\nu(S_1 \setminus S_o) \leq \epsilon$ such that the restriction to S_o of the map $s \to 1_K(s,-) \in L^1(T,\mu)$ is continuous.

Lemma 2. Let $f: S \times T \to E$ be a function and $K \in \mathcal{C}_f$. Assume the restriction to $S_o \in \mathcal{K}(S)$ of the map $s \to 1_K(s,-) \in L^1(\mu)$ is continuous, and that $f_s = f(s,-)$ is μ-summable for all $s \in S_o$. Then the map $s \to \int 1_K(s,t)\, f(s,t) d\mu(t)$ from S_o to E is continuous.

Assume this result then from lemma 1 and 2 it immediately follows that $s \to \int f_n(s,t) d\mu(t) = \int 1_{K_n}(s,t)\, f(s,t) d\mu(t)$ is ν-measurable. From

$$\int d\nu(s)\ \left| \int f_n(s,t) d\mu(t) \right|_B \leq \int d\nu(s) \int |f_n(s,t)|_B\, d\mu(t) = \int |f_n| dm < + \infty$$

it follows that the same function is totally summable and finally, applying for instance linear forms, it is clear that $\int d\nu(s) \int f_n(s,t) d\mu(t) = \int f_n\, dm$.

Thus all that remains is the Proof of Lemma 2. Replacing T by the projection of K we may assume without loss of generality that T is compact. Let $K_o = K \cap S_o \times T$. For $s \in S_o$ $1_K(s,-) = 1_{K_o}(s,-)$. Thus we may replace K by K_o and assume $K \subset S_o \times T$.

Put $I(s) = \int 1_K(s,t)\, f(s,t) d\mu(t)$ for $s \in S_o$. To prove that I is continuous we first consider a particular case:

Case 1 $E = \mathbb{R}$. Let $\hat{f}$ be a continuous extension of f/K to $S_o \times T$. Then if $M = \sup_{(s,t) \in K} |f(s,t)|$ and $s_o \in S_o$ we have

$$|I(s) - I(s_o)| \leq M \int |1_K(s,t) - 1_K(s_o,t)|\, d\mu(t)$$

$$+ \int 1_K(s_o,t) |\, \hat{f}(s,t) - \hat{f}(s_o,t) |\, d\mu(t)$$

and this is small for s near s_o, (the first part by hypothesis).

General case: E locally convex. $|\ |$ now stands for a continuous seminorm. We may write $I(s) = \int 1_K(s,t)\, f_K(s,t) d\mu(t)$ where $f_K = 1_K\, f$.

$$|\, I(s) - I(s_o) | \leq M \int |1_K(s,t) - 1_K(s_o,t)|\, d\mu(t)$$

$$+ \int 1_K(s_o,t)\ |f_K(s,t) - f_K(s_o,t)| d\mu(t) \quad ;$$

it suffices to prove that the second part is small for s near s_o.

Replacing T by K_{s_o} this becomes $R(s) = \int |f_K(s,t) - f_K(s_o,t)| d\mu(t)$ while $(s_o,t) \in K$ for all $t \in T$. Since $R(s_o) = 0$ it is sufficient to prove that R is a continuous function of s. Let $F(s,t) = |f_K(s,t) - f_K(s_o,t)|$. Then F/K and F/K^c

are continuous (for $(s,t) \in K^c$ $F(s,t) = |f_K(s_o,t)|$).

$$R(s) = \int 1_K(s,t)\, F(s,t)d\mu(t) + \int 1_{K^c}(s,t)\, F(s,t)d\mu(t).$$

The first part is continuous by the proof of case 1. The second part equals

$$\int 1_{K^c}(s,t)\, |f_K(s_o,t)|d\mu(t) = \int |f_K(s_o,t)|d\mu(t) - \int 1_K(s,t)\, |f_K(s_o,t)|d\mu(t)$$

which is also continuous in s.

Thus lemma 2 is proved and this terminates the proof of theorem 2.

Next we mention the theorem on superposition of integrals which, were it is not for a technical restriction, would be a generalization of the preceding theorem:

Theorem 3. Let T be a topological Hausdorff space, the compact subsets of which are metrisable. Let μ, μ_s for $s \in S$, be Radon measurable on T, ν a Radon measure on S (all moderated) and assume

$$\mu(A) = \int \mu_s(A)\, d\nu(s) \qquad \forall \ A \in B(T).$$

Then if $f: T \to E$ is totally μ-summable, f is totally μ_s-summable for ν-a.a s, $s \to \int f\, d\mu_s$ is totally ν-summable and

$$\int f\, d\mu = \int d\nu(s) \int f\, d\mu_s \ .$$

We only sketch the proof: Here again it suffices to prove that for $K \in \mathcal{C}_f$

i) $s \to \int_K f\, d\mu_s$ is totally ν-summable.

ii) $\int d\nu(s) \int_K f\, d\mu_s = \int_K f\, d\mu$. The main point is again to prove that K is metrisable makes it possible to conclude that the map $s \to \mu'_s = \mu_s/\mathcal{L} \in M(K)$ is ν-measurable with respect to the topology $\sigma(M(K), \mathcal{C}(K))$. One can then apply the following lemma to conclude:

Lemma. Let $(\mu'_s)_{s \in S_o}$ be a family of Radon measures on a compact set K, continuous in the weak dual topology $\sigma(M(K), C(K))$. Let $f: K \to E$ be continuous. Then $s \to \int f\, d\mu'_s$ is continuous on S_o.

We shall leave the proof of this lemma and the details of the proof of theorem 3 to the reader.

Finally we turn to the Lebesgue differentiation theorem: "$f(s) = \frac{d}{ds}\int_a^s f(t)dt$ ae" and its generalizations. This theorem is known to fail for general summable functions $f: [a,b] \to E$ (9) .

(9) See for instance Munroe [6] and Phillips [8] p.144.

Even if we assume $\int|f(s)|_p ds < +\infty$ for all continuous semi-norms (e.g. E a separable Hilbert space with its weak topology in Philips example), the problem is that although (by Bochner's generalization of Lebesgue's theorem) for each continuous semi-norm p, $\lim_{h\to o} |f(s) - \frac{1}{h}\int_s^{s+h} f(t)dt|_p = 0$ a.e, the exceptional nul set N_p depends on P in general, and $\bigcup_P N_p$ need not be negligeable (indeed can be the whole interval)(10). It is remarkable that if f: [a,b] → E is bounded and measurable this difficulty does not occur, nor does it arise if f is totally summable.

Let $S = R^m$, let μ be any Radon measure on S. Let $|\ |$ be any norm on $S = R^m$ and put $B(s,\epsilon) = \{t: |s-t| \leq \epsilon\}$. Recall that $s \in \operatorname{supp}\mu$ iff $\mu(B(s,\epsilon)) > 0$ for all $\epsilon > 0$.

<u>Theorem 4.</u> Let f: S → E be totally μ-summable. Then for almost all $s \in \operatorname{supp}\mu$

$$(*)\quad \lim_{\epsilon\downarrow 0} \frac{1}{\mu(B(s,\epsilon))} \int_{B(s,\epsilon)} |f(s) - f(t)|_p \, d\mu(t) = 0 \qquad \forall p$$

In particular:

$$f(s) = \lim_{\epsilon\downarrow 0} \frac{1}{\mu(B(s,\epsilon))} \int_{B(s,\epsilon)} f(t)\, d\mu(t) \qquad \mu \text{ a.e.}$$

<u>Definition:</u> A point $s \in \operatorname{supp}\mu$ for which (*) holds for all continuous semi-norms p, will be called a Lebesgue point of f (relative to μ). Thus the theorem states that μ a.e. $s \in S$ is a Lebesgue point.

<u>Remark:</u> If $\int|f|_B \, d\mu < +\infty$ relation (*) does not necessarily hold for μ a.e.s, with $|\ |_B$ instead of $|\ |_p$. Indeed it is easy to see, if μ is Lebesgue measure, that this would be true if and only if $f:S \to E_B$ (defined a e) were Bochner integrable. An explicit counter example: let $E = M[0,1]$ with the topology $\sigma(M[0,1],C[0,1])$ and let $f(t) = \delta_{(t)}$ the unit mass at t. Then f:(0,1) → E is continuous, so every point is a Lebesgue point. But $\frac{1}{h}\int_s^{s+h} \|\delta_{(s)} - \delta_{(t)}\|dt = 2$ for all $h > 0$, $\|\ \|$ being the norm of total variation. Any bounded subset of E is bounded in norm also; consequently $\varliminf_{h\downarrow 0} \frac{1}{h}\int_s^{s+h} |f(s) - f(t)|_B dt > 0$ for any bounded B, and all $s \in (0,1)$.

<u>Remark.</u> In theorem 4 it is clearly sufficient that f be locally totally summable (i.e. totally μ-summable over every compact subset of R^m)

<u>Proof.</u> It is known that if ρ is a real μ-integrable function, then μ a.e. point of suppμ is a Lebesgue point for ρ. This follows from the maximal inequality.

(10) If E is metrisable this difficulty does not arise, but then, as has been shown by Grothendieck [4] the above hypothesis implies that f:S → E is Bochner integrable.

$$\mu\{s: \rho^*(s) > \lambda\} \leq \frac{c}{\lambda} \int |\rho| d\mu \qquad \text{where}$$

$$\rho^*(s) = \sup_{\epsilon > 0} \frac{1}{\mu(B(s,\epsilon))} \int_{B(s,\epsilon)} |\rho(t)| d\mu(t)$$

which can be proved using the Besicovitch covering lemma in the case of the Euclidean norm and the Morse covering lemma in the case of other norms(11).

In particular, for any compact $K \subset S$, μ a.e. point of K is a density point

$$\lim_{\epsilon \downarrow 0} \frac{\mu(K \cap B(s,\epsilon))}{\mu(B(s,\epsilon))} = 1 \quad \mu \quad \text{a.a.} s \in K .$$

Equivalently: $$\lim_{\epsilon \downarrow 0} \frac{\mu(K^c \cap B(s,\epsilon))}{\mu(B(s,\epsilon))} = 0 \quad \mu \quad \text{a.a.} \quad s \in K.$$

We know (proposition 5) that any totally summable function is equal a.e. to a function ρg where $\rho \in L^1(\mu)$, $g: S \to E$ is μ-measurable and bounded.

The proof of the theorem now follows from the following two lemmas:

Lemma 1. Assume $f(t) = \rho(t)g(t)$ μ a.e. ρ and g as above. If s is a Lebesgue point for both ρ and g and if $f(s) = \rho(s)g(s)$, s is a Lebesgue point for f.

Proof.

$$\int_{B(s,\epsilon)} |f(s) - f(t)|_p d\mu(t) = \int_{B(s,\epsilon)} |\rho(s)g(s) - \rho(t)g(t)| d\mu(t)$$

$$\leq \int_{B(s,\epsilon)} |\rho(s)g(s) - \rho(s)g(t)|_p d\mu(t) + \int_{B(s,\epsilon)} |\rho(s)g(t) - \rho(t)g(t)|_p d\mu(t)$$

$$\leq |\rho(s)| \int_{B(s,\epsilon)} |g(s) - g(t)| d\mu(t) + \sup_t |g(t)|_p \int_{B(s,\epsilon)} |\rho(s) - \rho(t)| d\mu(t).$$

Dividing by $\mu(B(s,\epsilon))$ for $s \in \operatorname{supp}\mu$, and letting ϵ tend to zero, gives the result.

Lemma 2. Let $g: S \to E$ be bounded μ-measurable. Let g/K be continuous and let $s \in K$ be a density point of K. Then s is a Lebesgue point of g.

Proof. Let p and $\eta > 0$ be given. Then for $t \in K$ $|g(s) - g(t)|_p \leq \eta$ provided, $|s-t| \leq \delta$. For $0 < \epsilon \leq \delta$ we have

$$\int_{B(s,\epsilon)} |g(s) - g(t)|_p d\mu(t) = \int_{K \cap B(s,\epsilon)} |g(s)-g(t)|_p d\mu(t) + \int_{K^c \cap B(s,\epsilon)} |g(s)-g(t)| d\mu(t)$$

$$\leq \eta\, \mu(B(s,\epsilon)) + 2\|g\|_{p,\infty} \mu(K^c \cap B(s,\epsilon))$$

$$\leq 2\eta \; \mu(B(s,\epsilon)) \quad \text{for } \epsilon \text{ sufficiently small. (where } \|g\|_{p,\infty} = \sup_t |g(t)|_p).$$

(11) M. de Guzman [5] pp.6 and 37-41.

It remains only to observe that $S = N + \sum_{n=1}^{\infty} K_n$ with $\mu(N) = 0$, g/K_n continuous, and that μ a.e. point of each K_n is a density point, in order to conclude that μ a.e. point of S is a Lebesgue point of g; lemma 1 shows that μ a.e. point is a Lebesgue point of f.

Remark. One proves similarly that if $f: R \to E$ is (locally) totally summable with respect to Lebesgue measure, then for a.a $s \in R$.

$$\lim_{h \downarrow 0} \frac{1}{h} \int_s^{s+h} |f(s) - f(t)|_p dt = 0 \quad \text{for all } p .$$

In particular $f(s) = \frac{d}{ds} \int_a^s f(t)dt$ a.e.

Remark. Lemma 2 is an adaptation of a theorem due to Denjoy (1915): a point of approximate continuity of a bounded measurable function is a Lebesgue point.

Remark. (concerning the previous article). Throughout this article we might have made the less stringent assumption that f is μ-measurable relative to the weak topology on E (which is nevertheless more restrictive than scalar or so called weak measurability). But the conclusions would also have been weaker: for instance one would only have $f(s) = \lim_{h \to 0} \frac{1}{h} \int_s^{s+h} f(t)dt$ weakly, almost everywhere. On the other hand all those 'more general' results can be deduced from the previous ones by replacing E by its weak completion. In most cases E is a Frechet space, or a Suslin space, or a weak space, in which case both measurability conditions amount to the same thing.

References

[1] S. Bochner: Integration von Funktionen, derer Werte die Elemente eines Vectorraumes sind. Fund. Math. 20 pp.262-276 (1938).

[2] N. Bourbaki: Integration Chapter V, Paris Hermann.

[3] ________ : Integration Chatper IX, Paris Hermann.

[4] A. Grothendieck: Produits Tensoriels Topologiques et Espaces Nulceaires. A.M.S. Memoir n° 16 (1966).

[5] M. de Guzman: Differentiation of Integrals in R^n, Springer Lecture notes 481.

[6] M.E. Munroe: A note on weak differentiability of Pettis integrals. Bulletin A.M.S. 52 (1946) pp.167-174.

[7] B.J. Pettis: On integration in Vector Spaces. Trans. Amer. Math. Soc. 44, pp.277-304 (1938).

[8] R.S. Phillips: Integration in a convex linear topological space. Trans. Amer. Math. Soc. 47 pp.114-145 (1940).

[9] A. Pietsch: Nuclear Locally Convex Spaces, Springer-Verlag, Berlin 1972.

[10] L. Schwartz: Radon measures on arbitrary topological spaces. Tata Institute Publication, Bombay 1975.

[11] G.E.F. Thomas: The Lebesgue-Nikodym theorem for vector valued Radon measures. A.M.S. Memoir 139.

[12] ________ : Integration of functions with values in locally convex Suslin spaces. Transactions Amer.Math. Soc. 212 (1975) pp.61-81.

<u>Addendum</u> On some negative properties of the Pettis integral.

In the previous article we mentioned that the Pettis integral does not share certain properties with the Bochner integral. Below are some of these negative results:

Let E be a Banach space, which we assume to be separable to avoid the futile discussion of measurability. The space of Pettis integrable functions has the natural norm

$$N(f) = \sup_{|g| \leq 1} \| \int g f d\mu \| = \sup_{\|x'\| \leq 1} \int |< f,x'>| d\mu .$$

<u>Theorem 1</u> Let μ be Lebesgue measure on [0,1]. Let E be a separable infinite dimensional Banach space. Then the normed space of all Pettis μ-integrable functions $f: [0,1] \to E$ is incomplete.

<u>Proof.</u> In a previous paper [11][(1)] we proved that there exists a summable sequence $(x_n)_{n\in N}$ in E and functions $f_n \in \mathcal{L}^1[0,1]$ with $\int |f_n| dt \leq 1$ such that the vector measure $A \to \mu(A) = \sum_{n=1}^{\infty} \int_A f_n dt\, x_n$ does not posses a Pettis integrable density with respect to Lebesgue measure (and such that $\sum_{n=1}^{\infty} f_n x_n$ does not converge in measure). Let $F_n = \sum_{i=1}^{n} f_i x_i$. Then F_n is seen to be a Cauchy sequence in the Pettis norm, because of the relation $\lim_{N \to \infty} \sup_{\|x'\| \leq 1} \sum_{n=N}^{\infty} |< x_n, x' >| = 0$ (cf. Pettis [7]). On the other hand, there is no Pettis integrable function $F: [0,1] \to E$ such that $\lim_{n \to \infty} \int_A F_n dt = \int_A F\, dt$; a fortiori F_n is not convergent in the norm.

<u>Remark.</u> In this theorem Lebesgue measure can be replaced by any non atomic measure.

<u>Theorem 2.</u> Let E be any infinite dimensional Banach space. Then there exists a function $f: [0,1]^2 \to E$ Pettis integrable with respect to the two dimensional Lebesgue measure, such that the set of t for which $s \to f(s,t)$ fails to be Pettis integrable is not a nul set.

<u>Proof.</u> Let $[0,1] = \sum_{n=1}^{\infty} B_n$ be an infinite partition in sets of positive measure.

Proof. Let $f(s,t) = \frac{1}{\mu(B_n)} f_n(t) x_n$ for $s \in B_n$. Then f is easily seen to be Pettis integrable, but $s \to f(s,t)$ is integrable only if $\sum_n f_n(t) x_n$ converges (unconditionally) which is not the case for almost all t.

Concerning differentiation it seems likely that for every infinite dimensional Banach space E there is $f: [0,1] \to E$ Pettis summable such that $\sup_{h > 0} \| \frac{1}{h} \int_s^{s+h} f(t) dt \| = +\infty$ for a set {s} of positive measure.

(1) The numbers refer to the list at the end of the previous article.

Ein Nicht-Standard-Beweis für die Existenz eines Liftings

Von

B. Eifrig

Im folgenden wird mittels der Nicht-Standard-Analysis ein Beweis für die Existenz eines Liftings im Falle einer separablen Maßalgebra und bei vorgegebener unterer Dichte ohne die Restriktion der Separabilität gegeben. Die Integrationstheorie in Nicht-Standard-Modellen stößt auf große Schwierigkeiten [1], z.B. existiert noch kein eigenständiger Beweis für den Satz von Egerov.

Sei (Ω,F,P) ein vollständiger Wahrscheinlichkeitsraum, N die Menge der P-Nullmengen. Durch $A \Delta B \in N, A,B \in F$, wird auf F eine Äquivalenzrelation $\equiv$ definiert. Eine Abbildung $\theta: F \to F$, welche die folgenden Bedingungen (I)-(V) erfüllt, heißt Lifting.

(I) $\theta(A) \equiv A$;

(II) $A \equiv B$ impliziert $\theta(A) = \theta(B)$;

(III) $\theta(\emptyset) = \emptyset$, $\theta(\Omega) = \Omega$;

(IV) $\theta(A \cap B) = \theta(A) \cap \theta(B)$;

(V) $\theta(A \cup B) = \theta(A) \cup \theta(B)$.

Falls eine Abbildung $\theta: F \to F$ nur (I)-(IV) erfüllt, nennt man θ eine untere Dichte [2].

Im folgenden werde nun $L_2(\Omega,F,P)$ in ein Nicht-Standard-Modell [3] von geeigneter Saturiertheit [4] (z.B. $\varkappa > \aleph_1$) eingebettet.

Es gilt [2]:

THEOREM. Aus einer unteren Dichte θ läßt sich ein Lifting konstruieren.

B e w e i s . Auf dem Produkt der Menge der Mengen vom Maße 1 mit sich erzeugt die Inklusion eine offensichtlich konkurrente [3] binäre Relation. Es gibt also eine interne Menge $\Omega_\infty \subset {}^*\Omega$ mit ${}^*P(\Omega_\infty) = 1$, ${}^*\theta(\Omega_\infty) = {}^*\Omega$ und $\Omega_\infty \cap {}^*N = \emptyset$ für alle $N \in N$. Zu $\omega \in \Omega$ sei

(1) $F_\omega = \{A: \omega \in \theta(A)\}$.

Wegen (III) ist $F_\omega \neq \emptyset$, zudem besitzt F_ω die endliche Durchschnittseigenschaft. Es gibt demnach eine *-meßbare Menge $B^\omega_\infty \subset {}^*\Omega$ von positivem *-Maß mit

(2) a) $B^\omega_\infty \subset A$; $A \in F_\omega$; b) $\omega \in {}^*\theta(B^\omega_\infty)$;

$P(B^\omega_\infty \cap \Omega_\infty) > 0$ liefert die Existenz eines $y_\omega \in B^\omega_\infty \cap \Omega_\infty$. $\omega \in \theta(A)$ impliziert also $y_\omega \in {}^*A \cap \Omega_\infty$.

Sei

(3) $\hat{\theta}(A) = \{\omega : y_\omega \in {}^*A \cap \Omega_\infty\} = \{\omega : y_\omega \in {}^*A\}$.

Die Abbildung $\hat{\theta}: F \to \text{Pot}\Omega$ hängt nur von der Klasse $[A]$ bezüglich $\equiv$ ab, da wegen der Wahl von Ω_∞ für $N \in N$ stets $\Omega_\infty \cap {}^*N = \emptyset$ gilt. Das heißt $\hat{\theta}$ erfüllt (II), (3) liefert (III),(IV),(V).

Außerdem hat man

(4) $\theta(A) \subseteq \hat{\theta}(A), \quad \theta(CA) \subseteq \hat{\theta}(CA)$.

(4) liefert (I) für $\hat{\theta}$, denn

$$A \equiv \theta(A) \subseteq \hat{\theta}(A) = C\hat{\theta}(CA) \subseteq C\theta(CA) \equiv CCA = A. \qquad \text{q.e.d.}$$

Ist F separabel, so existiert eine Folge von Zerlegungen

$$Z_n = (A_{n,1}, A_{n,2}, \ldots, A_{n,m_n}), \; n = 1,2, \ldots$$

mit:

(5) a) $P(A_{n,k}) > 0$, $n = 1,2, \ldots, 1 \leq k \leq m_n$;

b) $A_{n,k} \cap A_{n,k'} = \emptyset$, $k \neq k'$, $1 \leq k, k' \leq m_n$;

c) $\bigcup_{k=1}^{m_n} A_{n,k} = \Omega$;

d) Z_{n+1} verfeinert Z_n;

e) $\underset{n\to\infty}{\text{l.i.m.}}\; T_n f = f$ mit

(6) $$(T_n f)(\omega) = \sum_{i=1}^{m_n} \frac{\int_{A_{n,i}} f dP}{P(A_{n,i})} c_{A_{n,i}}(\omega),$$

$\omega \in \Omega$, $f \in L_2(\Omega, F, P)$.

Für eine unendlich große Zahl ν aus der Monade eines schnellen (rapide) Choquetschen Ultrafilters [5],[6] gilt für $f \in L_\infty$

(7) $(T_\nu f)(\omega) = {}_1 f(\omega)$ für P-fast alle $\omega \in \Omega$.

LEMMA. Es gilt:

(8) $\theta(A) = \{\omega : (T_\nu c_A)(\omega) = {}_1 1\}$, $A \in F$

ist eine untere Dichte.

B e w e i s . (I),(II),(III) sind klar. Sei $A_{\nu,m(\omega)}$ diejenige Menge, die ω enthält. Wegen

$${}^*P(A \cap B \cap A_{\nu,m(\omega)}) \leq \min\{{}^*P(A \cap A_{\nu,m(\omega)}), {}^*P(B \cap A_{\nu,m(\omega)})\}$$

gilt: $\theta(A \cap B) \subset \theta(A) \cap \theta(B)$. $\omega \in \theta(A) \cap \theta(B)$ zieht $\omega \notin \theta(CA) \cup \theta(CB)$

nach sich.

Wertet man die Zerlegung $\{A, CA\} \wedge \{B, CB\}$ vermöge (6) aus, ergibt sich

$$\theta(A \cap B) \supset \theta(A) \cap \theta(B). \qquad \text{q.e.d.}$$

L i t e r a t u r v e r z e i c h n i s

[1] R.F. TAYLOR, On Some Properties of Bounded Internal Functions. In: Appl.of Model Theory to Algebra, Analysis and Probability. Amsterdam 1969

[2] A.-C. IONESCU-TULCEA, Topics in the Theory of Lifting.NewYork 1969.

[3] A. ROBINSON, Non-Standard-Analysis. Amsterdam 1966.

[4] W.A.J. LUXEMBURG, A General Theory of Monads. In: Appl.of Model Theory to Algebra, Analysis and Probability. Amsterdam 1969.

[5] G. MOKOBODZKI, Seminaire Brelot-Choquet-Deny 67-68.

[6] B. EIFRIG, Proceedings of the Oberwolfach Conference on Nonstandard Analysis.

Anschrift des Autors:

B. Eifrig
Institut für Angewandte Mathematik
Universität Heidelberg
69 H e i d e l b e r g
Im Neuenheimer Feld 5

RELEVEMENTS SUR UNE ALGEBRE D'ENSEMBLES

Jacques Gapaillard

Dans ce qui suit, on étudie systématiquement, au moyen de techniques élémentaires, les relations existant entre divers types de relèvements.

1. Notations et définitions

De façon générale, si $A, B \subset E$, on pose :

$$A - B = A \cap \complement_E B \quad \text{et} \quad A \Delta B = (A - B) \cup (B - A) .$$

Définition 1. On appelle couple de relèvement sur un ensemble non vide E tout couple $(\mathcal{a},\mathfrak{I})$ où $\mathcal{a}$ est une algèbre de parties de E et $\mathfrak{I}$ un idéal propre de $\mathcal{a}$.

Le couple de relèvement $(\mathcal{a},\mathfrak{I})$ est dit complet si

$$(I \in \mathfrak{I} , A \subset I) \Rightarrow A \in \mathfrak{I} .$$

Définition 2. On désigne par $R(\mathcal{a},\mathfrak{I})$, ou simplement R, l'ensemble des relèvements relatifs au couple de relèvement $(\mathcal{a},\mathfrak{I})$, c'est-à-dire l'ensemble des applications $L : \mathcal{a} \longrightarrow \mathcal{a}$ telles que :

(i) $\forall A \in \mathcal{a} , \ L(A) \Delta A \in \mathfrak{I}$;

(ii) $(A, B \in \mathcal{a}, \ A \Delta B \in \mathfrak{I}) \Rightarrow L(A) = L(B)$.

Dans R on distingue :

(a) RM, ensemble des relèvements monotones :

$$(A, B \in \mathcal{a}, A \subset B) \Rightarrow L(A) \subset L(B) \ ;$$

(b) SDI (resp. SDS), ensemble des sur-densités inférieures (resp. sous-densités supérieures) :

$$\forall A, B \in \mathcal{a}, \ L(A \cap B) \supset L(A) \cap L(B) \quad (\text{resp. } L(A \cup B) \subset L(A) \cup L(B)) \ ;$$

(c) $DI = SDI \cap RM$ (resp. $DS = SDS \cap RM$), ensemble des densités inférieures (resp. densités supérieures) ;

(d) $RL = DI \cap DS$, ensemble des relèvements linéaires ;

(e) $RD^{(n)}$, $n \in \mathbb{N}^*$, ensemble des relèvements disjoints d'ordre n :

$$(A_i \in \mathcal{a}, \ i = 1,\dots,n , \ \bigcap_{i=1}^{n} A_i \in \mathfrak{I}) \Rightarrow \bigcap_{i=1}^{n} L(A_i) = \emptyset \ ;$$

(f) $RD^{(\omega)} = \bigcap_{n=1}^{\infty} RD^{(n)}$, ensemble des relèvements totalement disjoints ;

(g) RsS (resp. RSS) , ensemble des relèvements sous-soustractifs (resp. sur-soustractifs) :

$$\forall A, B \in \mathcal{A}, \ L(A - B) \subset L(A) - L(B) \quad (\text{resp. } L(A - B) \supset L(A) - L(B));$$

(h) RS = RsS $\cap$ RSS , ensemble des relèvements soustractifs ;

(i) RsA (resp. RSA) , ensemble des relèvements sous-additifs (resp. sur-additifs) :

$$\forall A, B \in \mathcal{A}, \ L(A \,\Delta\, B) \subset L(A) \,\Delta\, L(B) \quad (\text{resp. } L(A \,\Delta\, B) \supset L(A) \,\Delta\, L(B)) ;$$

(j) RA = RsA $\cap$ RSA , ensemble des relèvements additifs.

La notion de relèvement monotone (voir [4]) a été introduite par D. KÖLZOW [10].

Définition 3. Soit $L \in R$. On appelle relèvement conjugué de L le relèvement L' défini par $L'(A) = E - (L(E - A))$ pour tout $A \in \mathcal{A}$.

Définition 4. On munit R de l'ordre $L_1 \leq L_2 \Leftrightarrow \forall A \in \mathcal{A}, \ L_1(A) \subset L_2(A)$, qui lui confère une structure de treillis.

Notations. $R_0 = \{L \in R \ ; \ L(\emptyset) = \emptyset\}$ $(= RD^{(1)})$, $R_1 = \{L \in R \ ; \ L(E) = E\}$, $R_r = R_0 \cap R_1$: ensemble des relèvements réguliers , $\underline{R} = \{L \in R \ ; \ L \leq L'\}$, $\overline{R} = \{L \in R \ ; \ L' \leq L\}$, $R^* = \underline{R} \cap \overline{R}$.

2. Relèvements soustractifs

Théorème 1. RsS = $\underline{RM}$.

. Démonstration. Soit $L \in$ RsS . Alors, d'une part :

$$A \subset B \Rightarrow L(A) = L(B - (B - A)) \subset L(B) - L(B - A) \subset L(B) ,$$

et d'autre part :

$$L(E - A) \subset L(E) - L(A) \subset E - L(A) \quad , \text{d'où } L(A) \subset E - L(E - A) .$$

Donc $L \in \underline{RM}$.

Réciproquement, si $L \in \underline{RM}$, on a :

$$L(A - B) \cap L(B) \subset L(E - B) \cap L(B) \subset (E - L(B)) \cap L(B) = \emptyset$$

d'où , $L(A - B) \subset L(A) \cap (E - L(B)) = L(A) - L(B)$.

Lemme 1. Si $L \in$ RSS et si $L(\emptyset) \subset L(A)$ pour tout $A \in \mathcal{A}$, alors $L \in$ RM .

. Démonstration. Si $A \subset B$, $L(A) - L(B) \subset L(A - B) = L(\emptyset) \subset L(B)$ d'où $L(A) \subset L(B)$.

Théorème 2. $RSS_0 = DS_0$.

. Démonstration. Soit $L \in RSS_0$. Alors $L \in RM$ d'après le lemme 1.

D'où
$$\begin{aligned} L(A) \cup L(B) &= L((A \cup B) - (B - A)) \cup L(B) \\ &\supset (L(A \cup B) - L(B - A)) \cup L(B) \\ &= L(A \cup B) \cup L(B) = L(A \cup B) . \end{aligned}$$

Donc $L \in SDS \cap RM = DS$.

Si $L \in DS$, il vient, compte tenu de $DS \subset RM$,
$$\begin{aligned} L(A) - L(B) &= L((A - B) \cup (A \cap B)) - L(B) \\ &= (L(A - B) \cup L(A \cap B)) - L(B) \\ &= L(A - B) - L(B) \subset L(A - B) . \end{aligned}$$

Théorème 3. $RS = \underline{DS} = \underline{RL} = RL_0$.

. Démonstration. Remarquons d'abord que $\underline{RM} \subset RM_0$ et que, si $L \in RS$, il vient $L(\emptyset) = L(\emptyset - \emptyset) = L(\emptyset) - L(\emptyset) = \emptyset$.

D'où, compte tenu des théorèmes 1 et 2 et de $DS \subset RM$:
$$RS = RS_0 = RsS \cap RSS_0 = \underline{RM} \cap DS_0 = \underline{RM} \cap DS = \underline{DS} .$$

D'autre part, on voit facilement que $RL_0 = \underline{RL}$.

Enfin, si $L \in RS$, on a :
$$L(A \cap B) = L(A - (A - B)) = L(A) - (L(A) - L(B)) = L(A) \cap L(B) ,$$

d'où $RS \subset DI$, et par suite :
$$RS = \underline{DS} \cap DI = \underline{RL} .$$

3. Relèvements additifs

Théorème 4. $RA \cap RM = RsA \cap RM = RSA \cap \underline{RM} = RL_0$.

. Démonstration. On procède par étapes :

(a) $DS \cap RD^{(2)} \subset RA$.

Soit $L \in DS \cap RD^{(2)}$ et soit $A, B \in \mathcal{a}$.

Si $A \cap B = \emptyset$, il vient, puisque $L(A) \cap L(B) = \emptyset$:
$$L(A \Delta B) = L(A \cup B) = L(A) \cup L(B) = L(A) \Delta L(B) .$$

Si A et B sont quelconques, on a alors :
$$\begin{aligned} L(A \Delta B) &= L((A \Delta (A \cap B)) \Delta (B \Delta (A \cap B))) \\ &= L(A \Delta (A \cap B)) \Delta L(B \Delta (A \cap B)) \\ &= (L(A \Delta (A \cap B)) \Delta L(A \cap B)) \Delta (L(B \Delta (A \cap B)) \Delta L(A \cap B)) \\ &= L(A) \Delta L(B) . \end{aligned}$$

(b) $RsA \cap RM \subset DS$.

Car si $L \in RsA \cap RM$, on peut écrire :

$$L(A \cup B) = L((A - B) \cup B) = L((A - B) \Delta B) \subset L(A - B) \Delta L(B)$$
$$\subset L(A - B) \cup L(B) \subset L(A) \cup L(B) \subset L(A \cup B) .$$

(c) $RsA \cap RM \subset RD^{(2)}$.

On considère $L \in RsA \cap RM$ et $A, B \in \mathcal{A}$, $A \cap B = \emptyset$. Il vient :

$$L(A) \cap L(B) = (L(A) \cup L(B)) - (L(A) \Delta L(B)) \subset (L(A) \cup L(B)) - L(A \Delta B)$$
$$= (L(A) \cup L(B)) - L(A \cup B) = \emptyset .$$

(d) $RA \cap RM \subset DI$.

D'après (c) , $RA \cap RM \subset RsA \cap RM \subset RD^{(2)}$.

D'où, pour $L \in RA \cap RM$:

$$L(A \cap B) \Delta (L(A) \cap L(B)) = (L(A \cap B) \Delta L(A)) \cap L(B)$$
$$= L(A - B) \cap L(B) = \emptyset .$$

(e) $RSA \cap \underline{RM} \subset RL_0$.

Car pour $L \in RSA \cap \underline{RM}$, il vient :

$$L(B) - L(A) \subset L(B) - L(A \cap B) = L(B) \Delta L(A \cap B)$$
$$\subset L(B \Delta (A \cap B)) = L(B - A) \subset L(B) - L(A)$$

la dernière inclusion résultant du théorème 1.

De sorte que $RSA \cap \underline{RM} \subset RS = RL_0$ (théorème 3).

(f) Démonstration du théorème 4.

Les résultats ci-dessus permettent d'écrire :

$$RL_0 \subset DS \cap RD^{(2)} \subset RA \cap RM \subset RsA \cap RM$$
$$\subset DS \cap RD^{(2)} \subset RA \cap RM \cap DS \subset DI_0 \cap DS = RL_0$$
$$\subset RA \cap \underline{RM} \subset RSA \cap \underline{RM} \subset RL_0$$

en remarquant que $RL_0 \subset RD^{(2)}$ et $RA \subset R_0$.

Corollaire 4.1. $RsA \cap DI = RA \cap DI = RsA \cap DS = RA \cap DS = RL_0$.

Le résultat $RA \cap DI = RL_0$ est établi et utilisé dans [13] .

Corollaire 4.2. $L \in RL_0$ si et seulement si, pour $A, B \in \mathcal{A}$, on a :

$$L(A \Delta B) = \begin{cases} L(A) - L(B) & \text{si } B \subset A \\ L(B) - L(A) & \text{si } A \subset B \\ L(A) \Delta L(B) & \text{autrement.} \end{cases}$$

Théorème 5. $RsA^* = RSA^* = RA^* = RA_r = RsA_r$.

. Démonstration. On établit successivement :

(a) $RsA^* \subset RA$.

Car, pour $L \in RsA^*$, il vient :

$$L(A) \Delta L(B) = L'(A) \Delta L(B) = E - (L(E - A) \Delta L(B)) \subset E - L((E - A) \Delta B) = L'(A \Delta B) = L(A \Delta B) \subset L(A) \Delta L(B) .$$

(b) $RSA^* \subset RA$.

Pour $L \in RSA^*$, on a :

$$L(A) \Delta L(B) \subset L(A \Delta B) = L'(A \Delta B) = E - L((E - A) \Delta B) \subset E - (L(E - A) \Delta L(B)) = L'(A) \Delta L(B) = L(A) \Delta L(B) .$$

(c) RsA_r $(= RsA_1) \subset R^*$.

Car si $L \in RsA_1$, on a :

$$L'(A) \Delta L(A) = E - (L(E - A) \Delta L(A)) \subset E - L((E - A) \Delta A) \subset E - L(E) = \emptyset .$$

(d) $RA^* \subset RA_r$.

Evident puisque $RA \subset R_0$.

(e) Démonstration du théorème 5.

Des résultats ci-dessus, et puisque $RA = RsA \cap RSA$, il vient :

$$RsA^* = RSA^* = RA^* \subset RA_r \subset RsA_r \subset RsA^* .$$

4. Existence de relèvements réguliers

Ce paragraphe développe certains résultats déjà annoncés dans [2].

Le lemme suivant est immédiat :

Lemme 2. Soit $(\mathfrak{a},\mathfrak{J})$ un couple de relèvement et $\underline{\underline{\mathcal{H}}}$ l'ensemble des parties non vides $\mathcal{H}$ de $\mathfrak{a}$ telles que :

(a) $(H \in \mathcal{H}, A \in \mathfrak{a}, A - H \in \mathfrak{J}) \Rightarrow A \in \mathcal{H}$,

(b) $H \in \mathcal{H} \Rightarrow E - H \notin \mathcal{H}$.

Alors, si $\underline{\underline{\mathfrak{J}}}$ est l'ensemble des idéaux propres de $\mathfrak{a}$ contenant $\mathfrak{J}$, on a :

(1) $\underline{\underline{\mathfrak{J}}} \subset \underline{\underline{\mathcal{H}}}$,

(2) Tout élément de $\underline{\underline{\mathcal{H}}}$ est contenu dans un élément maximal et un tel élément $\mathcal{H}^*$ vérifie

$$(A \in \mathfrak{a}, A \notin \mathcal{H}^*) \Rightarrow E - A \in \mathcal{H}^* ,$$

(3) Tout idéal maximal dans $\underline{\underline{\mathfrak{J}}}$ est maximal dans $\underline{\underline{\mathcal{H}}}$.

Théorème 6. $RM \neq \emptyset$ implique $RM^* \neq \emptyset$. Plus précisément, si $L_0 \in \underline{RM}$ et si $\mathcal{H} \in \underline{\underline{\mathcal{H}}}$, il existe $L_1 \in RM^*$, vérifiant $L_1 \geq L_0$, et coïncidant avec L_0 sur $\mathcal{H}$.

. Démonstration. Si $L \in RM$, on a $L_o = L \wedge L' \in \underline{RM}$.
Si $\mathcal{H} \in \underline{\mathcal{H}}$, $\mathcal{H}$ est contenu dans un élément maximal $\mathcal{H}^*$ (lemme 2).
Soit $L_1 : \mathcal{A} \longrightarrow \mathcal{A}$ telle que $L_1(A) = L_o(A)$ si $A \in \mathcal{H}^*$ et $L_1(A) = L_o'(A)$ si $A \notin \mathcal{H}^*$.
On a $L_1(A) \Delta A \in \mathcal{J}$ pour tout $A \in \mathcal{A}$. De plus :

$$(A, B \in \mathcal{A}, A \Delta B \in \mathcal{J}) \Rightarrow (A, B \in \mathcal{H}^* \text{ ou } A, B \notin \mathcal{H}^*)$$
$$\Rightarrow L_1(A) = L_1(B) \quad ,$$

d'où $L_1 \in R$.

Soit $A, B \in \mathcal{A}$, $A \subset B$. Puisque $L_o, L_o' \in RM$, $L_1(A) \subset L_1(B)$ est acquis si $A, B \in \mathcal{H}^*$ ou $A, B \notin \mathcal{H}^*$. Et si $A \in \mathcal{H}^*$, $B \notin \mathcal{H}^*$, il vient, compte tenu de $L_o \in \underline{RM}$:

$$L_1(A) = L_o(A) \subset L_o'(A) \subset L_o'(B) = L_1(B) \quad ,$$

d'où $L_1 \in RM$.

De plus, pour $A \in \mathcal{A}$, on a, si $A \in \mathcal{H}^*$

$$L_1'(A) = E - L_1(E - A) = E - L_o'(E - A) = L_o(A) = L_1(A)$$

puisque $E - A \notin \mathcal{H}^*$; et si $A \notin \mathcal{H}^*$

$$L_1'(A) = E - L_1(E - A) = E - L_o(E - A) = L_o'(A) = L_1(A)$$

puisque $E - A \in \mathcal{H}^*$ (lemme 2).

Donc $L_1 \in RM^*$. Et il est évident que L_1 majore L_o et coïncide avec L_o sur $\mathcal{H}$.

Corollaire 6.1 (voir [10]). $RM \neq \emptyset$ implique $RM_r \neq \emptyset$.
Car il est immédiat que $\overline{RM} \subset RM_1$, d'où $\underline{RM} \subset RM_o$ et par suite $RM^* \subset RM_r$.

Corollaire 6.2. Soit $L \in R$. Alors $L \in RM^*$ si et seulement si L est maximal dans $\underline{RM}$.

. Démonstration. L maximal dans $\underline{RM}$ implique $L \in RM^*$ d'après le théorème 6 ; tandis que $L \in RM^*$ et $L \leqslant L_o \in \underline{R}$ entraîne $L_o = L$ puisque $L \leqslant L_o \leqslant L_o' \leqslant L' = L$.

Théorème 7 (J. von Neumann, M.H. Stone [13]). $RL \neq \emptyset$ implique $RL_r = RL^* = DI^* = DS^* \neq \emptyset$. Plus précisément, si $L_o \in RL_o = \underline{RL}$ et si $\mathcal{J} \in \underline{\mathcal{J}}$, il existe $L_1 \in RL_r$ vérifiant $L_1 \geqslant L_o$ et coïncidant avec L_o sur $\mathcal{J}$.

. Démonstration. Si $L \in RL$, l'application $L_o : \mathcal{A} \longrightarrow \mathcal{A}$ définie par $L_o(A) = L(A) - L(\emptyset)$, vérifie $L_o \in RL_o = \underline{RL}$. Alors, d'après le lemme 2 et le théorème 6, si $\mathcal{J}^*$ est un idéal maximal (contenant $\mathcal{J}$) et si $L_1 : \mathcal{A} \longrightarrow \mathcal{A}$ est définie par $L_1(A) = L_o(A)$ si $A \in \mathcal{J}^*$ et $L_1(A) = L_o'(A)$ si $A \notin \mathcal{J}^*$, on

a $L_1 \in RM^*$ avec $L_1 \geqslant L_o$.
De $RL_o = \underline{RL}$ résulte $RL_1 = \overline{RL}$ d'où $RL_r = RL^* = DI^* = DS^*$.
Il suffit donc de prouver $L_1 \in DI$. Comme $L_o \in \underline{DS}$, il vient :

$$\forall B \in \mathcal{A},\ L_o(E) \cap L_o'(B) = (L_o(B) \cup L_o(E - B)) \cap L_o'(B) = L_o(B),$$

d'où $\forall A, B \in \mathcal{A}$, $L_o(A) \cap L_o(B) = L_o(A) \cap L_o'(B)$
et $L_1 \in DI$ en résulte facilement puisque les seuls cas possibles pour $A, B \in \mathcal{A}$, sont $\{A \cap B, A, B\} \subset \mathcal{J}^*$ ou $\{A \cap B, A, B\} \cap \mathcal{J}^* = \emptyset$ ou enfin $\{A \cap B, A\} \subset \mathcal{J}^*$ et $B \notin \mathcal{J}^*$.

<u>Corollaire 7.1</u>. Soit $L \in R$. Alors $L \in RL_r = RL^*$ si et seulement si L est maximal dans $RL_o = \underline{RL}$.

<u>Théorème 8</u>. $DI \neq \emptyset$ implique $DI_r \neq \emptyset$. Plus précisément, si $L_o \in DI_o$ et si $\mathcal{J} \in \underline{\underline{\mathcal{J}}}$, il existe $L_1 \in DI_r$ vérifiant $L_1 \geqslant L_o$ et coïncidant avec L_o sur $\mathcal{J}$.

. <u>Démonstration</u>. Si $L \in DI$ on lui associe $L_o \in DI_o$ comme dans la démonstration du théorème 7, et si $\mathcal{J}^*$ est un idéal maximal contenant $\mathcal{J}$, on considère $L_1 : \mathcal{A} \to \mathcal{A}$ définie par $L_1(A) = L_o(A)$ si $A \in \mathcal{J}^*$ et $L_1(A) = L_o(A) \cup (E - L_o(E))$ si $A \notin \mathcal{J}^*$. Alors $L_1 \in R_r$ et l'égalité $L_1(A \cap B) = L_1(A) \cap L_1(B)$ est claire si $A \cap B, A, B$ sont conjointement dans $\mathcal{J}^*$ ou hors de $\mathcal{J}^*$. Dans le cas $A \cap B, A \in \mathcal{J}^*$, $B \notin \mathcal{J}^*$, il vient, compte tenu de la monotonie de L_o :

$$\begin{aligned} L_1(A \cap B) = L_o(A \cap B) = L_o(A) \cap L_o(B) &= L_o(A) \cap (L_o(B) \cup (E - L_o(E))) \\ &= L_1(A) \cap L_1(B). \end{aligned}$$

<u>Lemme 3</u>. Si $RSA \neq \emptyset$, il existe $L_o \in RSA_o$ tel que $L_o(A) \subset L_o(E)$ pour tout $A \in \mathcal{A}$.

. <u>Démonstration</u>. Si $L \in RSA$, définissons L_o par

$$L_o(A) = L(A) \cap L(E) \cap (E - L(\emptyset))$$

pour tout $A \in \mathcal{A}$. Alors $L_o \in R_o$ et, pour tout $A \in \mathcal{A}$

$$L_o(A) \subset L(E) \cap (E - L(\emptyset)) = L_o(E).$$

Enfin, pour $A, B \in \mathcal{A}$, il vient :

$$\begin{aligned} L_o(A) \,\Delta\, L_o(B) &= (L(A) \,\Delta\, L(B)) \cap L(E) \cap (E - L(\emptyset)) \\ &\subset L(A \,\Delta\, B) \cap L(E) \cap (E - L(\emptyset)) = L_o(A \,\Delta\, B). \end{aligned}$$

<u>Théorème 9</u>. $RSA \neq \emptyset$ (resp. $RA \neq \emptyset$) implique $RSA_r \neq \emptyset$ (resp. $RA_r \neq \emptyset$).

. <u>Démonstration</u>. Si $RSA \neq \emptyset$, il existe $L_o \in RSA_o$ tel que $L_o(A) \subset L_o(E)$ pour tout $A \in \mathcal{A}$ (lemme 3). Alors, si $\mathcal{J}^*$ est un idéal maximal contenant $\mathcal{J}$, on pose, pour $A \in \mathcal{A}$, $L_1(A) = L_o(A)$ si $A \in \mathcal{J}^*$ et $L_1(A) = L_o(A) \cup (E - L_o(E))$

si $A \notin \mathcal{J}^*$.

Alors $L_1 \in R_r$. De plus, $L_1(A) \Delta L_1(B) \subset L_1(A \Delta B)$ est évident si $A, B, A \Delta B \in \mathcal{J}^*$.

Si $A \in \mathcal{J}^*$ et $B, A \Delta B \notin \mathcal{J}^*$, il vient :

$$\begin{aligned} L_1(A \Delta B) &= L_o(A \Delta B) \cup (E - L_o(E)) \supset (L_o(A) \Delta L_o(B)) \cup (E - L_o(E)) \\ &= L_o(A) \Delta L_o(B) \Delta (E - L_o(E)) = L_o(A) \Delta (L_o(B) \cup (E - L_o(E))) \\ &= L_1(A) \Delta L_1(B) . \end{aligned}$$

Enfin, si $A, B \notin \mathcal{J}^*$ et $A \Delta B \in \mathcal{J}^*$, on a :

$$\begin{aligned} L_1(A \Delta B) &= L_o(A \Delta B) \supset L_o(A) \Delta L_o(B) \\ &= L_o(A) \Delta (E - L_o(E)) \Delta L_o(B) \Delta (E - L_o(E)) \\ &= (L_o(A) \cup (E - L_o(E))) \Delta (L_o(B) \cup (E - L_o(E))) \\ &= L_1(A) \Delta L_1(B) . \end{aligned}$$

Le cas $RA \neq \emptyset$ se traite de façon analogue.

5. Relèvements disjoints

Lemme 4.

(1) Soit un entier $n \geq 2$. Alors si $L \in R$, on a $L \in RD^{(n)}$ si et seulement si

$$(A_i, B \in \mathcal{a}, i=1,\dots,n-1 , \bigcap_{i=1}^{n-1} A_i \subset B) \Rightarrow \bigcap_{i=1}^{n-1} L(A_i) \subset L'(B) .$$

(2) $RD^{(2)} \subset \underline{R}$, $\underline{RM} = RD^{(2)} \cap RM$, $\overline{RD}^{(2)} = RD^{(2)*} = RM^*$, et, si $n \geq 3$:

$\overline{RD}^{(n)} = RD^{(n)*} = \overline{RD}^{(\omega)} = RD^{(\omega)*} = RL^* = RL_r$.

. Démonstration. (1) est immédiat et implique $RD^{(2)} \subset \underline{R}$, $\underline{RM} \subset RD^{(2)}$ et $\overline{RD}^{(2)} \subset RM$, d'où les premiers résultats de (2). Enfin (1) entraîne $\overline{RD}^{(3)} \subset SDI$ et, comme $\overline{RD}^{(3)} \subset RD^{(2)} = RM^*$, il vient, pour $n \geq 3$:

$$RL^* \subset \overline{RD}^{(\omega)} \subset \overline{RD}^{(n)} \subset \overline{RD}^{(3)} \subset SDI \cap RM^* = DI^* = RL^* .$$

Lemme 5. Si $(\mathcal{a},\mathcal{J})$ est complet, pour tout $i = 2, 3, \dots, \omega$, et pour tout $L \in RD^{(i)}$, il existe $L_o \in RD^{(i)} \cap RM$ ($\subset \underline{RM}$) tel que $L \leq L_o \leq L'$.

. Démonstration. Soit un entier $n \geq 2$ et $L \in RD^{(n)}$. L'application $L_o : \mathcal{a} \to \mathcal{P}(E)$ définie pour tout $A \in \mathcal{a}$ par $L_o(A) = \bigcup\{L(B) ; B \in \mathcal{a}, B \subset A\}$ vérifie $L(A) \subset L_o(A) \subset L'(A)$ (d'après le lemme 4). Comme $(\mathcal{a},\mathcal{J})$ est complet, il en résulte $L_o \in R$ et $L \leq L_o \leq L'$. De plus, L_o est évidemment monotone. Enfin, si $A_1, \dots, A_n \in \mathcal{a}$ avec $A_1 \cap \dots \cap A_n \in \mathcal{J}$, il vient :

$$\begin{aligned} L_o(A_1) \cap \dots \cap L_o(A_n) &= \bigcap_{i=1}^{n} \bigcup\{L(B) ; B \in \mathcal{a}, B \subset A\} \\ &= \bigcup\{ \bigcap_{i=1}^{n} L(B_i) ; B_i \subset A_i , 1 \leq i \leq n\} = \emptyset . \end{aligned}$$

Théorème 10. Si $(\mathcal{A},\mathcal{I})$ est complet et pour $n = 2, 3, \ldots, \omega$, $RD^{(n)} \neq \emptyset$ implique $RD_r^{(n)} \neq \emptyset$. Plus précisément, si $L_0 \in RD^{(n)} \cap RM$ et si $\mathcal{J} \in \underline{\mathcal{I}}$, il existe $L_1 \in RD_r^{(n)}$ coïncidant avec L_0 sur $\mathcal{J}$ et tel que $L_1 \geq L_0$.

. Démonstration. Si $RD^{(n)} \neq \emptyset$, il existe $L_0 \in RD^{(n)} \cap RM$ (lemme 5). On définit alors L_1 comme dans la démonstration du théorème 8. On obtient ainsi $L_1 \in R_r$. Considérons $A_i \in \mathcal{A}$, $i = 1, \ldots, n$ avec $\bigcap_{i=1}^{n} A_i \in \mathcal{J}$ et remarquons qu'il est impossible que $A_i \notin \mathcal{J}^*$ pour $i = 1, \ldots, n$; en effet, ceci impliquerait $E - A_i \in \mathcal{J}^*$ pour $i = 1, \ldots, n$ d'où :

$$\bigcup_{i=1}^{n} (E - A_i) = E - \bigcap_{i=1}^{n} A_i \in \mathcal{J}^*$$

et par suite $E \in \mathcal{J}^*$. En réindexant éventuellement les A_i, on peut donc affirmer qu'il existe k entier, $1 \leq k \leq n$ tel que $A_i \in \mathcal{J}^*$ pour $1 \leq i \leq k$ et $A_i \notin \mathcal{J}^*$ pour $k+1 \leq i \leq n$. D'où :

$$\bigcap_{i=1}^{n} L_1(A_i) = \bigcap_{i=1}^{k} L_0(A_i) \cap \Big(\bigcap_{i=k+1}^{n} L_0(A_i) \cup (E - L_0(E))\Big)$$

$$= \Big(\bigcap_{i=1}^{n} L_0(A_i)\Big) \cup \Big(\bigcap_{i=1}^{k} L_0(A_i) \cap (E - L_0(E))\Big)$$

et les deux termes de cette réunion sont vides puisque $L_0 \in RD^{(n)} \cap RM$.

Lemme 6. Si $(\mathcal{A},\mathcal{I})$ est complet et si L est maximal dans $RD^{(\omega)}$, alors $L \in RM_r$. Ce lemme 6 est une conséquence immédiate du lemme 5 et du théorème 10, et le résultat suivant se déduit des lemmes 4 et 5.

Théorème 11. Supposons $(\mathcal{A},\mathcal{I})$ complet.

(1) $RM \neq \emptyset \iff RD^{(2)} \neq \emptyset$.

(2) $L \in RM^* \iff L$ maximal dans $\underline{RM} \iff L$ maximal dans $RD^{(2)}$.

Le lemme suivant est immédiat.

Lemme 7. Soit $(\mathcal{A},\mathcal{I})$ un couple de relèvement et $\underline{\mathcal{K}}$ l'ensemble des parties non vides $\mathcal{K}$ de $\mathcal{A}$ telles que :

(a) $(K \in \mathcal{K},\ A \in \mathcal{A},\ A \,\Delta\, K \in \mathcal{I}) \Rightarrow A \in \mathcal{K}$,

(b) $(A \in \mathcal{A},\ A \subset \bigcup_{i=1}^{n} K_i,\ K_i \in \mathcal{K},\ i=1,\ldots,n) \Rightarrow E - A \notin \mathcal{K}$.

Alors, si $\underline{\mathcal{I}}$ est l'ensemble des idéaux propres de $\mathcal{A}$ contenant $\mathcal{I}$, on a :

(1) $\underline{\mathcal{I}} \subset \underline{\mathcal{K}}$,

(2) Pour tout $\mathcal{K} \in \underline{\mathcal{K}}$, il existe $\mathcal{J} \in \underline{\mathcal{I}}$ tel que $\mathcal{K} \subset \mathcal{J}$.

Théorème 12 (voir [1]). Si $(\mathcal{A},\mathfrak{I})$ est complet, $RD^{(\omega)} \neq \emptyset$ implique $RL \neq \emptyset$. Plus précisément, si $L_o \in RD^{(\omega)}$, il existe $L_1 \in RL_r$ tel que $L_o \leq L_1 \leq L'_o$.

. Démonstration. Montrons que $RD^{(\omega)}$, supposé non vide, est inductif pour l'ordre habituel. Soit $(L_i)_{i \in I}$ une famille totalement ordonnée d'éléments de $RD^{(\omega)}$ et soit $i_o \in I$. Comme $RD^{(\omega)} \subset \underline{R}$, pour chaque $i \in I$ on a $L_i \leq L_{i_o} \leq L'_{i_o}$ ou $L_i \leq L'_i \leq L'_{i_o}$ selon que $L_i \leq L_{i_o}$ ou $L_{i_o} \leq L_i$ respectivement. Alors si $L^* : \mathcal{A} \longrightarrow \mathcal{P}(E)$ est définie par $L^*(A) = \cup\{L_i(A)\ ;\ i \in I\}$, il vient $L_{i_o}(A) \subset L^*(A) \subset L'_{i_o}(A)$ pour tout $A \in \mathcal{A}$, d'où $L^* \in R$ puisque $(\mathcal{A},\mathfrak{I})$ est complet. De plus, $L^* \in RD^{(\omega)}$ car si $A_k \in \mathcal{A}$, $k = 1, 2, \ldots, n$, il vient :

$$\bigcap_{k=1}^{n} L^*(A_k) = \bigcap_{k=1}^{n} \bigcup_{i\in I} L_i(A_k) = \bigcup_{i\in I} \bigcap_{k=1}^{n} L_i(A_k)$$

puisque $(L_i)_{i \in I}$ est totalement ordonnée, d'où il résulte que $L^* \in RD^{(n)}$.

Alors, par l'axiome de Zorn, tout $L_o \in RD^{(\omega)}$ est majoré par un élément maximal L_1 de $RD^{(\omega)}$. Il reste à prouver que $L_1 \in RL$ et, d'après le lemme 4, il suffit d'établir $L_1 = L'_1$.

Supposons donc $L_1 \neq L'_1$, ce qui impose $L_1 < L'_1$. Alors, il existe $A_1 \in \mathcal{A}$ et $x_1 \in E$ tels que $x_1 \in E - (L_1(A_1) \cup L_1(E - A_1)) \in \mathfrak{I}$, d'où $\{x_1\} \in \mathfrak{I}$ puisque $(\mathcal{A},\mathfrak{I})$ est complet. D'autre part, on a :

$$\mathcal{K} = \{A\ ;\ A \in \mathcal{A},\ x_1 \in L_1(E - A)\} \in \underline{\mathcal{K}}$$

avec $\mathcal{K} \neq \emptyset$ (lemme 6). Soit $\mathcal{J}^* \in \underline{\mathfrak{J}}$ un idéal maximal contenant $\mathcal{K}$ (lemme 7). Comme on peut toujours supposer $E - A_1 \in \mathcal{J}^*$, l'application $\tilde{L}_1 : \mathcal{A} \longrightarrow \mathcal{A}$ définie par $\tilde{L}_1(A) = L_1(A)$ si $A \in \mathcal{J}^*$ et $\tilde{L}_1(A) = L_1(A) \cup \{x_1\}$ si $A \notin \mathcal{J}^*$, est un relèvement et $L_1 < \tilde{L}_1$ puisque $A_1 \notin \mathcal{J}^*$ et $x_1 \notin L_1(A_1)$ implique $L_1(A_1) \subset \tilde{L}_1(A_1)$ strictement.

Cependant $\tilde{L}_1 \in RD^{(\omega)}$. En effet, si $n \geq 2$ et $B_1, \ldots, B_n \in \mathcal{A}$ avec $\bigcap_{i=1}^{n} B_i \in \mathfrak{I}$, il existe un entier k, $1 \leq k \leq n$ tel que, après réindexation des B_i, on ait $B_i \in \mathcal{J}^*$ pour $i = 1, \ldots, k$ et $B_i \notin \mathcal{J}^*$ pour $i = k+1, \ldots, n$. (voir démonstration du théorème 10).

D'autre part, $x_1 \in L_1(A)$ implique $E - A \in \mathcal{K} \subset \mathcal{J}^*$, d'où $A \notin \mathcal{J}^*$. En particulier, $x_1 \notin L_1(B_1)$, d'où :

$$\bigcap_{i=1}^{n} \tilde{L}_1(B_i) = \Big(\bigcap_{i=1}^{n} L_1(B_i)\Big) \cup \Big(\bigcap_{i=k+1}^{n} (L_1(B_i) \cup \{x_1\})\Big)$$

$$\subset \Big(\bigcap_{i=1}^{n} L_1(B_i)\Big) \cup (L_1(B_1) \cap \{x_1\}) = \emptyset .$$

Donc, L_1, élément maximal de $RD^{(\omega)}$, vérifie nécessairement $L'_1 = L_1$, d'où $L_1 \in RL^* = RL_r$, et le théorème est démontré car $L_o \leq L_1 = L'_1 \leq L'_o$.

Corollaire 12.1. Si $(\mathcal{A},\mathcal{J})$ est complet, SDI $\neq \emptyset$ implique RL $\neq \emptyset$ (a fortiori, DI $\neq \emptyset$ implique RL $\neq \emptyset$). Plus précisément, si $L_0 \in SDI_0$, il existe $L_1 \in RL_r$ tel que $L_0 \leq L_1 \leq L_0'$.

Remarques.

1. Les relèvements appartenant à $RD_r^{(\omega)} \cap RM$ correspondent aux prédensités considérées par K. Bichteler [1] qui montre, dans un cadre particulier, que l'existence d'une prédensité implique celle d'un relèvement linéaire.
2. Le corollaire 12.1 est connu pour les densités inférieures, au moins dans le cas des espaces mesurés (voir [1], [3], [6], [7], [8], [12], [15], [17]).
3. S. Graf a signalé une démonstration plus simple du théorème 12 par la considération d'ultrafiltres plutôt que d'idéaux maximaux.

Définition 5. Un relèvement $L \in RM$ est dit modéré par $L_0 \in R$ si :

$$(A_i \in \mathcal{A},\ i = 1, \ldots, n) \Rightarrow \bigcap_{i=1}^{n} L(A_i) \subset L_0(\bigcap_{i=1}^{n} A_i) .$$

On désignera par RMM l'ensemble des relèvements monotones modérés.

Théorème 13. Si $(\mathcal{A},\mathcal{J})$ est complet, RMM $\neq \emptyset$ implique RL $\neq \emptyset$. De façon précise, si $L \in RM$ est modéré par $L_0 \in R$, il existe $L_1 \in DI$ tel que $L \leq L_1 \leq L_0$.

. Démonstration. Pour $A \in \mathcal{A}$ on définit :

$$I(A) = \{(A_i)_{1\leq i\leq n}\ ;\ n \in \mathbb{N}^*,\ A_i \in \mathcal{A}_i,\ i = 1, \ldots, n,\ \bigcap_{i=1}^{n} A_i = A\}$$

et si $L \in RM$ est modéré par L_0, on pose :

$$L_1(A) = \bigcup_{(A_i)_{1\leq i\leq n} \in I(A)} \bigcap_{i=1}^{n} L(A_i)$$

d'où $L(A) \subset L_1(A) \subset L_0(A)$. Comme $(\mathcal{A},\mathcal{J})$ est complet, l'application $L_1 : A \longrightarrow L_1(A)$ est un relèvement. De plus, si $A, B \in \mathcal{A}$:

$$L_1(A) \cap L_1(B) = \bigcup_{(A_i) \in I(A),(B_j) \in I(B)} (\bigcap_i L(A_i)) \cap (\bigcap_j L(B_j))$$

$$\subset \bigcup_{(C_k) \in I(A \cap B)} \bigcap_k L(C_k) = L_1(A \cap B) .$$

Donc, $L_1 \in SDI$ et la première partie du théorème est établie d'après le corollaire 12.1.

En fait, $L_1 \in DI$. Pour cela, il suffit que $L_1 \in RM$. Or, si $A, B \in \mathcal{A}$, $A \subset B$, et si $C = B - A$, on a :

$$(A_i)_{1\leq i\leq n} \in I(A) \Rightarrow (A_i \cup C)_{1\leq i\leq n} \in I(B)$$

et comme $L(A_i) \subset L(A_i \cup C)$ pour $i = 1, \ldots, n$, il vient :

$$L_1(A) = \bigcup_{(A_i)\in I(A)} \bigcap_i L(A_i) \subset \bigcup_{(A_i)\in I(A)} \bigcap_i L(A_i \cup C)$$

$$\subset \bigcup_{(B_j)\in I(B)} \bigcap_j L(B_j) = L_1(B) .$$

Corollaire 13.1.

(1) Si $(\mathcal{A},\mathfrak{J})$ est complet et si $L \in RM$ et $L_o \in SDI$ avec $L \leqslant L_o$, il existe $L_1 \in DI$ vérifiant $L \leqslant L_1 \leqslant L_o$.

(2) Si, de plus, $L \in RM^*$ et $L_o \in SDI_o$, on a $L \in RL^*$.

Corollaire 13.2.

(1) Si $(\mathcal{A},\mathfrak{J})$ est complet, RL, DI, DS, SDI, SDS, $RD^{(\omega)}$ et RMM sont conjointement vides ou non vides.

(2) Si $L \in R$, on a $L \in RL^*$ si et seulement si L est maximal dans l'un quelconque des ensembles DI_o, DS_1, SDI_o, SDS_1, $RD^{(\omega)}$.

6. Décompositions

Définition 6. Si $(\mathcal{A},\mathfrak{J})$ est un couple de relèvement et $\mathfrak{j}$ un idéal de $\mathcal{A}$ non contenu dans $\mathfrak{J}$, on pose $\mathfrak{J}_o = \mathfrak{j} \cap \mathfrak{J}$ et $\mathfrak{j}_+ = \mathfrak{j} - \mathfrak{J}_o$.
On appelle alors $\mathfrak{j}$-décomposition de $(\mathcal{A},\mathfrak{J})$ toute partie $\mathcal{Z}$ de $\mathfrak{j}_+$ telle que :

a) $(Z_1, Z_2 \in \mathcal{Z}, Z_1 \neq Z_2) \Rightarrow Z_1 \cap Z_2 = \emptyset$,

b) $\forall J \in \mathfrak{j}_+, \exists Z \in \mathcal{Z}, J \cap Z \in \mathfrak{j}_+$.

Cette notion de décomposition généralise directement la notion de décomposition d'un espace mesuré chez D. Kölzow [9], [10] (voir aussi la notion d'espace strictement localisable dans [8] et celle de somme directe d'espaces mesurés chez I. G. Segal [16]).

Théorème 14. Soit $(\mathcal{A},\mathfrak{J})$ un couple de relèvement, $\mathfrak{j}$ un idéal de $\mathcal{A}$ non contenu dans $\mathfrak{J}$ et $L \in R$ tel que :

$$(J_1, J_2 \in \mathfrak{j}_+, J_1 \cap J_2 \in \mathfrak{J}_o) \Rightarrow L(J_1) \cap L(J_2) = \emptyset .$$

Alors $(\mathcal{A},\mathfrak{J})$ possède une $\mathfrak{j}$-décomposition.

. Démonstration. C'est celle de D. Kölzow dans [10] pour obtenir une décomposition d'un espace mesuré à partir d'un relèvement linéaire.
On considère $\Sigma = \{\sigma \,;\, \sigma \subset \mathfrak{j}_+ , (J_1, J_2 \in \sigma, J_1 \neq J_2) \Rightarrow J_1 \cap J_2 \in \mathfrak{J}_o\}$ et l'axiome de Zorn montre que Σ admet un élément maximal σ_o. Alors, à cause de l'hypothèse faite sur L, il est évident que $\mathcal{Z} = \{J \cap L(J) \,;\, J \in \sigma_o\}$ est une $\mathfrak{j}$-décomposition de $(\mathcal{A},\mathfrak{J})$.

Corollaire 14.1. Soit $(\mathfrak{A},\mathfrak{I})$ un couple de relèvement tel que $RM \neq \emptyset$ et $\mathfrak{J}$ un idéal de $\mathfrak{A}$ non contenu dans $\mathfrak{I}$. Alors $(\mathfrak{A},\mathfrak{I})$ possède une $\mathfrak{J}$-décomposition.

7. Relèvements dans un espace mesuré

Les notations suivantes sont celles de D. Kölzow dans [10].

Soit $(E,\mathfrak{M},\varphi)$ un espace mesuré, où $\mathfrak{M}$ est une tribu (σ-algèbre) de parties de E et φ une mesure positive (finie ou non) sur $\mathfrak{M}$.

On pose $\mathcal{S} = \{S \; ; \; S \in \mathfrak{M} \, , \varphi(S) < +\infty\}$, $\mathcal{S}_+ = \{S \; ; \; S \in \mathcal{S} \, , \; \varphi(S) > 0\}$,

$$\mathfrak{N} = \{N \; ; \; N \in \mathfrak{M} \, , \; \varphi(N) = 0\} \quad , \; \mathfrak{N}_{loc} = \{N \; ; \; N \in \mathfrak{M}, \; \forall S \in \mathcal{S} \, , \; N \cap S \in \mathfrak{N}\}.$$

On rappelle que l'espace mesuré $(E,\mathfrak{M},\varphi)$ coïncide avec son extension de Carathéodory si et seulement si

a) φ est complète,

b) $\mathfrak{M} = \{M \; ; \; M \subset E \, , \; \forall S \in \mathcal{S} \, , \; M \cap S \in \mathfrak{M}\}$.

Pour un tel espace, $(N \in \mathfrak{N}_{loc}, A \subset N)$ implique $A \in \mathfrak{N}_{loc}$.

D'autre part, on suppose toujours dans la suite que l'espace mesuré considéré $(E,\mathfrak{M},\varphi)$ est <u>non trivial</u> en ce sens que $E \notin \mathfrak{N}_{loc}$. Alors $(\mathfrak{M},\mathfrak{N}_{loc})$ est un couple de relèvement, complet en particulier si l'espace coïncide avec son extension de Carathéodory.

Un relèvement $L \in R(\mathfrak{M}, \mathfrak{N}_{loc})$ est dit <u>borné</u> si $L(\mathcal{S}) \subset \mathcal{S}$ et on notera R_b l'ensemble des relèvements bornés, d'où les notations évidentes RL_b , RM_{ob} , etc...

L'espace $(E,\mathfrak{M},\varphi)$ étant supposé non trivial, on peut envisager des $\mathcal{S}$-décompositions de $(\mathfrak{M}, \mathfrak{N}_{loc})$ que nous appellerons simplement <u>décompositions de l'espace mesuré</u> $(E,\mathfrak{M},\varphi)$.

Théorème 15. Tout espace mesuré $(E,\mathfrak{M},\varphi)$ tel que $RM(\mathfrak{M}, \mathfrak{N}_{loc}) \neq \emptyset$ possède une décomposition.

C'est un cas particulier du corollaire 14.1. Ce résultat figure déjà dans [2] et [4] avec des hypothèses restrictives. Il a aussi été obtenu indépendamment, dans le cadre de l'intégrale de Stone, par W. Strauss [18]. On en déduit aussitôt l'énoncé suivant qui constitue une réponse affirmative à une question posée par D. Kölzow dans [10] (à ce sujet, voir aussi [18]).

Corollaire 15.1. Si $(E,\mathfrak{M},\varphi)$ coïncide avec son extension de Carathéodory, $RM(\mathfrak{M}, \mathfrak{N}_{loc}) \neq \emptyset$ implique $RL(\mathfrak{M}, \mathfrak{N}_{loc}) \neq \emptyset$.

. Démonstration. On utilise que, dans un tel espace, l'existence d'une décomposition entraîne celle d'un relèvement linéaire (voir [10] ou la démonstration du théorème 16 ci-dessous).

Remarquons que cette démonstration s'appuie essentiellement sur le théorème fondamental de D. Maharam et J. von Neumann d'existence d'un relèvement linéaire pour un espace mesuré où la mesure est finie et complète (voir entre autres [6], [11], [15], [17], [19]).

Corollaire 15.2. Soit $(E,\mathfrak{M},\varphi)$ un espace mesuré coïncidant avec son extension de Carathéodory et tel que $RM(\mathfrak{M},\mathfrak{N}_{loc}) \neq \emptyset$.
Alors, si $L \in RM(\mathfrak{M},\mathfrak{N}_{loc})$, il existe $L_o \in RM_{ob}(\mathfrak{M},\mathfrak{N}_{loc})$ tel que $L_o \leq L$ et il existe une décomposition $\mathcal{Z}_o$ de l'espace mesuré telle que $L_o(Z) = Z$ pour tout $Z \in \mathcal{Z}_o$.

. Démonstration. D'après le théorème 15, il existe une décomposition $\mathcal{Z}$ de l'espace mesuré. D'autre part, $L_1(M) = L(M) - L(\emptyset)$ définit $L_1 \in RM_o(\mathfrak{M},\mathfrak{N}_{loc})$, $L_1 \leq L$.
Alors, en posant $L_o(M) = \cup\{L_1(M \cap Z) \cap Z \; ; \; Z \in \mathcal{Z}\}$ pour tout $M \in \mathfrak{M}$, on montre facilement que l'on définit ainsi $L_o \in RM_o(\mathfrak{M},\mathfrak{N}_{loc})$ avec $L_o \leq L_1 \leq L$.
Ensuite, en utilisant que, pour chaque $S \in \mathcal{S}_+$, $\{Z \; ; \; Z \in \mathcal{Z}, \; S \cap Z \in \mathcal{S}_+\}$ est dénombrable ([10], lemme 6), on obtient aisément que L_o est borné.
Enfin, il est immédiat que $\mathcal{Z}_o = L_o(\mathcal{Z})$ est aussi une décomposition.

L'énoncé suivant complète le théorème d'équivalence de D. Kölzow ([10], théorème 21) (voir aussi [6]).

Théorème 16. Pour un espace mesuré $(E,\mathfrak{M},\varphi)$ coïncidant avec son extension de Carathéodory, les conditions suivantes sont équivalentes :

(1) $(E,\mathfrak{M},\varphi)$ possède une décomposition ;

(2) $RL_b(\mathfrak{M},\mathfrak{N}_{loc}) \neq \emptyset$;

(3) $RM(\mathfrak{M},\mathfrak{N}_{loc}) \neq \emptyset$;

(4) $SDI(\mathfrak{M},\mathfrak{N}_{loc}) \neq \emptyset$;

(5) $SDS(\mathfrak{M},\mathfrak{N}_{loc}) \neq \emptyset$;

(6) $RD^{(\omega)}(\mathfrak{M},\mathfrak{N}_{loc}) \neq \emptyset$;

(7) $RD^{(2)}(\mathfrak{M},\mathfrak{N}_{loc}) \neq \emptyset$;

(8) Il existe $L \in R(\mathfrak{M},\mathfrak{N}_{loc})$ tel que :

$$(S_1, S_2 \in \mathcal{S}_+, \; S_1 \cap S_2 \in \mathfrak{N}) \Rightarrow L(S_1) \cap L(S_2) = \emptyset .$$

. Démonstration. (1) ⇒ (2). Il suffit de remarquer que le relèvement linéaire L obtenu par D. Kölzow [10], théorème 7, à partir d'une décomposition, est borné. Rappelons que L s'obtient de la façon suivante : si $Z \in \mathcal{Z}$, on considère le sous-espace mesuré $(Z,\mathfrak{M} \cap Z, \varphi_Z)$ et, comme φ_Z est finie et complète, il existe sur $\mathfrak{M} \cap Z$ un relèvement linéaire L_Z tel que $L_Z(\emptyset) = \emptyset$; on pose alors

$L(M) = \bigcup\{L_Z(M \cap Z) \; ; \; Z \in \mathcal{Z}\}$ pour chaque $M \in \mathfrak{M}$ et on montre facilement que L est un relèvement linéaire et borné.

Les autres implications sont évidentes ou sont des conséquences immédiates de résultats antérieurs.

Signalons encore le résultat suivant, conséquence du théorème 7.

Théorème 17. Si $(E,\mathfrak{M},\varphi)$ est un espace mesuré tel que E ne s'écrive pas $S \cup N$ où $S \in \mathcal{S}$ et $N \in \mathfrak{N}_{loc} - \mathfrak{N}$, $RL_b(\mathfrak{M},\mathfrak{N}_{loc}) \neq \emptyset$ implique $RL_{rb}(\mathfrak{M},\mathfrak{N}_{loc}) \neq \emptyset$.

8. Relèvements dans un espace de Baire

Parmi les nombreuses définitions équivalentes des espaces de Baire, retenons celle-ci : un espace topologique $(E,\mathcal{O})$ est dit espace de Baire si, $\mathcal{C}_1$ désignant l'ensemble des parties de E de première catégorie (maigres), on a $\mathcal{O} \cap \mathcal{C}_1 = \{\emptyset\}$.

On dit qu'une partie B de E a la propriété de Baire si elle s'écrit $B = O \,\Delta\, N$ où $O \in \mathcal{O}$ et $N \in \mathcal{C}_1$. L'ensemble $\mathcal{B}$ de ces parties est une tribu, $\mathcal{C}_1$ est un idéal propre de $\mathcal{B}$ et $(\mathcal{B},\mathcal{C}_1)$ est un couple de relèvement complet.

Théorème 18. Soit $(B, \mathcal{C}_1)$ le couple de relèvement associé à un espace de Baire $(E,\mathcal{O})$.

(1) Il existe $\alpha \in DS_r(\mathcal{B}, \mathcal{C}_1)$ et $\beta \in DI_r(\mathcal{B}, \mathcal{C}_1)$ tels que, pour tout $B \in \mathcal{B}$, $\beta(B) = \widehat{\alpha(B)}$ et $\alpha(B) = \overline{\beta(B)}$;

(2) le couple (α,β) de densités jouissant des propriétés ci-dessus est unique ;

(3) $\alpha = \beta'$ (d'où $\beta = \alpha'$).

Nous omettrons la démonstration élémentaire de ce théorème qui n'est en fait qu'une façon d'exprimer la propriété bien connue selon laquelle, dans chaque classe d'équivalence de $\mathcal{B}$ selon $\mathcal{C}_1$, il existe un ouvert régulier (c'est-à-dire égal à l'intérieur de son adhérence) et un seul, l'ensemble des ouverts réguliers étant stable par intersections finies (voir [14]). De façon précise, si $O \,\Delta\, N$ est une représentation de $B \in \mathcal{B}$ ($O \in \mathcal{O}$ et $N \in \mathcal{C}_1$), les densités α et β sont définies par $\alpha(B) = \overline{O}$ et $\beta(B) = \mathring{\overline{O}}$.

Par application du corollaire 12.1, on obtient le résultat suivant démontré directement par S. Graf [5].

Théorème 19. Soit $(\mathcal{B}, \mathcal{C}_1)$ le couple de relèvement associé à un espace de Baire $(E,\mathcal{O})$.

Alors il existe $L \in RL_r(\mathcal{B}, \mathcal{C}_1)$ tel que, pour tout $B = O \,\Delta\, N \in \mathcal{B}$, avec $O \in \mathcal{O}$ et $N \in \mathcal{C}_1$, on ait $\mathring{\overline{O}} \subset L(B) \subset \overline{O}$.

Terminons par une remarque sur l'éventualité de l'existence, dans un espace topologique, d'une mesure de catégorie, c'est-à-dire d'une mesure (σ-additive) définie sur la tribu des ensembles ayant la propriété de Baire et pour laquelle les ensembles de mesure nulle sont exactement les ensembles de première catégorie. Dans [14], [5] et [6] on trouve des réponses à ce difficile problème. Ainsi, un espace topologique métrisable et séparable (par exemple métrisable compact) sans point isolé n'admet aucune mesure de catégorie finie. Par contre, un espace métrisable compact admet une mesure de catégorie simplement additive comme le montre le résultat suivant.

Théorème 20. Pour tout espace de Baire séparable, il existe une mesure de catégorie simplement additive et finie.

. Démonstration. Soit $(E,\mathcal{O})$ un espace de Baire, $D = \{x_n \ ; n \in \mathbb{N}\}$ une partie de E dénombrable et partout dense et $(a_n)_{n\in\mathbb{N}}$ une suite de réels strictement positifs telle que $\sum_{n=0}^{\infty} a_n < +\infty$.

Pour $A \subset E$, posons $\mathbb{N}(A) = \{n \ ; n \in \mathbb{N} , x_n \in A\}$.

Soit alors $L \in RL_r(\mathcal{B}, \mathcal{C}_1)$ et $\varphi : \mathcal{B} \longrightarrow \mathbb{R}$ définie par $\varphi(B) = \sum_{n\in\mathbb{N}(L(B))} a_n$ si $B \notin \mathcal{C}_1$ et $\varphi(B) = 0$ si $B \in \mathcal{C}_1$.

Alors il est immédiat que φ est la mesure de catégorie simplement additive annoncée.

BIBLIOGRAPHIE

[1] K. Bichteler, Integration theory, Lecture Notes 315, Berlin-Heidelberg-New York 1973.

[2] J. Gapaillard, Sur un théorème de Kölzow, C. R. Acad. Sci. Paris 271 (1970), 91-93.

[3] J. Gapaillard, Relèvements sur une algèbre de parties d'un ensemble, C. R. Acad. Sci. Paris 274 (1972), 1798-1800.

[4] J. Gapaillard, Relèvements monotones, Arch. Math. 24 (1973), 169-178.

[5] S. Graf, Lifting- und Fortsetzungsprobleme, insbesondere Hüllenkonstruktionen, Diplomarbeit, Erlangen 1972.

[6] S. Graf, Schnitte Boolescher Korrespondenzen und ihre Dualisierungen, Dissertation, Erlangen 1973.

[7] A. Ionescu Tulcea, On the lifting property (V), Ann. Math. Statist. 36 (1965), 819-828.

[8] A. Ionescu Tulcea and C. Ionescu Tulcea, Topics in the theory of lifting, Ergebn. Math. 48, Berlin-Heidelberg-New York 1969.

[9] D. Kölzow, Adaptions- und Zerlegungseigenschaften von Massen, Math. Z. 94 (1966), 309-321.

[10] D. Kölzow, Differentiation von Massen, Lecture Notes 65, Berlin-Heidelberg-New York 1968.

[11] D. Maharam, On a theorem of von Neumann, Proc. Amer. Math. Soc. 9 (1958), 987-994.

[12] J. von Neumann, Algebraische Repräsentanten der Funktionen "bis auf eine Menge vom Masse Null", J. Reine Angew. Math. 165 (1931), 109-115.

[13] J. von Neumann und M. H. Stone, The determination of representative elements in the residual classes of a Boolean algebra, Fund. Math. 25 (1935), 353-378.

[14] J. C. Oxtoby, Spaces that admit a category measure, J. Reine Angew. Math. 205 (1960), 156-170.

[15] J. Pellaumail, Une preuve de l'existence d'un relèvement. Application : Un théorème de Radon-Nikodym faible. Université de Rennes 1971.

[16] I. G. Segal, Equivalence of measure spaces, Amer. J. Math. 73 (1951), 275-313.

[17] M. Sion, A proof of the lifting theorem, University of British Columbia 1970.

[18] W. Strauss, Funktionalanalytische Fassung des Satzes von Radon-Nikodym I, J. Reine Angew. Math. 249 (1971), 92-132.

[19] T. Traynor, An elementary proof of the lifting theorem, Pacific J. Math. 53 (1974), 267-272.

ON THE EXISTENCE OF LOWER DENSITIES IN NONCOMPLETE MEASURE SPACES

Siegfried Graf and Heinrich von Weizsäcker

One of the main open problems concerning measure theoretic liftings is to decide whether every finite (not necessarily complete) measure space admits a lifting. In this note it is shown that the analogous problem for (lower) densities does have a positive solution. Some aspects of the non σ-finite case are also considered.

I. Finite measure spaces.

For the definition of liftings, linear liftings and lower (upper) densities see e.g. [4] or [10].

Theorem 1: Every finite measure space has a lower density (and hence also an upper density).

The proof consists in a slight modification of the two major steps in the Ionescu Tulcea proof [4] of the theorem of D. Maharam [5], see lemma 1 and lemma 2 below. Once these lemmas are established, the standard argument using Zorn's lemma completes the proof of the theorem.

Let $(X,\mathfrak{A},\mu)$ be a finite measure space. For $f \in \mathcal{L}^\infty(X,\mathfrak{A},\mu)$ and a σ-subalgebra $\mathfrak{B}$ of $\mathfrak{A}$ the symbol $E(f|\mathfrak{B})$ denotes a conditional expectation of f with respect to $\mathfrak{B}$.

The first lemma is essentially contained in Traynor [9]:

Lemma 1: Let $(\mathfrak{A}_n)_{n\in\mathbb{N}}$ be an increasing sequence of σ-subalgebras of $\mathfrak{A}$ and let $\mathfrak{A}_\infty$ be the σ-algebra generated by $\bigcup_{n=1}^{\infty}\mathfrak{A}_n$. For each $n \in \mathbb{N}$ let ϱ_n be a lower density of $(X,\mathfrak{A},\mu|\mathfrak{A}_n)$ such that $\varrho_n|\mathfrak{A}_m = \varrho_m$ for all $m \in \mathbb{N}$ with $m \leq n$. Then

$$\varrho_\infty(A) = \bigcap_{k=1}^{\infty}\bigcup_{n=1}^{\infty}\bigcap_{m=n}^{\infty}\varrho_m(\{x\in X:\ E(1_A|\mathfrak{A}_m)(x) \geq 1-\tfrac{1}{k}\}) \qquad (A\in\mathfrak{A}_\infty)$$

defines a lower density for $(X,\mathfrak{A}_\infty,\mu|\mathfrak{A}_\infty)$ satisfying $\varrho_\infty|\mathfrak{A}_n = \varrho_n$ for all $n\in\mathbb{N}$.

The formulation of the induction step given in lemma 2 apparently has been discovered independently by several people, but we know of no reference except [3], where part a) has been formulated for general Boolean algebras. Let $\mathfrak{F}$ be a σ-subalgebra of $\mathfrak{A}$ and let A be an element of $\mathfrak{A}\setminus\mathfrak{F}$. Let $\mathfrak{L}$ denote the σ-algebra generated by $\mathfrak{F}\cup\{A\}$. Choose elements B and C of $\mathfrak{F}$ such that

$$B = \text{ess inf}\,\{F\in\mathfrak{F}:\ A\subset F\}$$

and

$$C = \text{ess inf}\,\{F \in \mathcal{F}: A' \subset F\}$$

where A' denotes the complement of A. Then we have

Lemma 2: a) If ϱ is a lower density (resp. upper density, lifting) of $(X,\mathcal{F},\mu|\mathcal{F})$, then the formula

$$\varrho_0((D\cap A)\cup(E\cap A')) = (A\cap\varrho((B\cap D)\cup(B'\cap E)))\cup(A'\cap\varrho((C\cap E)\cup(C'\cap D))) \qquad (D,E\in\mathcal{F})$$

defines a lower density (resp. upper density, lifting) ϱ_0 of $(X,\mathcal{L},\mu|\mathcal{L})$ such that $\varrho_0|\mathcal{F} = \varrho$.

b) If $l: \mathcal{L}^\infty(X,\mathcal{F},\mu|\mathcal{F}) \to \mathcal{L}^\infty(X,\mathcal{F},\mu|\mathcal{F})$ is a linear lifting (resp. lifting), then the formula

$$l_0(f\cdot 1_A + g\cdot 1_{A'}) = l(f\cdot 1_B + g\cdot 1_{B'})\cdot 1_A + l(f\cdot 1_{C'} + g\cdot 1_C)\cdot 1_{A'} \qquad (f,g\in\mathcal{L}^\infty(X,\mathcal{F},\mu|\mathcal{F}))$$

defines a linear lifting (resp. lifting) $l_0: \mathcal{L}^\infty(X,\mathcal{L},\mu|\mathcal{L}) \to \mathcal{L}^\infty(X,\mathcal{L},\mu|\mathcal{L})$ such that $l_0|\mathcal{L}^\infty(X,\mathcal{F},\mu|\mathcal{F}) = l$.

The proof of this lemma is straightforward: First show (using the definition of B and C) that the right hand side does not depend on the representation of the argument on the left hand side; then verify the required properties of ϱ_0 and l_0 using the corresponding properties of ϱ and l.

Remarks: a. Lemma 1 is a simple generalization of the Lebesgue density theorem. If X is a second countable topological space and $\mathcal{B}$ is the Borel-σ-algebra of X, then lemma 1 shows that a lower density φ of $(X,\mathcal{B},\mu)$ can be chosen such that $\varphi(B)$ is at most of the third Baire class for each $B\in\mathcal{B}$.

b. The formulas in lemma 2 can be simplified considerably if one is not interested in the extension property $\varrho_0|\mathcal{F} = \varrho$ resp. $l_0|\mathcal{L}^\infty(X,\mathcal{F},\mu|\mathcal{F}) = l$ (cf. [6], p. 196).

c. Using part b) of lemma 2 it is possible to show that a finite measure space has a linear lifting, if the intersection of $2^{\aleph_0}$ sets of measure zero is measurable. This remark is due to Fremlin.

d. Theorem 1 once more seems to indicate that the existence of a lower density is a much weaker statement than the existence of a linear lifting. Another hint in this direction is the following observation: In Solovay's model of Zermelo-Fraenkel set theory without axiom of choice [8] one has $(L^\infty(\mu))' = L^1(\mu)$ for any finite measure space (cf. p. 2 in [8] . In this model every set of reals has the Baire property and is Lebesgue measurable. The proof that this implies $(L^\infty)' = L^1$ is omitted in [8], but it is contained in Ch. 5 of [2].) But then von Neumann's argument ([4], p. 96) showing the non-existence of a linear lifting in the spaces $L^p([0,1])$ for $p<\infty$ carries over to the case $p=\infty$. So even in the case of Lebesgue measure any proof for the existence of a linear lifting necessarily relies on some nonconstructive tool like the axiom of choice. For a similar statement concerning (multiplicative) liftings Christensen ([2], p. 111) has given a different argument which does not make use of set theoretic models. In the context of general Boolean algebras the gap between lower densities and linear liftings is illustrated in [10].

II. Nonfinite measure spaces.

Let $(X,\mathfrak{A},\mu)$ be a measure space such that $\mu(X) > 0$ and

$$\mu(A) = \sup\{\mu(B): B\in\mathfrak{A},\ B \subset A,\ \mu(B) < \infty\}$$

for each $A \in \mathfrak{A}$. A partition $\mathfrak{Z}$ of $(X,\mathfrak{A},\mu)$ is a subset of $\mathfrak{A}$ satisfying

(i) $0 < \mu(Z) < \infty$ $\qquad \forall Z \in \mathfrak{Z}$

(ii) $Z \neq Z' \implies Z \cap Z' = \emptyset$ $\qquad \forall Z,Z' \in \mathfrak{Z}$

(iii) $(\mu(A \cap Z) = 0 \quad \forall Z \in \mathfrak{Z}) \implies \mu(A) = 0$ $\qquad \forall A \in \mathfrak{A}$.

Call $(X,\mathfrak{A},\mu)$ strictly localizable, if there is a partition $\mathfrak{Z}$ of $(X,\mathfrak{A},\mu)$ such that

$$\Big((X\setminus \bigcup_{Z\in\mathfrak{Z}} Z) \cap A \in \mathfrak{A} \text{ and } Z \cap A \in \mathfrak{A}\ \forall Z \in \mathfrak{Z}\Big) \implies A \in \mathfrak{A} \qquad \forall A \subset X.$$

From theorem 1 it is easy to derive

Theorem 2: Every strictly localizable measure space has a lower density.

For Radon measures we get

Theorem 3: Let X be a locally compact space and let μ be a Radon measure on X with corresponding essential measure $\mu^\bullet$. Let $\mathfrak{B}$ (resp. $\mathfrak{B}_{loc}$) be the σ-algebra of all Borel (resp. locally Borel) sets in X. Then $(X,\mathfrak{B}_{loc},\mu^\bullet)$ has a lower density φ. If in addition X is metrizable, then one may choose φ such that $\varphi(\mathfrak{B}_{loc}) \subset \mathfrak{B}$.

Proof: 1. Using the existence of a μ-concassage ([7], p. 46) it is easy to see that $(X,\mathfrak{B}_{loc},\mu^\bullet)$ is strictly localizable, hence there is a lower density for $(X,\mathfrak{B}_{loc},\mu^\bullet)$ by theorem 2.

2. Assume that X is locally compact and metrizable. Then by a theorem of A.H. Stone X is paracompact and hence it is a direct sum of a family of σ-compact open subspaces X_i $(i \in I)$ (cf. [1], p. 96). Remark a. shows that for each i there is a lower density φ_i for $(X_i,\mathfrak{B}(X_i),\mu|\mathfrak{B}(X_i))$ such that the the Baire class of $\varphi_i(A)$ is at most 3 for all $A \in \mathfrak{B}(X_i)$. Then $\varphi(B) = \bigcup_{i\in I} \varphi_i(B \cap X_i)$ defines a lower density such that $\varphi(B) \in \mathfrak{B}$ for all $B \in \mathfrak{B}_{loc}$.

Problems: 1. a) Does $(X,\mathfrak{A},\mu)$ have a density, if it has a partition ?
b) Does every Radon measure on a locally compact space X admit a Borel density ?

2. Does there exist a countable ordinal α with the following property: For all compact Radon measure spaces $(X,\mathcal{B},\mu)$ there is a density φ such that all the sets $\varphi(B)$ $(B\in\mathcal{B})$ are at most of Baire class α in X ?

Note that "no" in 2. implies "no" in 1.; "yes" in 2. implies "yes" in 1.b), if X is paracompact.

References.

1. Bourbaki, N.: General topology 1. Paris; Hermann 1966.

2. Christensen, J.P.R.: Topology and Borel structure. Amsterdam etc.; North Holland 1974.

3. Graf, S.: Schnitte Boolescher Korrespondenzen und ihre Dualisierungen. Thesis. Erlangen 1973.

4. Ionescu Tulcea, A. and Ionescu Tulcea, C.: Topics in the theory of lifting. Berlin etc.; Springer 1969.

5. Maharam, D.: On a theorem of von Neumann. Proc. Amer. Math. Soc. 9 (1958), 987 - 994.

6. Meyer, P.A.: Probabilités et potentiel. Paris; Hermann 1966.

7. Schwartz, L.: Radon measures on arbitrary topological spaces and cylindrical measures. London; Oxford University Press 1973.

8. Solovay, R.: A model of set theory in which every set of reals is Lebesgue measurable. Ann. Math. 92 (1970), 1 - 56.

9. Traynor, T.: An elementary proof of the lifting theorem. Pac. J. Math. 53 (1974), 267 - 272. (Abstract in this volume)

10. Weizsäcker, H. v.: Some negative results in the theory of lifting. In this volume.

SOME NEGATIVE RESULTS IN THE THEORY OF LIFTING

Heinrich von Weizsäcker

We show that the following classical theorems in the theory of lifting in some sense are best possible.

A. The theorem of A. and C. Ionescu Tulcea [5] on the existence of a lifting commuting with the left translations of a locally compact group.

B. The theorem of von Neumann and Stone [11] on the construction of a lifting from a lower density in an abstract Boolean algebra.

Accordingly, these notes have two parts which can be read independently. Our results in the first part are based on a general lemma on automorphisms of complete Boolean algebras (A.1). It permits to infer from the existence of a lifting commuting with a set of transformations a strong necessary condition on the fixpoints of these transformations (A.2). As a consequence in a connected locally compact group with left Haar measure the group of left translations is maximal with respect to the property of admitting a commuting lifting (A.3).

In the second part we prove a property of the one-point compactification of an uncountable discrete space (B.1) which shows that in the theorem of von Neumann and Stone the completeness assumptions on the ideal in general cannot be weakened (B.3). At the same time this result gives a nice illustration for the importance of the metrizability conditions in some of the topological selection theorems of Michael (B.4). The connection between these problems is given by an extension of the Stone duality theory for Boolean algebras, as it has been used by Graf in [2] (B.2).

<u>General notations</u>. For the finite operations and relations in a Boolean algebra we use the usual set theoretic symbols. A^c denotes the complement of A and $A \triangle B$ denotes the symmetric difference $(A \setminus B) \cup (B \setminus A)$ of A and B. We repeat the definition of a lifting since often it is given

only for measure spaces. Let $\mathfrak{A}$ be a Boolean algebra with zero element Θ and unit element $\underline{1}$. Let $\mathfrak{N}$ be an ideal in $\mathfrak{A}$. We write $A \equiv_{\mathfrak{N}} B$ or just $A \equiv B$ if $A \triangle B \in \mathfrak{N}$. A map $\rho : \mathfrak{A} \to \mathfrak{A}$ is called a lifting (with respect to $\mathfrak{N}$), if it satisfies the following conditions

I.	$\rho(\Theta) = \Theta,\ \rho(\underline{1}) = \underline{1}$	
II.	$A \equiv B \implies \rho(A) = \rho(B)$	$A \in \mathfrak{A},\ B \in \mathfrak{A}$
III.	$A \equiv \rho(A)$	$A \in \mathfrak{A}$
IV.	$\rho(A \cap B) = \rho(A) \cap \rho(B)$	$A \in \mathfrak{A},\ B \in \mathfrak{A}$
V.	$\rho(A \cup B) = \rho(A) \cup \rho(B)$	$A \in \mathfrak{A},\ B \in \mathfrak{A}.$

If ρ satisfies I, II, III and IV (I, II, III and V), then it is called a lower (upper) density. For a definition of a lifting using the $\mathcal{L}^\infty$-spaces see the remarks preceding theorem B.3.

Part A. Liftings commuting with point transformations

This is an outline of slight generalizations of the main results of [12]. The crucial argument is in the following lemma which perhaps is of independent interest.

Lemma A.1. Let Σ be a finite set of automorphisms of the complete Boolean algebra $\mathfrak{A}$. If there is a maximal filter $\mathfrak{F}$ in $\mathfrak{A}$ which is Σ-invariant, i.e. $\sigma F \in \mathfrak{F}$ for all $\sigma \in \Sigma$ and $F \in \mathfrak{F}$, then there is an element $A \neq \Theta$ of $\mathfrak{A}$ such that $\sigma C = C$ whenever $C \in \mathfrak{A}$, $C \subset A$ and $\sigma \in \Sigma$.

Proof: For any automorphism σ of $\mathfrak{A}$ there are pairwise disjoint elements $A^\sigma, B_1^\sigma, B_2^\sigma, B_3^\sigma$ of $\mathfrak{A}$ such that

a)	$A^\sigma \cup B_1^\sigma \cup B_2^\sigma \cup B_3^\sigma = \underline{1}$	
b)	$B_i^\sigma \cap \sigma B_i^\sigma = \Theta$	$1 \leq i \leq 3$
c)	$\sigma C = C$	$C \in \mathfrak{A},\ C \subset A^\sigma.$

In fact let $A^\sigma = \sup\{D \in \mathfrak{A} : \sigma E = E$ whenever $E \in \mathfrak{A}$ and $E \subset D\}$. For B_1^σ choose a maximal element of $\{B \in \mathfrak{A} : B \cap \sigma B = \Theta\}$ which exists by Zorn's lemma. For B_2^σ and B_3^σ take σB_1^σ and $(A^\sigma \cup B_1^\sigma \cup B_2^\sigma)^c$ respectively. Then B_1^σ, B_2^σ and B_3^σ satisfy b). (For B_3^σ use the maximality of B_1^σ and the fact that σ^{-1} exists).

Now let $\mathfrak{F}$ be a Σ-invariant maximal filter in $\mathfrak{A}$. Then one of the atoms of the finite subalgebra generated by the A^σ and B_i^σ ($\sigma \in \Sigma$, $1 \leq i \leq 3$) must be an element of $\mathfrak{F}$. But in view of a), b) and the Σ-invariance of $\mathfrak{F}$ the only possible candidate is $\bigcap\{A^\sigma : \sigma \in \Sigma\}$. Hence we have $\bigcap\{A^\sigma : \sigma \in \Sigma\} \neq \Theta$ which by c) proves the lemma.◇

In the sequel let $\mathfrak{K}$ be an algebra of subsets of a set Ω and let $\mathfrak{N}$ be an ideal in $\mathfrak{K}$ such that $\mathfrak{K}/\mathfrak{N}$ is a complete Boolean algebra. This is true e.g. if

- $(\Omega,\mathfrak{K},\mu)$ is a localizable measure space and $\mathfrak{N} = \{N \in \mathfrak{K} : \mu(N) = 0\}$,
- Ω is a Baire topological space (e.g. Polish or locally compact), $\mathfrak{N}$ is the system of all subsets of Ω which are locally of first category and $\mathfrak{K}$ is the σ-algebra of all sets with the Baire property (i.e. $K \in \mathfrak{K}$ iff there is an open subset U of Ω such that $K \equiv_{\mathfrak{N}} U$).

Further let G be a set of bijections $s : \Omega \to \Omega$ which are 'bi-measurable and null-set preserving', i.e. $s\mathfrak{K} = \mathfrak{K}$ and $s\mathfrak{N} = \mathfrak{N}$ for all s in G. A lifting ρ is said to <u>commute</u> with G if $s^{-1}\rho(B) = \rho(s^{-1}B)$ holds for all $s \in G$ and $B \in \mathfrak{K}$. For $S \subset G$ denote by Fix S the set of fixpoints $\{\omega : s\omega = \omega \text{ for all } s \in S\}$.

For the next theorem consider the following regularity conditions.

i) $\mathfrak{K}$ is a σ-algebra, $\mathfrak{N}$ is a σ-ideal and $\mathfrak{K}$ has a countable subset which separates the points of Ω.

ii) Ω is a Hausdorff space, the elements of G are continuous, $\mathfrak{K}$ contains the open sets and the implication

$$N \cap U_i \in \mathfrak{N} \text{ for all } i \in I \implies N \cap \Big(\bigcup_{i\in I} U_i\Big) \in \mathfrak{N}$$

holds for all $N \subset \Omega$ and all families $(U_i)_{i\in I}$ of open sets in Ω.

<u>Theorem A.2.</u> Assume that either i) or ii) holds. If there is a lifting commuting with G, then for every finite subset S of G Fix $S \neq \emptyset$ implies Fix $S \in \mathfrak{K}\setminus\mathfrak{N}$.

Proof: Assume that ρ is a lifting commuting with G and that $S \subset G$ is finite. S induces a finite set Σ of automorphisms of the complete Boolean algebra $\mathfrak{K}/\mathfrak{N}$. If there is an element ω_0 of Fix S, then $\{\pi B : \omega_0 \in \rho(B)\}$

is a Σ-invariant maximal filter in $\mathfrak{R}/\mathfrak{N}$. Here π denotes the projection of $\mathfrak{R}$ onto $\mathfrak{R}/\mathfrak{N}$. So by the lemma there is an element A of $\mathfrak{R}\setminus\mathfrak{N}$ such that $sC \equiv C$ whenever $C \in \mathfrak{R}$, $C \subset A$ and $s \in S$. Using either i) or ii) it is not difficult to show that Fix $S \in \mathfrak{R}$ and $A \cap (\text{Fix } S)^c \in \mathfrak{N}$. Because of $A \in \mathfrak{R}\setminus\mathfrak{N}$ this completes the proof.◇

Corollary A.3. Let Ω be a connected locally compact group with left Haar measure μ, let $\mathfrak{R}$ be the σ-algebra of all μ-measurable sets and let $\mathfrak{N}$ be the ideal of all μ-null-sets. If G is a set of continuous bimeasurable and null-set preserving bijections on Ω which is strictly larger than the set of all left translations, then there is no lifting commuting with G.

Proof: Assume that there is a lifting commuting with all left translations and some other $s \in G$. For $y \in \Omega$ denote by t_y the left translation $x \longmapsto yx$. Consider the map $r : x \longmapsto s(x)x^{-1}$. Because of the equality $r^{-1}(\{y\}) = \text{Fix}(t_{y^{-1}} \circ s)$ we can apply the theorem A.2 to get

$$r(\Omega) = \{ y : \mu(r^{-1}(\{y\})) > 0 \}.$$

Thus, μ being σ-finite, $r(\Omega)$ is countable and - by the continuity of r - connected. So it is a singleton. Hence s must be a left translation which is a contradiction.◇

Using the ideas of [4] we can give the following partial converse to theorem A.2.

Proposition A.4. Assume that $\mathfrak{R}$ is a σ-algebra and that $\mathfrak{N}$ is a σ-ideal which is complete in the sense that it is also an ideal in the power set of Ω. If G is a finitely generated abelian group such that for every finite subset S of G Fix $S \neq \emptyset$ implies Fix $S \in \mathfrak{R}\setminus\mathfrak{N}$, and if there is a lower density, then there is a lifting commuting with G.

Proof: For every subgroup H of G denote the set $\{ \omega : H = \{s : s\omega = \omega\} \}$ by $\Omega(H)$. On $\Omega(H)$ the factor group G/H acts faithfully by $(sH)(\omega) = s\omega$. The same arguments as in the proof of Theorem 1 and Theorem 2 in [4] show that there is a lifting ρ_H commuting with G/H (and hence with G) on $\Omega(H)$, whenever $\Omega(H) \in \mathfrak{R}\setminus\mathfrak{N}$. So on these sets we can define a lifting by

(1) $\rho(A) \cap \Omega(H) = \rho_H(A \cap \Omega(H))$ if $\Omega(H) \in \mathfrak{K}\setminus\mathfrak{N}$.

Now let H be a subgroup of G such that $\emptyset \neq \Omega(H) \in \mathfrak{N}$. Since H is finitely generated, we have Fix H $\in \mathfrak{K}\setminus\mathfrak{N}$ and therefore there is a (not necessarily unique) subgroup $\overline{H}$ of G such that $H \subset \overline{H}$ and $\Omega(\overline{H}) \in \mathfrak{K}\setminus\mathfrak{N}$. Now choose from any G-orbit in $\Omega(H)$ exactly one $\overline{H}$-suborbit and let N be the union of these suborbits. If ω_o is any point in $\Omega(\overline{H})$, define ρ on $\Omega(H)$ by

(2) $\rho(A) \cap \Omega(H) = \bigcup \{sN : s\omega_o \in \rho_{\overline{H}}(A \cap \Omega(\overline{H})),\ s \in G\}$, if $\Omega(H) \in \mathfrak{N}$.

The sets $\Omega(H)$ form a countable measurable partition of Ω and therefore (1) and (2) together give the definition of a lifting commuting with G on the whole of Ω.◇

Problems.1. Find necessary conditions on G for the existence of linear liftings commuting with G.

2. Find (necessary or sufficient) conditions for the existence of liftings commuting with a non-injective map $s : \Omega \to \Omega$ satisfying $s^{-1}\mathfrak{K} \subset \mathfrak{K}$ and $s^{-1}\mathfrak{N} \subset \mathfrak{N}$. In particular, does A.1 have an analogue for non-invertible endomorphisms of complete Boolean algebras?

Part B. On the theorem of von Neumann and Stone

Let us first collect some topological definitions and notations. For a toplogical space Y denote by $C_b(Y)$ the Banach space of all real bounded continuous functions on Y with the sup-norm, by M(Y) the space of all bounded signed Borel measures in Y endowed with the topology $\sigma(M(Y), C_b(Y))$ and by P(Y) the subset of M(Y) consisting of all proba bility measures in M(Y). Let $\mathfrak{F}(Y)$ be the set of all closed subsets of Y. For a subset Z of $\mathfrak{F}(Y)$ a mapping $\sigma : Z \to Y$ is called a selection if $\sigma(z) \in z$ for all $z \in Z$. We call a map $m : Z \to P(Y)$ a randomized selection if supp $m(z) \subset z$ for all $z \in Z$. Let X be another toplogical space. For any $f : X \to \mathfrak{F}(Y)$ and any subset A of Y denote by $f^{-1}(A)$ the set $\{x \in X : f(x) \subset A\}$. f is upper semicontinuous if $f^{-1}(U)$ is open for all open subsets U of Y. f is lower semicontinuous if $f^{-1}(F)$ is closed for all closed subsets F of Y. The Vietoris topology (or exponential topology) on a subset Z of $\mathfrak{F}(Y)$ may be defined as the coarsest topology on Z for which the inclusion $Z \hookrightarrow \mathfrak{F}(Y)$ is upper and lower semicontinuous.

The basic construction. Let $k = \aleph_\alpha$ be an infinite cardinal, let I be a set of cardinality $\aleph_{\alpha+1}$ and let ω be any point such that $\omega \notin I$. Denote $I \cup \{\omega\}$ by Y. On Y consider two topologies: the topology s of the one-point compactification of the discrete space I and on the other hand the topology t which induces on I the discrete topology but for which the neighbourhoods of ω are the sets of the form $Y \setminus J$ where card $J < k$ and $J \subset I$. Thus we have $s = t$ if $k = \aleph_0$ and $s \subsetneqq t$ if $k > \aleph_0$. For n = 2 and n = 3 let Y(n) be the n-fold symmetric product of Y, i.e. the set of all non-void subsets of Y with at most n elements. Consider Y(n) as a subset of $\mathfrak{F}(Y,t)$ and let t(n) be the corresponding Vietoris topology on Y(n). Then t(n) is nothing but the quotient topology of the product topology on $(Y,t)^n$. We now have the following

Theorem B.1. a) There is no continuous selection $\sigma: (Y(2),t(2)) \longrightarrow (Y,s)$.
b) There is no continuous randomized selection $m : (Y(3),t(3)) \longrightarrow P(Y,s)$.

Proof: 1. We start by proving the following assertion. Let J be a set of cardinality $\aleph_{\alpha+1}$, let $\tau : J(2) \longrightarrow J$ be a selection and for $x \in J$ let M(x) be the set $\{ y \in J : \tau(\{x,y\}) = x \}$. Then there is an element x of J such that card $M(x) \geq k$. In fact let H be any subset of J such that card H = k. We may assume card $M(y) < k$ for all $y \in H$. For $K = \bigcup \{ M(y) : y \in H \}$ we get card $K \leq k \cdot k = k$ and hence $J \setminus K \neq \emptyset$. By the definition of K we have $\tau(\{x,y\}) = x$ for all $y \in H$ and therefore card $M(x) \geq$ card $H = k$ for $x \notin K$.
2. Now let us establish for any continuous randomized selection $z \longmapsto p_z$, $p : (Y(2),t(2)) \longrightarrow P(Y,s)$ the following estimation

$$(1) \qquad \text{card} \{ y \in I : p_{\{y,\omega\}}(\{y\}) \neq \tfrac{1}{2} \} \leq k.$$

Obviously, (1) implies part a) of the theorem. For the proof of (1) consider the function $f : Y \times Y \longrightarrow \mathbb{R}$, $(x,y) \longmapsto p_{\{x,y\}}(\{y\})$. f satisfies

(2) $\quad 0 \leq f(x,y) = 1 - f(y,x) \qquad x \in Y,\ y \in Y$

and

(3) $\quad \lim_{y \longrightarrow_t \omega} f(x,y) = f(x,\omega) \qquad x \in I.$

Here $\longrightarrow_t$ denotes convergence in the topology t. Let $J_1 = \{ y \in I : f(y,\omega) < \frac{1}{2} \}$ and $J_2 = \{ y \in I : f(y,\omega) > \frac{1}{2} \}$. Then there is a selection $\tau : J_1(2) \longrightarrow J_1$ such that $p_{\{x,y\}}(\{\tau(\{x,y\})\}) \geq \frac{1}{2}$ for all $x \in J_1$, $y \in J_1$. If card $J_1 = \aleph_{\alpha+1}$, then by 1. there is an element x of J_1 for which card $\{ y : f(x,y) \geq \frac{1}{2} \} \geq k$. But then (3) and the definition of the topo logy t give $f(x,\omega) \geq \frac{1}{2}$ in contradiction to $x \in J_1$. Thus we get card $J_1 \leq k$ and by symmetry also card $J_2 \leq k$. This proves (1) and part a) of the theorem.

3. In order to prove part b) we first improve the estimation (1): For every $\varepsilon > 0$ we have

(4) $\quad$ card $K_\varepsilon < k$, where $K_\varepsilon = \{ y \in I : |p_{\{y,\omega\}}(\{y\}) - \frac{1}{2}| > \varepsilon \}$.

Let f be as in 2. For every y in I let N(y) be the set of all $x \in I$ for which $|f(y,x) - f(y,\omega)| > \frac{\varepsilon}{2}$. Because of (3) we have card $N(y) < k$ for all $y \in I$. (1) implies card $K_\varepsilon \leq k$ and hence card $N \leq k \cdot k = k$ for $N = \bigcup \{ N(y) : y \in K_\varepsilon \}$. Then by (1) again there is an element x of $I \setminus N$ such that $f(x,\omega) = \frac{1}{2}$. For this x and any y in K_ε we have $x \notin N(y)$ and $|f(y,\omega) - \frac{1}{2}| > \varepsilon$ and therefore $|f(x,y) - f(x,\omega)| = |f(x,y) - \frac{1}{2}| = |f(y,x) - \frac{1}{2}| \geq |f(y,\omega) - \frac{1}{2}| - |f(y,\omega) - f(y,x)| > \varepsilon - \frac{\varepsilon}{2} = \frac{\varepsilon}{2}$. This implies $K_\varepsilon \subset N(x)$ and a fortiori card $K_\varepsilon \leq$ card $N(x) < k$.

4. Now assume, if possible, that there is a continuous randomized selection $m : (Y(3), t(3)) \longrightarrow P(Y,s)$. (1) implies card $J = \aleph_{\alpha+1}$ for the set $J = \{ y \in I : m_{\{y,\omega\}}(\{y\}) = \frac{1}{2} \}$. Fix $y \in J$ and $\varepsilon > 0$. By the continuity of m there is a set J(y) such that $\omega \in J(y)$, $y \notin J(y)$ and the following conditions are fulfilled:

(5) $\quad$ card $(Y \setminus J(y)) < k$

(6) $\quad |m_{\{y,x,z\}}(\{y\}) - \frac{1}{2}| \leq \varepsilon \qquad x \in J(y),\ z \in J(y).$

For $x \in J(y)$ and $z \in J(y)$ we define a probability measure $p^y_{\{x,z\}}$ by

(7) $$p^y_{\{x,z\}}(A) = (m_{\{y,x,z\}}(\{x,z\}))^{-1} \cdot m_{\{y,x,z\}}(A \cap \{x,z\}).$$

Since $J(y)$ with the relative toplogy is homeomorphic to Y and since $p^y_{\{x,z\}}$ is continuous in $\{x,z\}$ we can apply (4) to get

(8) $$\text{card } (J(y) \setminus L(y)) < k$$

for the set $L(y) = \{ x \in J(y) : |p^y_{\{x,\omega\}}(\{x\}) - \frac{1}{2}| \leq \varepsilon \}$. (6) and (7) imply

(9) $$|m_{\{y,x,\omega\}}(\{x\}) - \tfrac{1}{4}| \leq 2\varepsilon + \varepsilon^2 \qquad x \in L(y).$$

Now let $L(y)$ be choosen in this way for all y in J. For ε sufficiently small (e.g. $\varepsilon = \frac{1}{13}$) (6) and (9) show that for $x \in J$ and $y \in J$ either $x \notin L(y)$ or $y \notin L(x)$ holds. Thus there is a selection $\tau : J(2) \longrightarrow J$ such that $x \notin L(y)$ if $\tau(\{x,y\}) = y$. But then 1. implies for some $y \in J$ that $\text{card}\,\{x \in J : x \notin L(y)\} \geq k$ in contradiction to (5) and (8). So m cannot exist and the proof of the theorem is complete.◇

Remark. 1. In part b) of the theorem we cannot substitute Y(3) by Y(2) since the map $\{x,y\} \longmapsto \frac{1}{2}(\varepsilon_x + \varepsilon_y)$ (where ε_x denotes the Dirac measure in x) from $(Y(2),t(2))$ to $P(Y,s)$ is continuous.

2. Young [13] has given an example of a compact totally disconnected space X for which there is no continuous selection $\sigma : X(2) \longrightarrow X$. As we shall see in the proof of B.4 part a) of the theorem shows that for $X = (Y,s)$ and $k > \aleph_0$ there is not even a Baire measurable selection $\sigma : X(2) \longrightarrow X$.

The translation procedure. The following theorem enables us to translate problems concerning the existence of liftings into topological problems and vice versa. B.2 a) and b)i) are partially contained in Halmos [3] and (in an extended form) in Graf [2], p. 32. Part b)ii) is an easy consequence of a). For a discussion of the first statement in b)i) in the context of Riesz spaces see Nagel [10]. If $\mathfrak{C}$ is a set of subsets of a set X, denote by $\mathcal{L}^\infty(\mathfrak{C})$ the closed linear hull of the set $\{1_C : C \in \mathfrak{C}\}$ in the Banach space of all real bounded functions on X with the sup-norm.

Theorem B.2. Let $\mathfrak{A}$ be the Boolean algebra of all clopen (i.e. simultaneously closed and open) subsets of a compact totally disconnected space Y. Let $\mathfrak{B}$ be an algebra of subsets of a set X.

a) A map $\varphi : \mathfrak{A} \longrightarrow \mathfrak{B}$ is a $\cap$-homomorphism satisfying $\varphi(Y) = X$ if and only if there is a map $\psi : X \longrightarrow \mathfrak{F}(Y)$ such that $\varphi = \psi^{-1}$. ψ is uniquely determined by $\psi(x) = \bigcap\{A \in \mathfrak{A} : x \in \varphi(A)\}$ for all $x \in X$.

b) Let φ and ψ be given as in a).

i) A map $\rho : \mathfrak{A} \longrightarrow \mathfrak{B}$ is an algebra-homomorphism if and only if there is a map $\sigma : X \longrightarrow Y$ such that $\rho = \sigma^{-1}$ We have $\varphi(A) \subset \rho(A)$ for all $A \in \mathfrak{A}$ if and only if $\sigma(x) \in \psi(x)$ holds for all $x \in X$

ii) Let $T : C(Y) \longrightarrow \mathcal{L}^{\infty}(\mathfrak{B})$ be a positive linear operator satisfying $T(1_Y) = 1_X$ and let T' be the adjoint map from X to P(Y) defined by $\langle T'x, f\rangle = Tf(x)$ $(x \in X, f \in C(Y))$. Then $T(1_A) \geq 1_{\varphi(A)}$ holds for all $A \in \mathfrak{A}$ if and only if supp $T'x \subset \psi(x)$ holds for all $x \in X$.

Application to liftings in Boolean algebras. The following definitions and elementary facts are in complete analogy to the measure theoretic case (cf. [9],p.34ff). Let $\mathfrak{B}$ be an algebra of subsets of a set X. Let $\mathfrak{N}$ be an ideal in $\mathfrak{B}$. A linear map $T : \mathcal{L}^{\infty}(\mathfrak{B}) \longrightarrow \mathcal{L}^{\infty}(\mathfrak{B})$ satisfying

I'.	$T(1_X) = 1_X$	
II'.	$f - g \in \mathcal{L}^{\infty}(\mathfrak{N}) \implies Tf = Tg$	$f \in \mathcal{L}^{\infty}(\mathfrak{B})$, $g \in \mathcal{L}^{\infty}(\mathfrak{B})$
III'.	$f - Tf \in \mathcal{L}^{\infty}(\mathfrak{N})$	$f \in \mathcal{L}^{\infty}(\mathfrak{B})$
IV'.	$f \geq 0 \implies Tf \geq 0$	$f \in \mathcal{L}^{\infty}(\mathfrak{B})$
V'.	$T(f\cdot g) = (Tf)\cdot(Tg)$	$f \in \mathcal{L}^{\infty}(\mathfrak{B})$, $g \in \mathcal{L}^{\infty}(\mathfrak{B})$

is also called a lifting. If T satisfies I',II',III' and IV', then it is called a linear lifting. For any lifting $\rho : \mathfrak{B} \longrightarrow \mathfrak{B}$ there is a unique lifting $T_\rho : \mathcal{L}^{\infty}(\mathfrak{B}) \longrightarrow \mathcal{L}^{\infty}(\mathfrak{B})$ such that $T_\rho(\sum_{i=1}^{n} a_i 1_{B_i}) = \sum_{i=1}^{n} a_i 1_{\rho(B_i)}$ for all $a_i \in \mathbb{R}$, $B_i \in \mathfrak{B}$, $n \in \mathbb{N}$. If $\mathfrak{B}$ is a σ-algebra and T is a linear lifting, then $\varphi_T(B) = \{x \in X : T(1_B)(x) = 1\}$ ($B \in \mathfrak{B}$) defines a lower density such that $T(1_B) \geq 1_{\varphi_T(B)}$ for all $B \in \mathfrak{B}$. If in addition T is a lifting, then φ_T is also a lifting and $\rho = \varphi_{T_\rho}$ holds for all liftings ρ If φ is a lower density and if ρ is a lifting, then $\varphi(B) \subset \rho(B)$ holds for all $B \in \mathfrak{B}$ if and only if $T_\rho(1_B) \geq 1_{\varphi(B)}$ holds for all $B \in \mathfrak{B}$.

The following theorem is a classical tool for proofs of the existence of liftings. A lattice V is called [conditionally] k-complete if every subset W of V of cardinality $\leq k$ [which has lower and upper bounds in V] has a supremum and an infimum in V Let $\aleph_\alpha$ be a fixed infinite cardinal.

Theorem (von Neumann and Stone [11],Th.18). Let $\mathfrak{N}$ be an ideal in a Boolean algebra $\mathfrak{B}$ such that $\mathfrak{N}$ is conditionally k'-complete for all cardinals $k' < \aleph_\alpha$. If card $\mathfrak{B}/\mathfrak{N} \leq \aleph_\alpha$, then for any lower density $\varphi : \mathfrak{B} \longrightarrow \mathfrak{B}$ there is a lifting $\rho : \mathfrak{B} \longrightarrow \mathfrak{B}$ such that $\varphi(B) \subset \rho(B)$ for all $B \in \mathfrak{B}$.

Our aim is to show that in the above theorem the condition card $\mathfrak{B}/\mathfrak{N} \leq \aleph_\alpha$ cannot generally be weakened to card $\mathfrak{B}/\mathfrak{N} \leq \aleph_{\alpha+1}$ even if one is interested only in the existence of linear liftings:

Theorem B.3. Assume $\sup_\gamma k_\gamma < \aleph_\alpha$ for all families $(k_\gamma)_{\gamma\in\Gamma}$ of cardinals such that card $\Gamma < \aleph_\alpha$ and $k_\gamma < \aleph_\alpha$ for all $\gamma \in \Gamma$. Then for i = 1 and i = 2 there are a set algebra $\mathfrak{B}_i$, an ideal $\mathfrak{N}_i$ in $\mathfrak{B}_i$ and a lower density φ_i with respect to $\mathfrak{N}_i$ such that the following is true.

1) $\mathfrak{N}_i$ is k'-complete for all $k' < \aleph_\alpha$.

2) $\mathfrak{B}_1/\mathfrak{N}_1$ and $\mathfrak{B}_2/\mathfrak{N}_2$ are isomorphic with card $\mathfrak{B}_i/\mathfrak{N}_i = \aleph_{\alpha+1}$.

3) There is no linear lifting $T : \mathcal{L}^\infty(\mathfrak{B}_1) \longrightarrow \mathcal{L}^\infty(\mathfrak{B}_1)$ such that $T(1_B) \geq 1_{\varphi_1(B)}$ for all $B \in \mathfrak{B}_1$.

4) There is a linear lifting $T : \mathcal{L}^\infty(\mathfrak{B}_2) \longrightarrow \mathcal{L}^\infty(\mathfrak{B}_2)$ such that $T(1_B) \geq 1_{\varphi_2(B)}$ for all $B \in \mathfrak{B}_2$, but there is no lifting with this property (or equivalently: there is no lifting $\rho : \mathfrak{B}_2 \longrightarrow \mathfrak{B}_2$ such that $\varphi_2(B) \subset \rho(B)$ for all $B \in \mathfrak{B}_2$).

Note that the hypothesis on $\aleph_\alpha$ cannot be ommitted, since if it were not satisfied then by 1) $\mathfrak{N}_i$ would be $\aleph_\alpha$-complete and so the theorem of von Neumann and Stone could be used to show that 3) and 4) are not possible.

Proof of B.3: Choose Y and Y(n) (n = 2,3) as in B.1 for $k = \aleph_\alpha$. For any $L \subset I$ of cardinality less than $\aleph_\alpha$ we define an equivalence relation $\sim_L$ on Y(n) by

$$z \sim_L z' \iff (z \cap L = z' \cap L \text{ and } (z \subset L \iff z' \subset L)).$$

Consider the algebras

$$\mathfrak{G}_L(n) = \{ Z \subset Y(n) : (z \in Z, z \sim_L z') \implies z' \in Z \}$$

and

$$\mathfrak{G}(n) = \bigcup \{ \mathfrak{G}_L(n) : L \subset I \text{ and } \operatorname{card} L < \aleph_\alpha \}.$$

Then $\mathfrak{G}(n)$ is closed under k'-fold unions and intersections for all cardinals $k' < \aleph_\alpha$. Also it is easy to see that $\mathfrak{G}(n)$ is contained in the set of all clopen subsets of (Y(n),t(n)).

Let $\mathfrak{B}$ be the algebra of all clopen sets in (Y,s) and let $\mathfrak{B}' = j\mathfrak{B}$ where $j : Y \longrightarrow Y(n)$, $y \longmapsto \{y\}$ is the canonical inclusion. Then we define $\mathfrak{B}_1$ and $\mathfrak{N}_1$ by $\mathfrak{N}_1 = \{ N \in \mathfrak{G}(3) : N \cap jY = \emptyset \}$ and $\mathfrak{B}_1 = \{ B \cup N : B \in \mathfrak{B}', N \in \mathfrak{N}_1 \}$, similarly $\mathfrak{N}_2 = \{ N \in \mathfrak{G}(2) : N \cap jY = \emptyset \}$ and $\mathfrak{B}_2 = \{ B \cup N : B \in \mathfrak{B}', N \in \mathfrak{N}_2 \}$. 1) and 2) are satisfied. The lower densities φ_1 are given by $\varphi_1(jB \cup N)) = \{ z \in Y(3) : z \subset B \}$ and $\varphi_2(jB \cup N) = \{ z \in Y(2) : z \subset B \}$. Let now $T : \mathcal{L}^\infty(\mathfrak{B}_1) \longrightarrow \mathcal{L}^\infty(\mathfrak{B}_1)$ be a linear lifting such that $T(1_B) \geq 1_{\varphi_1(B)}$ for all $B \in \mathfrak{B}_1$. Then there is a positive linear map T^* from C(Y,s) to $\mathcal{L}^\infty(\mathfrak{B}_1)$ such that $Tf = T^*(f|_{jY} \circ j)$ for all $f \in \mathcal{L}^\infty(\mathfrak{B}_1)$ and $T^*(1_B) \geq 1_{\varphi_1(jB)}$ for all $B \in \mathfrak{B}$. By B.2 a) we have $\varphi_1 \circ j = \psi^{-1}$ where $\psi : z \longmapsto \bigcap \{ B \in \mathfrak{B} : z \subset B \}$, i.e. $\psi(z) = z$. Thus by B.2 b)ii) the adjoint map $z \longmapsto (f \longmapsto T^*f(z))$ from Y(3) into P(Y,s) is a randomized selection which is continuous with respect to t(3) because of $\mathcal{L}^\infty(\mathfrak{B}_1) \subset \mathcal{L}^\infty(\mathfrak{G}(3)) \subset C_b(Y(3),t(3))$. But this is a contradiction to part b) of theorem B.1. Hence 3) is established. Similarly one can prove 4) using part a) of B.1 and the subsequent remark.◇

Note that in the example given in the above proof there is a trivial lifting defined by

$$\rho(jB \cup N) = \begin{cases} jB & \text{if } \omega \notin B \\ jB \cup (Y(n) \setminus jY) & \text{if } \omega \in B. \end{cases}$$

<u>Comparison with topological selection theorems</u>. Theorem B.1 implies the following theorem which illustrates the importance of the metrizability conditions on the range in three theorems of Michael ([7],Th.2, [8],Th.1.2, [9],Th.1.1. A weaker version of these metrizability conditions has been given recently by Fakhoury [1].)

<u>Theorem B.4.</u> There exist a totally disconnected compact space Z, a compact subset X of a locally convex linear space E and an open map π from X onto Z such that for the set valued map $\Phi : z \longmapsto \pi^{-1}(z)$ the following is true.

1) Φ is upper and lower semicontinuous.

2) For all $z \in Z$, $\Phi(z)$ is a compact convex triangle (possibly degenerated to an interval or to a point).

3) Φ has no Baire measurable selection (i.e. there is no Baire measurable map $f : Z \longrightarrow X$ such that $f(z) \in \Phi(z)$ for all $z \in Z$).

4) There is no linear map T from C(X) to the space $B_o(Z)$ of all bounded real Baire measurable functions on Z such that

(1) $$\inf_{x \in \pi^{-1}(z)} g(x) \leq Tg(z) \leq \sup_{x \in \pi^{-1}(z)} g(x)$$

for all $z \in Z$ and $g \in C(X)$.

Proof: Choose in theorem B.1 $k \geq \aleph_1$. For Z take the set Y(3) endowed with the Vietoris topology s(3) induced by the Alexandrov topology s on Y. Let $X = \{\mu \in P(Y,s) : \text{supp}\, \mu \in Z\}$ and define π by $\pi(\mu) = \text{supp}\, \mu$ for all $\mu \in X$. Then X is a compact subset of M(Y,s) and π is onto. Property 1) of Φ and hence the openness of π follows from the definition of the topology s(3). The extreme points of the triangle $\Phi(z)$ are the Dirac measures ε_y ($y \in z$).

For a proof of 3) it is in view of theorem B.1 b) sufficient to show that any Baire measurble map from Z into a completely regular space M is continuous with respect to the topology t(3). Consider the set

$$\mathfrak{B} = \{ \{z \in Z : z \subset U\} \cap \{z \in Z : z \cap U' \neq \emptyset\} : U, U' \in \mathfrak{B}\} ,$$

where

$\mathfrak{B} = \{ U \subset Y : \text{either } U \text{ or } Y \setminus U \text{ is finite and does not contain } \omega \}$.

Then $\mathfrak{B}$ is a base of clopen sets for the topology $s(3)$. Hence $\mathfrak{B}$ generates the σ-algebra of all Baire sets in Z. Since $k \geq \aleph_1$ the system $\mathfrak{G}(3)$ defined in the proof of theorem B.3 is a σ-algebra containing $\mathfrak{B}$. Thus (cf. the proof of B.3) every Baire set in Z is open for the topology $t(3)$. This completes the proof of our assertion since the topology in M has a base consisting of Baire sets.

For the proof of 4) assume that $T : C(X) \longrightarrow B_o(Z)$ is linear and satisfies (1). Consider the map $R : C(Y,s) \longrightarrow C(X)$, $h \longmapsto (\mu \longmapsto \mu(h))$ and define for each $z \in Z$ a functional $m(z)$ on $C(Y,s)$ by $m(z)(h) = T(R(h))(z)$. Because of (1) we get

$$\inf_{y \in z} h(y) = \inf_{\mu \in \pi^{-1}(z)} R(h)(\mu) \leq m(z)(h) \leq \sup_{\mu \in \pi^{-1}(z)} R(h)(\mu) = \sup_{y \in z} h(y)$$

for all $z \in Z$ and all $h \in C(Y,s)$. This implies $m(z) \in P(Y,s)$ and $\operatorname{supp} m(z) \subset z$ for all $z \in Z$. Since $B_o(Z) \subset C_b(Y(3),t(3))$ we get that $m : (Y(3),t(3)) \longrightarrow P(Y,s)$ is continuous in contradiction to theorem B.1.

◇

<u>Problems</u>: 1. Find an ideal in a Boolean algebra for which a lower density exists but no lifting does.

Problem 1 should not be too difficult. In the following problems Y is a compact space and $\mathfrak{F}(Y)$ is the space of all closed subsets of Y endowed with the Vietoris topology. Using theorem B.2 it is easy to see that problem 1 is equivalent to problem 1'.

1'. Find a totally disconnected Y such that $\{ \{y\} : y \in Y \}$ is not a retract of $\mathfrak{F}(Y)$.

2. Does there exist a continuous selection $\sigma : \mathfrak{F}(Y) \longrightarrow Y$, if Y is hyperstonean ? ("Yes" would imply the existence of Borel liftings).

3. Does there always exist a Borel measurable selection $\sigma : \mathfrak{F}(Y) \longrightarrow Y$?

R e f e r e n c e s

1. Fakhoury,H.: Sélections continues dans les espaces uniformes. C.R. Acad. Sci. Paris Sér. A 280 (1975) 213 - 216.

2. Graf,S.: Schnitte Boolescher Korrespondenzen und ihre Dualisierung. Thesis. Erlangen 1973.

3. Halmos,P.: Algebraic Logic. New York: Chelsea 1962.

4. Ionescu Tulcea,A.: On the lifting property V. Ann. Math. Stat. 36 (1965) 819-828.

5. Ionescu Tulcea,A. and Ionescu Tulcea,C.: On the existence of a lifting commuting with the left translations of an arbitrary locally compact group. 'Proceedings Fifth Berkeley Symposium of Math. Stat. and Probability' 63-67. University of California Press 1967.

6. Ionescu Tulcea,A. and Ionescu Tulcea,C.: Topics in the theory of liftings. Berlin etc.: Springer 1969.

7. Michael,E.: Selected selection theorems. Am. Math. Monthly 63 (1956) 233-238.

8. Michael,E.: A selection theorem. Proc. Am. Math. Soc. 17 (1966) 1404-1406.

9. Michael,E.: A linear mapping between function spaces. Proc. Am. Math. Soc. 15 (1964) 407-409.

10. Nagel,R.: Darstellung von Verbandsoperatoren auf Banach-Verbänden. Publ. Acad. de Ciencias, Zaragoza XXVII (1972) No. 3.

11. von Neumann,J. and Stone,M.H.: The Determination of Representative Elements in the Residual Classes of a Boolean Algebra. Fund. Math. 25 (1935) 353-376.

12. v.Weizsäcker,H.: Eine notwendige Bedingung für die Existenz masztheoretischer Liftings. To appear in Arch. Math.

13. Young,G.S.: Representation of Banach spaces. Proc. Am. Math. Soc. 13 (1962) 667-668.

DIFFERENTIATION OF MEASURES

S.D. Chatterji

§1. INTRODUCTION.

The main aim of this paper is to indicate that most of the results of Chatterji [2(b)] extend to vector-valued measures if the vector space concerned satisfies a certain property (called the Radon-Nikodym property and defined in section 2). We note also, as in [2(b)], that these very general differentiation theorems (of section 3) include, on the one hand, the usual martingale convergence theory and on the other, they permit us to prove generalizations of the classical differentiation theorem of Lebesgue in $\mathbb{R}^d$ (cf. section 4) without any appeal to Vitali type covering theorems. In the final section (section 5), we state a conjecture concerning the validity of a generalization of Lebesgue's differentiation theorem to abstract spaces.

§2. NOTATION AND PRELIMINARIES.

Let (Ω,Σ) be a Borel space (i.e. Σ is a σ-algebra of subsets of the abstract set Ω), E a (real or complex) Banach space and $\mathcal{A}$ an arbitrary subalgebra of Σ. If μ: $\mathcal{A} \to E$ is an additive E-valued set function defined on $\mathcal{A}$, the formula

$$|\mu|(A) = \sup\{ \sum_{j=1}^{n} \|\mu(A_j)\| \;|A_j \in \mathcal{A}, \bigcup_{j=1}^{n} A_j = A, A_j \text{ disjoint}, n \geq 1\}$$

defines a non-negative additive, possibly infinite, set function on $\mathcal{A}$; $|\mu|$ is called the total variation set function of μ. If $\|\mu\| = |\mu|(\Omega) < \infty$, we shall say that μ is an E-valued additive set function of bounded variation. By $M(\mathcal{A}, E)$, we shall note the Banach space of all E-valued additive set functions on $\mathcal{A}$ which are of bounded variation, equipped with the total variation norm $\|\mu\|$. By $M^\sigma(\mathcal{A}, E)$, we note the subspace of $M(\mathcal{A}, E)$ which consists of all σ-additive E-valued set functions of bounded variation. By $M(\mathcal{A})$ and $M^\sigma(\mathcal{A})$, we shall denote the corresponding spaces of scalar valued set functions and by $M_+(\mathcal{A})$, $M_+^\sigma(\mathcal{A})$ their non-negative elements respectively.

Let $A_1 \subset A_2 \subset \cdots \subset \Sigma$ be an increasing sequence of algebras. Clearly $A_\infty = \bigcup_{n=1}^{\infty} A_n$ is an algebra. If $\mu: A_\infty \to E$ is an additive E-valued set function, $\Pi_n\mu$ will denote the restriction of μ to A_n. It is evident that if $\mu \in M(A_\infty, E)$ then $\Pi_n\mu \in M(A_n, E)$ for $n \geq 1$ and $|\Pi_n\mu|(A) \leq |\Pi_{n+1}\mu|(A) \leq |\mu|(A)$ for any $A \in A_n$.

A Banach space E is said to have the Radon-Nikodym property (RNP) if any $\mu \in M^\sigma(\mathcal{B}, E)$ is of the form $\mu(A) = \int_A f(\omega)|\mu|(d\omega)$, $A \in \mathcal{B}$, where $\mathcal{B}$ is the class of Borel sets of the unit interval [0,1] and f is an E-valued strongly measurable function on [0,1] with $|\mu|$-a.e. separable range (i.e. f is $|\mu|$-Bochner measurable). Clearly, $\|f(\omega)\| = 1$ a.e.$|\mu|$. The RNP was introduced in Chatterji [2(a)], where it is shown that E has the RNP iff for any $\mu \in M^\sigma(\Sigma, E)$, for any arbitrary Borel space (Ω, Σ), and for any $\lambda \in M_+^\sigma(\Sigma)$ such that $\mu \ll \lambda$ (i.e. $\lambda(A) = 0$ implies $\mu(A) = 0$), one has that $\mu(A) = \int_A f(\omega)\, \lambda(d\omega)$ with $f \in L_E^1(\Omega,\Sigma,\lambda)$ = the Banach space formed from the λ-equivalence classes of E-valued Bochner integrable functions. It also follows from the considerations of [2(a)] that E has the RNP iff every separable subspace of E has the RNP. Several important classes of Banach spaces have the RNP. For example, any reflexive space or any separable dual space has the RNP. Several geometric characterizations of RNP have been given recently (cf. the related papers in this volume). However, certain important spaces like C[0,1] or c_0 or ℓ^∞ or L^1 (say over [0,1]) are known not to have the RNP. The following lemma concerning spaces E having the RNP will be useful in the sequel.

<u>LEMMA 2.1</u>: Let E have the RNP. Then any $\mu \in M(A, E)$ has separable range.

<u>PROOF</u>: In case μ is σ-additive, the representation (cf. supra) $\mu(A) = \int_A f\, d|\mu|$, along with the $|\mu|$-a.e. separable range property of f, gives the assertion of the lemma. In the general case, a passage to the Stone space connected with $B(\Omega, A)$ as in Dunford and Schwartz [3],p.312, proves the result. Q.E.D.

If $\mu \in M(A, E)$ and $\lambda: A \to [0,\infty]$ is a σ-additive, σ-finite measure then $\mu = \mu_1 + \mu_2$ where μ_1 is λ-continuous and μ_2 is λ-singular in the sense of Dunford and Schwartz [3], p.131. The existence and unicity of such a decomposition is easy to prove, using, say, the passage to a Stone space, as indicated in the proof of Lemma 2.1 (cf. Chatterji [2(b)], p.29). If, furthermore, E has the RNP, then

$$\mu(A) = \int_A f \, d\lambda + \mu_2(A)$$

where μ_2 is λ-singular and $f \in L^1_E(\Omega, \tilde{A}, \lambda)$, $\tilde{A}$ being the σ-algebra generated by A. The element f will be denoted by $D_\lambda\mu$ or more simply by $D\mu$. It is to be rememberd that $D\mu$ is $\tilde{A}$-measurable and that everything here depends on A as well as λ.

§3. SOME CONVERGENCE THEOREMS.

In this section, $\{A_n\}_{n>1}$ is a fixed increasing, sequence of algebras in the Borel space $(\Omega, \sum)$ and λ is a fixed, σ-additive, non-negative measure, defined at least on A_∞, such that $\Pi_1\lambda$ is σ-finite; i.e. there exist $A_n \in A_1$ with $\lambda(A_n) < \infty$ such that $\Omega = \bigcup_{n=1}^{\infty} A_n$.

PROPOSITION 3.1: Let E be a Banach space with the RNP. If $\mu \in M(A_\infty, E)$ then $D\Pi_n\mu \to D\mu$ strongly a.e. λ.

PROOF: Without loss of generality, we may take λ to be a finite or even a probability measure; hence, we shall suppose that $\lambda(\Omega) = 1$. Let us write $\mu = \alpha + \beta$ where α is a λ-integral (i.e. $\alpha(A) = \int_A f \, d\lambda$, $A \in A_\infty$) and β is λ-singular. This is possible since E has the RNP, as indicated in section 2; according to our notation, $f = D\mu$. Also, $D\Pi_n\mu = D\Pi_n\alpha + D\Pi_n\beta$. But, it is obvious that $D\Pi_n\alpha = E(f|\tilde{A}_n)$ = the λ-conditional expectation of f given $\tilde{A}_n$, the σ-algebra generated by A_n. A standard fact is that $D\Pi_n\alpha = E(f|\tilde{A}_n) \to E(f|\tilde{A}_\infty) = f = D\mu$ strongly a.e. λ. (cf. Chatterji [2(a)], p.23). It is enough, therefore, to show that $D\Pi_n\beta \to 0$ strongly a.e. λ. The argument for this, given in Chatterji [2(b)], p.5, in the scalar case, goes through with obvious modifications and completes the proof. Q.E.D.

REMARQUE 3.1: It is to be noted that the proof outlined above gurantees the a.e. converge of $D\Pi_n\mu$ in many cases, even when E does not have the RNP. For example, if μ is of the form $\mu(A) = \int_A f\, d\lambda + \theta(A)$ and θ is λ-singular then $D\Pi_n\mu \to f$ a.e. provided that $D\Pi_n\theta$ be well-defined. This latter condition is always valid if the algebras $\{A_n\}_{n\geq 1}$ are all purely atomic.

Exactly as in [2(b)], p.6-7, we can prove the following generalization of Prop.3.1.

PROPOSITION 3.2: Let E be a Banach space having the RNP. Let $\mu_n \in M(A_n, E)$ and let $\mu(A) = \lim_{n\to\infty} \mu_n(A)$ exist for all $A \in A_\infty$. Suppose that

(i) $\mu \in M(A_\infty, E)$

and

(ii) $|\mu_n - \Pi_n\mu| \leq \nu_n \in M_+(A_n)$

where $\nu_{n+1} \leq \nu_n$ and $\nu_n \to o$; then $D\mu_n \to D\mu$ a.e. (λ) strongly.

REMARK 3.2: As indicated in [2(b)], the convergence theorems above, contain the martingale and semi-martingale convergence theorems of classical probability theory. Also, as shown in [2(b)], the propositions of the above type can be used to characterize spaces E having the RNP.

§4. LESBESGUE DIFFERENTIATION THEOREM IN $\mathbb{R}^d$.

Exactly as in [2(b)], p.10, we can use prop.3.1 to prove the following differentiation theorem for E-valued set functions.

PROPOSITION 4.1: Let λ be the Lebesgue measure defined on the Borel sets Σ of $\mathbb{R}^d$ and let $\mu \in M(\Sigma, E)$ where E is a Banach space with the RNP. Then there exists a set $N \in \Sigma$ with $\lambda(N) = o$ such that for any $x \in (\mathbb{R}^d \setminus N)$ and any sequence of sets $A_n \in \Sigma$ with $o < \lambda(A_n) < \infty$, converging regularly (see [7] p.269) to x, $\lim_{n\to\infty} \mu(A_n)/\lambda(A_n) = f(x)$ exists and defines a Radon-Nikodym derivative ($f = D\mu$) of the λ-continuous part of μ with respect to λ.

REMARK: The proof of this is exactly as that in section 4 of [2(b)]; one only has to notice that the necessary separability is assured by our lemma 2.1. If μ is λ-singular, then in the above $\mu(A_n)/\lambda(A_n) \to o$ for $x \in (\mathbb{R}^d \setminus N)$, even without the assumption that E has the RNP; this can be derived from Remark 3.1. The interest of the present method of proof is that it avoids any use of Vitali type covering theorems.

§5. MISCELLANEOUS REMARKS.

As noted in [2(b)], the main theorems of section 3 can be generalised even to the case of a finitely additive scalar measure λ, provided that one replaces the derivatives involved by approximate derivatives and λ-a.e. convergence by a slightly modified notion (called quasi-a.e. convergence in section 3 of [2(b)]).

Two interesting problems which have been much studied in differentiation theory are as follows.

(1) Generalize proposition 4.1 by replacing λ there by a Haar measure on a suitable locally compact group (or, more generally, to some type of invariant measure on a nice homogeneous space) with a good choice of A_n's converging to points x.

(2) Generalize proposition 4.1 by such a choice of A_n's converging to x that the proposition remains valid for "arbitrary" non-negative λ.

For problem (2), Besicovitch [1] (cf. Federer [4], Guzmàn [5], Morse [6]) has given a remarkable solution. If the A_n's are balls centered at x (or, more generally, symmetric convex bodies centred at x) in $\mathbb{R}^d$, then prop.4.1 remains valid for any λ (with minor but obvious modifications to cover the non-σ-finite case). Much effort has been devoted recently to generalize this to some metric spaces other than $\mathbb{R}^d$. (cf. papers in these Lecture Notes). Obviously, if the space is locally euclidean, Besicovitch type theorems can be immediately deduced.

Incidentally, this covers the case of Lie groups, almost the only concrete instances where the theorems proved in the study of problem (1) apply; ironically, in these cases, Besicovitch's theorem gives a much stronger result in that the denominator measure is allowed to be arbitrary. It would seem that the Lebesgue or Besicovitch differentiation theory is possible in "finite-dimensional" spaces only. Although several interesting results have been proved which certainly confirm this point of view, the following conjecture remains open, to the best of my knowledge. If a locally compact (say, even, second countable) group is such that prop.4.1 remains valid, for suitable choices of $A_n \to x$, with λ a Haar measure, then the group is either o-dimensional or else a locally Euclidean (i.e. Lie) group. In this connexion cf. W.W. Comfort and H. Gordon, Trans. Amer. Math. Soc. 99(1961) p.83-90; R.E. Edwards and E. Hewitt, Acta Math. 113(1965) p.181-218.

REFERENCES

1. Besicovitch, A.S.
   A general form of the covering principle and relative differentiation of additive functions, Proc. Cambridge Philos. Soc.41 (1945), 103-110; also 42(1946), 1-10.

2. Chatterji, S.D.
   (a) Martingale convergence and the Radon-Nikodym theorem in Banach spaces, Math. Scand.22 (1968), 21-41.
   (b) Differentiation along algebras, Manuscripta Math.4 (1971), 213-224.

3. Dunford, N. and Schwartz, J.T.
   Linear operators I, Interscience, 1958.

4. Federer, H.
   Geometric measure theory. Grundlehren der mathematischen Wissenschaften, Bd.153, Springer-Verlag 1969.

5. Guzmàn, Miguel de
   Differentiation of integrals in $\mathbb{R}^n$, Lecture Notes in Mathematics, No 481, Springer-Verlag, 1975.

6. Morse, A.P.
Perfect blankets, Trans. Amer. Math. Soc.6 (1947), 418-442.

7. Munroe, M.E.
Introduction to measure and integration, Addison-Wesley, 1953.

Département de Mathématiques
Ecole Polytechnique Fédérale de Lausanne
61, avenue de Cour
1007 Lausanne
Switzerland

DIFFERENTIATION OF INTEGRALS IN R^n

Miguel de Guzmán.

The following is a short nontechnical description of some aspects of the theory. For technical details one can see the works by Hayes and Pauc [1970], Bruckner [1971] and Guzmán [1975].

The local theory of the differentiation of integrals starts with the classical theorem of Lebesgue: Let f be a real function in $L^1(R^n)$. Then, for almost every $x \in R^n$ we have, for each sequence of open Euclidean balls $B(x, r_k)$ centered at x such that $r_k \to 0$,

$$\lim \frac{1}{|B(x, r_k)|} \int_{B(x, r_k)} f(y)dy = f(x) \quad \text{as} \quad k \to \infty .$$

At the sight of this result one can ask several meaningful questions: Why Euclidean balls? What happens if the Euclidean balls are replaced, say, by rectangular parallelepipeds or by intervals, i.e. rectangular parallelepipeds whose edges are parallel to the coordinate axes? Why should f be in $L^1(R^n)$? What happens if $\int f$ is replaced by a more general measure or if one imposes something like $f \in L^p(R^n)$ with $p > 1$?

The surprising answers to questions like the ones we have formulated have been the main motivation for the development of this interesting theory. H. Bohr, in 1918 (cf. Caratheodory [1927], p. 689) and Banach [1924] showed that intervals in R^2 behave much worse than cubic intervals or Euclidean balls with respect to the covering theorem of Vitali, which was then fundamental for the proof of the theorem of Lebesgue. So it became a challenging problem to find out whether the replacement of Euclidean balls by intervals centered at the point x in the statement of the Lebesgue theorem would lead to a true proposition or not. The first result in this direction was the so-called strong density theorem, first proved by Saks [1933], stating that if the function f is the characteristic function of a measurable set, then Euclidean balls can be replaced by intervals. Later on Zygmund [1934] showed that this can also be done if f is in any space $L^p(R^p)$, with $1 < p \leq \infty$, and a year later Jessen, Marcinkiewicz and Zygmund [1935] proved that the same is valid if f is in $L(1 + \log^+ L)^{n-1}(R^n)$. On the other hand Saks [1934] proved that there exists a function g in $L(R^n)$ such that the Lebesgue statement is false for g if one take intervals instead of balls.

On the other hand, Zygmund (cf. Nikodym [1927], remark at the end) had previously observed, as a byproduct of a certain construction of Nikodym, that rectangles in R^2 behave still much worse than intervals with respect to differentiatio properties. Nikodym constructed a set N of full measure in the unit square Q of R, i.e. $|N| = 1$, such that for each point p in N there is a line $l(p)$ passing through p such that $l(p) \cap N = \{p\}$. It is easy to deduce from here that there exists a measurable set M in R such that the statement of the Lebesgue theorem is false if f is substituted by the characteristic function of M and the Euclidean balls are replaced by rectangles centered at the corresponding points.

Such facts motivated the introduction of more general differentiation bases by Busemann and Feller [1934] and by R. de Possel [1936] and the study of the problem of characterizing bases which would permit a differentiation theorem similar to that of Lebesgue.

A differentiation basis in R^n in the sense of Busemann and Feller is a collection $\mathcal{B}$ of open bounded sets of R^n such that for each $x \in R^n$ there is at least one sequence $\{R_k\}$ of sets of $\mathcal{B}$ contracting to x, i.e. containing x and such that $\delta(R_k) \to 0$. Let us call $\mathcal{B}(x)$ the family of those sets of

$\mathcal{B}$ that contain x. Then, for $f \in L_{loc}(R^n)$ and $x \in R^n$ one defines the upper derivative of the integral of f at the point x as

$$\bar{D}(\int f, x) = \sup \limsup_{k \to \infty} \frac{1}{|R_k|} \int_{R_k} f$$

where the sup is taken over all sequences $\{R_k\} \subset \mathcal{B}(x)$ contracting to x in the above sense. In a similar way one defines the lower derivative of $\int f$ at x and one says that $\mathcal{B}$ differentiates $\int f$ at x if $\bar{D}(\int f, x) = f(x) = \underline{D}(\int f, x)$. If Φ is a class of functions in $L_{loc}(R^n)$, one says for brevity that $\mathcal{B}$ differentiates Φ if, for each $f \in \Phi$, $\mathcal{B}$ differentiates $\int f$ at almost every $x \in R^n$.

Busemann and Feller [1934] gave a characterization of bases having the density property, i.e. differentiating the class of characteristic functions of measurable sets, and also a characterization of bases differentiating $L^1(R^n)$. Such characterizations were given in terms of certain "halo" conditions, which, in a veiled form, were in fact weak type conditions for the Hardy-Littlewood maximal operator relative to the basis $\mathcal{B}$. For $\mathcal{B}$ one can define its maximal operator M in the following way. If $f \in L^1_{loc}(R^n)$ and $x \in R^n$ one sets

$$Mf(x) = \sup \{\frac{1}{|R|} \int_R |f| : R \in \mathcal{B}(x)\}$$

The function Mf is measurable. Then one can say, for example, for a basis $\mathcal{B}$ invariant by homothecies, that $\mathcal{B}$ differentiates $L^1(R^n)$ if and only if M if of weak type (1, 1).

R. de Possel [1936] considered characterizations of differentiation properties of a basis by means of covering properties of the type of the Vitali theorem. One of his interesting results can be described as follows. For a set

$A \subset R^n$ we shall say that a collection T of sets of the family B is a Vitali cover of A when for each $x \in A$ there is a sequence of sets of T contracting to x. Then, the theorem of de Possel affirms that B differentiates $L^\infty(R^n)$ if and only if for each measurable set A, for each Vitali cover T of A and for each $\varepsilon > 0$, one can select a sequence $\{R_k\} \subset T$ such that: (i) $|A - \bigcup R_k| \leq \varepsilon$, (ii) $|\bigcup R_k - A| \leq \varepsilon$, (iii) $\int |\Sigma \chi_{R_k} - \chi_{\bigcup R_k}| \leq \varepsilon$. Condition (iii) affirms that, although the sets $\{R_k\}$ need not be disjoint, as in the Vitali theorem, their overlap is L^1 - small.

The theory of differentiation has grown around these types of results. It can be described as a study of the relation of the covering properties and weak type properties for the Hardy-Littlewood maximal operator with the differentiation properties of a basis. For a detailed account of such aspects and of many other interesting considerations of the theory one can see the works quoted at the beginning

The theory abounds with interesting open problems whose investigation may lead to important developments in the basic structures of the analysis.

REFERENCES.

BANACH, S. [1924], Sur un theorème de M. Vitali, Fund. Math. 5(1924), 130-136.

BRUCKNER, A.M. [1971], Differentiation of integrals, Amer. Math. Monthly 78(1971) (Slaught Memorial Paper, n° 12).

BUSEMANN, H. and FELLER, W. [1934], Zur Differentiation der Lebesgueschen Integrale Fund. Math. 22(1934), 226-256.

CARATHEODORY, C. [1927], Vorlesungen über reelle Fuktionen^2 (Leipzig, 1927).

de GUZMAN, M. [1975], Differentiation of Integrals in R^n (Springer Lecture Notes, Vol. 481, 1975).

HAYES, C.A. and PAUC, C.Y. [1970], Derivation and Martingales. (Springer, Berlin, 1970).

JESSEN, B., MARCINKIEWICZ, J. and ZYGMUND, A. [1935], Note on the differentiability of multiple integrals, Fund. Math. 25(1935), 217-234.

NIKODYM, O. [1927], Sur la mesure des ensembles plans dont tous les points sont rectilinéairement accessibles, Fund. Math. 10(1927), 116-168.

de POSSEL, R. [1936], Sur la dérivation abstraite des fonctions d'ensemble, J. Math. Pures Appl. 15(1936), 391-409.

SAKS, S. [1933], Theorie de l'Integrale (Warszawa, 1933).

SAKS, S. [1934], Remarks on the differentiability of the Lebesgue indefinite integral, Fund. Math. 22(1934), 257-261.

ZYGMUND, A. [1934], On the differentiability of multiple integrals, Fund. Math 23(1934), 143-149.

PACKINGS AND COVERINGS WITH BALLS IN FINITE DIMENSIONAL NORMED SPACES

Flemming Topsøe
University of Copenhagen

Introduction. Considering the development of measure theory with the present day sophisticated, and in most respects satisfactory theory, it is embarrasing that no natural Vitali type covering theorem seems to be known in infinite-dimensional Banach spaces. Probably, it is hopeless to find a satisfactory result in Banach spaces as $C([0,1])$ since if follows, by a result of Roy O. Davies [3], that there exist distinct probability measures on such spaces which agree on all balls (compare with the recent results [5] of Hoffmann-Jørgensen). But what about "nice" spaces, indeed, what is the situation in Hilbert space?

In section 2, we shall suggest a precise form of a covering theorem which could conceivably hold in "nice" spaces, but at present, this is no more than a working hypothesis. As the reader will have guessed from the title, all we can do is to contrast the general situation with the rather beautifull and profound results in finite dimensional spaces. However, we do hope that just pointing out the general problem is of interest. Actually, with the rescent progress in the geometry of Banach spaces - to a great part induced by another facet of differentiation theory, the Radon-Nikodym theorem - this seems to be the right time to hope for further progress.

1. General theory.

We shall work in a complete separable metric space (X,d), even though less will suffice for the results of this section. $d(A)$ denotes the diameter of $A \subseteq X$. A^δ denotes the δ-neighbourhood of A, i.e. the set of y with $d(y,x) < \delta$ for some $x \in A$. $B(x,\delta)$ denotes the open ball $\{x\}^\delta$, $B[x,\delta]$ the closed ball $\{y : d(y,x) \leq \delta\}$. The letter G, respectively F, indicates a subset of X which is assumed to be open, respectively closed.

$M_+^\infty(X)$ denotes the non-negative Radon measures on X. $\mu \in M_+^\infty(X)$ is locally finite $(\forall x\ \exists G \ni x : \mu G < \infty)$, and for $A \subseteq X$ we have

$$\mu^* A = \inf\{\mu G : G \supseteq A\}.$$

For some of the results below, measures which are "more infinite" than Radon measures could be allowed.

By a <u>Vitali system</u> $\mathcal{V}$, we shall here understand a class of pairs $(A,\mathcal{S})$ with A a subset of X and $\mathcal{S}$ a set of closed subsets of X, such that the following two conditions are satisfied

VS 1: $$(A,\mathcal{S}) \in \mathcal{V},\ B \subseteq A \Rightarrow (B,\mathcal{S}) \in \mathcal{V}.$$

VS 2: $$(A,\mathcal{S}) \in \mathcal{V},\ F \text{ closed} \Rightarrow (A \setminus F, \mathcal{S}_F) \in \mathcal{V} \quad \text{where} \quad \mathcal{S}_F = \{S \in \mathcal{S} : S \cap F = \emptyset\}.$$

All the general results mentioned below go through when VS 2 is replaced by the weaker condition

VS 2' $$(A,\mathcal{S}) \in \mathcal{V},\ F \text{ closed} \Rightarrow (A \setminus F^\delta, \mathcal{S}_F) \in \mathcal{V} \text{ for all } \delta > 0.$$

If $\mathcal{V}$ is a Vitali system and $\mathcal{F}$ a class of closed subsets of X, then we denote by $\mathcal{V}[\mathcal{F}]$ the Vitali system defined by

$$(A,\mathcal{S}) \in \mathcal{V}[\mathcal{F}] \Leftrightarrow (A,\mathcal{S}) \in \mathcal{V},\ \mathcal{S} \subseteq \mathcal{F}.$$

Let $\mathcal{V}$ be a Vitali system and let $\mu \in M_+^\infty(X)$. We shall say that the <u>packing theorem</u> holds for $(\mathcal{V},\mu)$ if, for any $(A,\mathcal{S}) \in \mathcal{V}$, there exists a sequence (S_n) of disjoint sets from $\mathcal{S}$ such that

$$\mu^*(A \setminus \cup S_n) = 0.$$

By a sequence we here mean an indexed set with indexset either $\mathbb{N} = \{1,2,\cdots\}$ or a finite section $\{1,2,\cdots,N\}$. And we shall say that the covering theorem holds if, to any $(A,\mathcal{S}) \in \mathcal{V}$ and any $\varepsilon > 0$, there exists a sequence (S_n) of sets in $\mathcal{S}$ such that

$$\cup S_n \supseteq A, \ \Sigma\mu S_n \leq \mu^*A + \varepsilon.$$

Let us collect several observations concerning the packing theorem, all of a rather straightforward nature, and probably all known to workers in the field (at least with VS 2 rather than VS 2'). Especially, we mention (vi) which, I think, is used - explicitely or implicitly - by every author in the field starting with Vitali and Banach. For the reader who is interested in the litterature, we mention M.de Guzmán [4] and E.M. Alfsen [1] as sources for further information and references.

(i): If $\mathcal{V}$ is a Vitali system, and μ a measure for which the packing theorem holds, then the packing theorem holds for any measure absolutely continuous w.r.t. μ.

(ii): If $\mathcal{V}$ is a Vitali system, if the packing theorem holds for all the measures μ_n; $n \geq 1$, and if $\Sigma\mu_n \in M_+^\infty(X)$, then the packing theorem holds for $\Sigma\mu_n$.

(iii): If $\mu \in M_+^\infty(X)$ and $(\mathcal{V}_n)$ is an increasing sequence of Vitali systems such that the packing theorem holds for μ w.r.t. each $\mathcal{V}_n$, then the packing theorem also holds for μ w.r.t. the Vitali system $\mathcal{V}$ defined by

$$(A,\mathcal{S}) \in \mathcal{V} \Leftrightarrow \exists A_n \uparrow A \ \forall n: (A_n,\mathcal{S}) \in \mathcal{V}_n.$$

(iv): It suffices to check sets of finite measure to establish the packing theorem, more precisely, if $\mathcal{V}$ is a Vitali system and $\mu \in M_+^\infty(X)$ is such that, for every $(A,\mathcal{S}) \in \mathcal{V}$ with $\mu^*(A) < \infty$, there exist disjoint sets (S_n) in $\mathcal{S}$ with $\mu^*(A \setminus \cup S_n) = 0$, then the packing theorem holds.

(v): If the Vitali system $\mathcal{V}$ is such that whenever $(A,\mathcal{S}) \in \mathcal{V}$ there exists a Borel set B with $B \supseteq A$ and $(B,\mathcal{S}) \in \mathcal{V}$, then, for a given measure μ, it suffices to check Borel sets of finite measure to establish the packing theorem.

(vi): If, for a Vitali system $\mathcal{V}$ and $\mu \in M_+^\infty(X)$, there exists a positive constant ρ such that, for every $(A,\mathcal{S}) \in \mathcal{V}$ with $\mu^*A < \infty$, there exist disjoint sets (S_n) in $\mathcal{S}$ for which

$$\mu(\cup S_n) \geqq \rho\ \mu^*A,$$

then the packing theorem holds.

To my mind, these results should be considered as important but simple results. Note that the packing theorem requires the construction of certain disjoint sets, but none of the properties (i)-(vi) above gives a clue to how this can in fact be achieved. The next result, the main idea of which is due to Besicovitch (cf. [2], Lemma 2), improves on the situation.

Lemma 1. Let $\mathcal{V}$ be a Vitali system for which there exists a natural number κ such that, for every $(A,\mathcal{S}) \in \mathcal{V}$, there exist sets (S_n) from $\mathcal{S}$ which covers A - i.e. $\cup S_n \supseteqq A$ - and for which the number of indices $i < n$ with $S_i \cap S_n \neq \emptyset$ is less than κ for each n. In short, the condition can be written as follows:

$$\underset{\kappa}{\exists}\ \underset{(A,\mathcal{S})\in\mathcal{V}}{\forall}\ \underset{\cup S_n\supseteqq A}{\exists}\ \underset{n}{\forall}\ \#\{i < n : S_i \cap S_n \neq \emptyset\} < \kappa.$$

Under this condition, the packing theorem as well as the covering theorem holds for every $\mu \in M_+^\infty(X)$.

Proof. If (S_n) is a sequence of sets such that, for all n, $\#\{i < n : S_i \cap S_n \neq \emptyset\} < \kappa$, then there exist sets E_ν; $\nu = 1,2,\cdots,\kappa$, such that each E_ν is a disjoint union of sets from (S_n) and such that $\cup_1^\kappa E_\nu = \cup S_n$.

To see this, define a mapping φ assigning to each index n in the indexset of (S_n), a natural number $\varphi(n) \leqq \kappa$ in such a way that $i \neq j$, $S_i \cap S_j \neq \emptyset \rightarrow \varphi(i) \neq \varphi(j)$; this can be done by induction. Then the sets $E_\nu = \cup\{S_n : \varphi(n) = \nu\}$; $\nu = 1,2,\cdots,\kappa$ have the desired properties.

Now it is easy to verify for any $\mu \in M_+^\infty(X)$, the condition from (vi) with $\rho = \kappa^{-1}$. Hence the packing theorem holds.

To verify the covering theorem, let $(A,\mathcal{S}) \in \mathcal{V}$, let $\varepsilon > 0$ and consider a $\mu \in M_+^\infty(X)$. We may assume that $\mu^*A < \infty$. Choose $G_1 \supseteqq A$ such that $\mu G_1 \leqq \mu^*A + \varepsilon$. By VS 2 (we simplify by working with this axiom rather than with VS 2'), we may assume that $S \subseteqq G_1$ for all $S \in \mathcal{S}$. Choose, by the packing theorem, (S_n) disjoint in $\mathcal{S}$ such that $\mu^*(A \setminus \cup S_n) = 0$. Put $N = A \setminus \cup S_n$ and choose $G_2 \supseteqq N$ such that $\mu G_2 < \varepsilon$. As $(N, \mathcal{S}) \in \mathcal{V}$, there exist (S_n') in $\mathcal{S}$ with $\cup S_n' \supseteqq A$ and

$$\#\{i < n : S_i' \cap S_n' \neq \emptyset\} < \kappa;\ n = 1,2,\cdots.$$

Now use (S_n') to construct the sets E_ν; $\nu = 1,\cdots,\kappa$ described in the beginning of the proof. We have

$$\cup S_n \cup \cup S_n' \supseteqq A$$

and

$$\begin{aligned} \Sigma\mu S_n + \Sigma\mu S_n' &= \mu(\cup S_n) + \Sigma_1^\kappa\ E_\nu \\ &\leqq \mu(G_1) + \kappa\ \mu(G_2) \\ &\leqq \mu^*A + \varepsilon + \kappa\varepsilon\ . \end{aligned}$$

This implies the desired result. ∎.

2. Some special Vitali systems.

Again, (X,d) is a complete separable metric space. For $0 \leqq \alpha < \infty$ we define a Vitali system $\mathcal{V}_\alpha$ by

$$(A,\mathcal{S}) \in \mathcal{V}_\alpha \Leftrightarrow \forall_{x \in A}\ \forall_{\delta > 0}\ \exists_{B[y,r] \in \mathcal{S}}\ d(x,y) \leqq \alpha r,\quad r \leqq \delta.$$

Of course, we may assume that $\mathcal{S}$ consists entirely of closed balls. The defining property of $\mathcal{V}_\alpha$ can be expressed in the equivalent form, that for each $\delta > 0$, the inclusion

(1) $$A \subseteqq \cup\{B[y,\alpha r] : B[y,r] \in \mathcal{S}\ ,\ r \leqq \delta\}$$

should hold.

$\mathcal{V}_0$ is the <u>centered</u> system $\mathcal{V}_\alpha$, with $\alpha > 0$ is the α-<u>skew</u> system, and the systems with $\alpha > 1$ could be called <u>hyper skew systems</u>.

Clearly, $\mathcal{V}_\alpha$ increases as α increases. In agreement with observation (iii) of the preceeding section, we define Vitaly systems

$\mathcal{V}_{1-}$ and $\mathcal{V}_{\infty}$ by

$$(A,\mathcal{S}) \in \mathcal{V}_{1-} \Leftrightarrow \underset{x\in A}{\forall}\ \underset{0\leq\alpha<1}{\exists}\ \underset{\delta>0}{\forall}\ \underset{B[y,r]\in\mathcal{S}}{\exists}\ d(x,y) \leq \alpha r,\quad r \leq \delta$$

and

$$(A,\mathcal{S}) \in \mathcal{V}_{\infty} \Leftrightarrow \underset{x\in A}{\forall}\ \underset{0\leq\alpha<\infty}{\exists}\ \underset{\delta>0}{\forall}\ \underset{B[y,r]\in\mathcal{S}}{\exists}\ d(x,y) \leq \alpha r,\quad r \leq \delta.$$

Naturally, we may replace $r \to \alpha\cdot r$ by a function of r and thus generalize the definition. To be precise, let $\alpha(\cdot)$ denote an increasing positive function defined for all $r > 0$ and define the Vitali system $\mathcal{V}_{\alpha(\cdot)}$ be replacing the requirement (1) with

$$A \subseteq \cup\{B[y,\alpha(r)] : B[y,r] \in \mathcal{S},\ r \leq \delta\}.$$

$\alpha(\cdot)$ is the <u>function of skewness</u>. If $\alpha(r) < r$ for all r, the skewness is <u>moderate</u>, and if $\alpha(r) \geq r$ for all r, $\mathcal{V}_{\alpha(\cdot)}$ is a <u>hyper skew</u> system.

Next, we define the <u>almost centered</u> system $\mathcal{V}_{0+}$ by

$$(A,\mathcal{S}) \in \mathcal{V}_{0+} \Leftrightarrow \underset{x\in A}{\forall}\ \underset{r>0}{\forall}\ \underset{0<r'\leq r}{\exists}\ \underset{B[x_n,r_n]\in\mathcal{S}}{\exists}\ x_n \to x,\quad r' \leq r_n \leq r \quad \text{for all } n.$$

We also wish to mention some Vitali systems which are of a general nature in that no restriction to balls occurs. $\mathcal{V}'$ and $\mathcal{V}''$ are defined by

$$(A,\mathcal{S}) \in \mathcal{V}' \Leftrightarrow \underset{x\in A}{\forall}\ \underset{\delta>0}{\forall}\ \underset{F\in\mathcal{S}}{\exists}\ d(F) < \delta,\quad x \in \overset{\circ}{F};$$

$$(A,\mathcal{S}) \in \mathcal{V}'' \Leftrightarrow \underset{x\in A}{\forall}\ \underset{\delta>0}{\forall}\ \underset{F\in\mathcal{S}}{\exists}\ d(F) < \delta,\quad x \in F.$$

To a given $\mu \in M_+^{\infty}(X)$ we define the Vitali system $\mathcal{V}_{\mu}$ by

$$(A,\mathcal{S}) \in \mathcal{V}_{\mu} \Leftrightarrow \underset{\delta>0}{\forall}\ \underset{(F_n)\in\mathcal{S}}{\exists}\ \mu^*(A \setminus \cup F_n) = 0,\quad d(F_n) < \delta \quad \text{for all } n.$$

In a way, for the study of the packing theorem, the Vitali systems $\mathcal{V}_{\mu}$ are the most general ones. Clearly, $\mathcal{V}' \subseteq \mathcal{V}_{\mu}$ for every μ.

If $\mu \in M_+^\infty(X)$ and $\mathcal{F}$ is a class of closed bounded subsets of X such that $\mu F > 0$ for all $F \in \mathcal{F}$ and such that, for every $F \in \mathcal{F}$ and $\delta > 0$, there exists $F' \in \mathcal{F}$ with $F \subseteq \mathring{F}' \subseteq F^\delta$, then it can be proved that the packing theorem for μ and each of the Vitali systems $\mathcal{V}'[\mathcal{F}]$, $\mathcal{V}''[\mathcal{F}]$ and $\mathcal{V}_\mu[\mathcal{F}]$ are equivalent.

Among the Vitali systems introduced, we have found that it is only the systems $\mathcal{V}_{0+}$, $\mathcal{V}'[\mathcal{F}]$ and $\mathcal{V}_\mu[\mathcal{F}]$ that lend themselves to a thorough analysis, in the sense that it is possible to formulate necessary and sufficient conditions for the packing theorem to hold, with conditions which appear to be very weak.

To formulate the conditions concerning $\mathcal{V}'[\mathcal{F}]$ and $\mathcal{V}_\mu[\mathcal{F}]$, we introduce some terminology. A <u>triangular array</u> is a system (F_{nk}) with indexset $\dot{N} \times \dot{N}$ such that, for each n, $F_{nk} = \emptyset$ eventually in k. By a <u>subsystem</u> of (F_{nk}) we understand a system (F_{nk}^*) with $F_{nk}^* = F_{nk}$ or $F_{nk}^* = \emptyset$ for each n,k. The triangular array (F_{nk}) is <u>asymptotically infinitesimal</u> if $\lim_n \sup_k d(F_{nk}) = 0$.

<u>Proposition 1</u>. <u>Given $\mu \in M_+^\infty(X)$, and $\mathcal{F}$ a class of closed bounded subsets of X. Then a necessary and sufficient condition that the packing theorem holds for μ and $\mathcal{V}_\mu[\mathcal{F}]$ is, that there exists $\rho > 0$ such that, for every asymptotically infinitesimal triangular array (F_{nk}) of sets in $\mathcal{F}$, one can find a subsystem (F_{nk}^*) consisting of pairwise disjoint sets such that</u>

$$\mu(\cup F_{nk}^*) \geqq \rho\, \mu(\bigcap_n \bigcup_k F_{nk}).$$

<u>If we replace the set $\cap\cup F_{nk}$, occuring on the right hand side, by the set $\cap\cup\mathring{F}_{nk}$, we obtain the necessary and sufficient condition for the packing theorem to hold for μ and $\mathcal{V}'[\mathcal{F}]$</u>

The proof of necessity is trivial, as $(A,\mathcal{P}) \in \mathcal{V}_\mu[\mathcal{F}]$ with $A = \cap\cup F_{nk}$, $\mathcal{P} = \{F_{nk}\}$, and the proof of sufficiency is carried out rather easily by appeal to (vi) of the preceeding section.

Many of the usual results are easy consequences of Proposition 1. As these results are not the main theme of this paper, we just mention, as a sample, that if, for some constant c, $\mu(F^{d(F)}) \leq c\,\mu F$ holds for all $F \in \mathcal{F}$, then the packing theorem for μ and $\mathcal{V}_\mu[\mathcal{F}]$ holds.

We now formulate the analogous result for the Vitali system $\mathcal{V}_{0+}$.

Proposition 2. Let $\mu \in M_+^\infty(X)$. A necessary and sufficient condition that the packing theorem holds for μ w.r.t. $\mathcal{V}_{0+}$ is, that there exists $\rho > 0$ such that, for any choice of sequences $r_1 > r_2 > \cdots \to 0$ and $\varepsilon_1 > \varepsilon_2 > \cdots \to 0$, and any family

$$B[x_{mnk}, r_{mnk}] \;;\; (m,n) \in \mathbb{N} \times \mathbb{N}, \quad k = 1,2,\cdots,N_{mn}$$

of closed balls with

$$r_{n+1} \leqq r_{mnk} \leqq r_n \quad \text{for all} \quad m,n,k,$$

there exists a subsystem $B^*[x_{mnk}, r_{mnk}]$ of $B[x_{mnk}, r_{mnk}]$ - i.e. $B^*[\cdot,\cdot] = B[\cdot,\cdot]$ or $B^*[\cdot,\cdot] = \emptyset$ for all m,n,k - consisting of pairwise disjoint sets such that

$$\mu\left(\bigcup_{mnk} B^*[x_{mnk}, r_{mnk}]\right) \geqq \rho\, \mu\left(\bigcap_{mn} \bigcup_{k} B[x_{mnk}, \varepsilon_m]\right).$$

As to the proof, we only wish to mention that necessity is trivial, and that sufficiency goes via an auxillary Vitaly system which satisfies VS 2' but not VS 2 (consider the definition of $\mathcal{V}_{0+}$ and take "$\forall x \in A$" as the third instead of as the first quantor - it turns out to be sufficient to establish the packing theorem for this auxillary Vitali system).

It may even be that the constant ρ can be allowed to depend on the choice of the sequences (r_n) and (ε_m).

The working hypothesis mentioned in the introduction is, that the packing theorem holds in l^2 for $_{0+}$ and "almost any" $\mu \in M_+^\infty(l^2)$, eg. for all Gaussian measures or all μ with $\mu(\partial B) = 0$ for every ball B. The "natural" hypothesis would concern the centered Vitali system $\mathcal{V}_0$, but this system is hard to analyze, and one can argue that if there at all exists a positive result, it surely can not matter if the balls are slightly off centered, hence a system like the almost centered system $\mathcal{V}_{0+}$ enters into the considerations.

3. Packing and covering theorems in finite dimensional spaces.

In this section, (X,d) is the space $\mathbb{R}^N$ provided with some norm $\|\ \|$.

Theorem 1. The packing theorem as well as the covering theorem holds for the Vitali system $\mathcal{V}_{1-}$ and any $\mu \in M_+^\infty(\mathbb{R}^N)$.

This beautiful result goes back to Besicovitch [2] and to Morse [6]. The paper by Morse contains the full result but in a very disguised form (cf. [6], Theorem 5.13) and has thus been unnoticed by many authors (including the present, cf. the acknowledgements).

The proof we shall now present is influenced by an unpublished manuscript by B. Jessen containing a proof of the packing theorem for $\mathcal{V}_0$.

Proof of Theorem 1. It suffices to establish the result for the Vitali systems $\mathcal{V}_\alpha$ with $\alpha < 1$. We shall verify the condition of Lemma 1 for $\mathcal{V}_\alpha$. Assume then that $(A, \mathcal{S}) \in \mathcal{V}_\alpha$. Consider (1) with $\delta = 1$. As A is a countable union of relatively compact sets, it is seen, that for $\alpha < \alpha' < 1$ we can find $B[x_k, r_k] \in \mathcal{S}$; $k = 1,2,\cdots$ (perhaps only finitely many) such that

$$A \subseteq \bigcup_k B[x_k, \alpha' r_k] ,$$

$$r_1 \geq r_2 \geq \cdots .$$

Assume, for the sake of simplicity, that $\alpha' = \alpha$.

Select indices $k_1, k_2, \cdots$ as follows: $k_1 = 1$. If k_i for $i < j$ have been selected, select k_j as the first index k for which

$$\|x_k - x_{k_i}\| > r_{k_i} - \alpha r_k \text{ for all } i < j,$$

unless no such index exists, in which case the construction stops.

We claim that

$$A \subseteq \bigcup_i B[x_{k_i}, r_{k_i}]. \tag{2}$$

Indeed, if $x \in A$, there exists k with $\|x - x_k\| \leq \alpha r_k$; if k is one of the selected indices, x clearly belongs to the right hand side of (2), and if not, we have $\|x_k - x_{k_i}\| \leq r_{k_i} - \alpha r_k$ for one of the selected indices k_i, hence $\|x - x_{k_i}\| \leq r_{k_i}$ and again, x is seen to belong to the right hand side of (2).

Changing the notation somewhat, replacing k_i by i, we see that what we have found, is a sequence of balls $B[x_i, r_i]$; $i = 1,2,\cdots$ in $\mathcal{S}$ such that

$$r_1 \geq r_2 \geq \cdots \tag{3}$$

(4) $\quad \|x_j - x_i\| > r_i - \alpha r_j \quad$ for $i < j$

(5) $\quad A \subseteq \cup_i \; B[x_i, r_i]$.

We shall prove that with this choice of sets from $\mathcal{S}$, and with a suitable κ, the condition of Lemma 1 is fulfilled. The condition $\cup S_n \supseteq A$ occuring in Lemma 1 is just (5).

To verify the essential condition of Lemma 1, we fix n and attempt to bound the number of "bad" indices, i.e. indices i with $i < n$ and $B[x_i, r_i] \cap B[x_n, r_n] \neq \emptyset$. For this purpose, we define y_i ; $i < n$ by

$$y_i = \frac{1}{r_i + r_n} \; (x_i - x_n).$$

We claim, for indices less than n, that:

(6) If $B[x_i, r_i]$ intersects $B[x_n, r_n]$, then $\|y_i\| \leq 1$;

(7) If $i \neq j$ and $B[x_i, r_i]$ and $B[x_j, r_j]$ both intersect $B[x_n, r_n]$, then $\|y_i - y_j\| \geq \frac{1-\alpha}{2}$.

(6) is obvious. To verify (7), assume $i < j$. We find

$$x_i - x_j = (r_j + r_n)(y_i - y_j) + (r_i - r_j)y_i$$

from which it follows, employing (3), (4) and (6) that

$$r_i - \alpha r_j \leq \|x_i - x_j\| \leq (r_j + r_n)\| y_i - y_j\| + (r_i - r_j)\|y_i\|$$

$$\leq 2 r_j \| y_i - y_j\| + r_i - r_j .$$

This implies the desired inequality $\| y_i - y_j\| \geq \frac{1-\alpha}{2}$.

(6) and (7) says that the points y_i corresponding to "bad" indices all lie in the unit ball and that any two of these points have a distance bounded below by $\frac{1-\alpha}{2}$. Clearly then, the number of such points is bounded above by some number depending only on α and the dimension N. ∎

<u>Theorem 2. The packing theorem holds for $\mathcal{V}_\infty$ and Lebesgue measure on R^N.</u>

This is in fact a simple consequence of Theorem 1. We need only consider a fixed $\mathcal{V}_\alpha$ $(\alpha > 1)$ and blow the balls up by some suitable factor to relate the problem to one for a Vitali system $\mathcal{V}_\beta$ with $\beta < 1$; due to the special properties of Lebesgue measure this only introduces a factor which can be taken care of via observation (vi) of section 1.

In connection with Theorem 2, it is interesting to ask for which functions of skewness $\alpha(\cdot)$, the packing theorem holds for $\mathcal{V}_{\alpha(\cdot)}$, and Lebesgue measure. Only the hyper skew case $\alpha(r) \geq r$ is of interest. Not to complicate the matter more than necessary, assume that $\alpha(\cdot)$ is strictly increasing and continuous so that it is clear what we mean by the inverse function $\alpha^{-1}(\cdot)$.

Conjecture. Under the above mentioned conditions on $\alpha(\cdot)$, the packing theorem holds for $\mathcal{V}_{\alpha(\cdot)}$ and Lebesgue measure in R^N, if and only if

$$\int_0^1 \frac{(\alpha^{-1}(r))^N}{r^{N+1}} \, dr = \infty .$$

At present, we only have a proof of this conjecture in one direction (necessity).

Acknowledgements. The basic problem concerning Vitali type theorems in infinite-dimensional spaces has been discussed by J. Hoffmann-Jørgensen and By Gunnar Andersen, and I heard about it from them. I also thank Hoffmann-Jørgensen and Gunnar Andersen for fruitful discussions on most of the meterial presented in this paper.

Until the Oberwolfach meeting, I thought that Theorem 1 was new. Hoffmann-Jørgensen had shown me a proof (essentially a simplified write-up using Ramsey theory of corresponding meterial from Federer's book on geometric measure thory), which covered the case $\mathcal{V}_\alpha$ for $\alpha \leq 2/3$, but I was not aware of the fact, pointet out to me by M. de Guzmán, that the result is contained in the paper [6] by A.P. Morse. The main interest, therefore, lies in the relative simplicity of the proff, and here I have profited by the access, Børge Jessen has kindly given me to an unpublished manuscript of his.

References.

[1] Alfsen, E.M.: Some coverings of Vitali type. Math. Ann. 159, 203-216 (1965).

[2] Besicovitch, A.S.: A general form of the covering principle and relative differentiation of additive functions. Proc. Cambridge Philos. Soc. 41, 103-110 (1945).

[3] Davies, R.O.: Measures not approximable or not specifiable by means of balls. Mathematika 18, 157-160 (1971).

[4] de Guzman, M.: Differentiation of integrals in R^n. To appear in the Springer Lecture Notes Series.

[5] Hoffmann-Jørgensen, J.: Measures which agree on balls. Århus University, preprint series 1974/75 no. 23.

[6] Morse, A.P.: Perfect blankets. Trans. Amer. Math. Soc. 61, 418-442 (1947).

ON THE RADON-NIKODYM THEOREM IN LOCALLY CONVEX SPACES*

G.Y.H. Chi
Department of Mathematics
University of Pittsburgh

§ 0. Introduction

In 1968, Rieffel [28] proved the fundamental Radon-Nikodym theorem for Banach spaces. This result generalized the classical Lebesgue-Nikodym theorem from R^n to arbitrary Banach spaces. Since then, various efforts have been made to extend Rieffel's Radon-Nikodym theorem to locally convex spaces (l.c.s.'s for short). Such extensions were motivated in part by the desire to prove the existence of conditional expectations for random distributions (see [10]). Metivier (1967,[26]) used the convergence of martingales to obtain the representation theorem for vector measures with values in a locally convex space. Rieffel, in the same paper, applied the Banach space Radon-Nikodym theorem to prove the representation theorem for vector measures with compact average range. Tweddle (1970,[38]) proved a Radon-Nikodym theorem for l.c.s.'s. However, the derivatives there may take values in the algebraic dual of the dual of the given l.c.s. Lewis (1971, [22]) proved the Radon-Nikodym theorem for Frechet spaces. The Lebesgue-Nikodym theorem for certain nuclear spaces were obtained independently by Chi [2], Lewis [22], and Thomas [37]. Kupka (1972,[20]) proved a general Radon-Nikodym theorem for normed linear spaces. A more general representation theorem for group valued measures was obtained by Sion (1973,[34]).

Recently, the related problem of characterizing Banach spaces with the (RNP) have been almost completely solved by Maynard [25], Davis and Phelps [7], Phelps [27], and Huff [12]. The corresponding problem for Frechet spaces have been partly solved by Chi [4] and independently by Saab [31, 32]. In [5], the present author was able to exhibit a wide class of l.c.s.'s having the (RNP). However, the basic problem of establishing the analogue of Rieffel's Radon-Nikodym theorem for l.c.s. still remains open.

The purpose of this paper is to establish the analogue of Rieffel's Radon-Nikodym theorem for a class of quasi-complete l.c.s.'s having the property (BM) (see Definition 2.1). This partially answers Problem 2 raised in [5]. This class of l.c.s.'s includes, for instance, the Frechet spaces, the (LF)-spaces (strict),

---

*This work was completed while the author was visiting at the University of Florida, and the University of Bucharest under a Fulbright Research Grant,1974-1975.

AMS(MOS) 1970 Subject Classification : Primary 28A45, 46G1o; Secondary 46A05.

Key Words and Phrases : Radon-Nikodym theorem, vector measures, locally convex spaces.

the Montel (DF)-spaces, the strong duals of metrizable Montel spaces, the strong duals of metrizable Schwartz spaces, the precompact duals of separable metrizable spaces, and the quasi-complete dual nuclear spaces.

The basic idea used here lies in the construction of a Banach space over the range of the vector measure in question and in the subsequent application of Rieffel's Radon-Nikodym theorem. This method of constructing a Banach space was essentially due to Larman and Rogers [21]. The approach is elementary. In § 1, some preliminary results are established. The proofs of Lemma 1.4, Lemma 1.5, Lemma 1.6, and Theorem 1.8 are given because these results are valid for arbitrary l.c.s's and also because these results are not as straight forward as they may appear to be. In § 2, various examples of l.c.s.'s with property (BM) will be given, and finally in § 3, the Radon-Nikodym theorem will be established for quasi-complete l.c.s.'s with property (BM).

§ 1. Preliminaries

Throughout this paper only standard terminologies in the measure theory and the theory of l.c.s.'s will be used.

Let (Ω,Σ,μ) be a fixed probability space, where Ω is an abstract set, Σ a σ-algebra of subsets of Ω, and μ a probability measure defined on Σ. Without loss of generality, one can assume that Σ is μ-complete. Let $\Sigma^+ = \{ S \in \Sigma \mid \mu(S) > 0 \}$.

Let E be a l.c.s. with 0-neighborhood base $\Theta(E)$ and with property (BM). See Definiton 2.1. For every $U \in \Theta$, let p_U denote the associated continuous seminorm. Let $m : \Sigma \to F$ be a vector measure. For every $U \in \Theta$, and $S \in \Sigma$, the U-variation of m over S is defined to be

$$V(m,U)(S) = \sup\{\sum_{i=1}^{n} p_U(m(S_i)) \mid S \quad S_i \in \Sigma, \text{ disjoint}, 1 \le i \le n\}.$$

$V(m,U)(\cdot)$ is an extended real-valued measure. m is said to have bounded variation iff $V(m,U)(\Omega) < \infty$ for every $U \in \Theta$. m is μ-continuous denoted by $m \ll \mu$ iff for every $U \in \Theta$, $V(m,U)(\cdot) \ll \mu$. Let $S \in \Sigma^+$, then the average range of m over S is the set

$$A_S(m) = \{ \frac{m(T)}{\mu(T)} \mid S \supset T \in \Sigma^+ \}.$$

A function $\phi : \Omega \to E$ is called simple iff ϕ is of the form

$$\phi = \sum_{i=1}^{n} x_i \chi_{S_i}, \qquad (\chi_S \text{ is the characteristic function of } S)$$

where $x_i \in E$, and $S_i \in \Sigma$, $1 \le i \le n$, disjoint. A function $f : \Omega \to E$ is strongly measurable iff there exists a sequence of simple functions ϕ_n such that $f = \lim_n \phi_n$ a.e.(μ). A function $f : \Omega \to E$ is Borel measurable iff

$f^{-1}(V) \in \Sigma$, for every V open in E. Simple functions are Borel measurable. However, a function which is the a.e. (μ) limit of a sequence of simple functions need not be Borel measurable in general, unless E is metrizable. For this and other very relevant and interesting results, see [14] and the talk presented by Masani [24] in this Conference.

Definition 1.1 A strongly measurable function $f : \Omega \to E$ is said to have the Egoroff property if there exists a sequence of simple functions, ϕ_n , such that $\lim_n \phi_n = f$ a.e.(μ), and for every $S \in \Sigma^+$, there exists $T \subset S$, $T \in \Sigma^+$, such that $\mu(S \setminus T) < \varepsilon$ and $\lim_n \phi_n = f$ uniformly on T .

Remark 1.1 Let f have Egoroff property. Then even though f need not be Borel measurable, $p_U(f)$ is still measurable, for every $U \in \Theta$.

Let $\Psi(\mu;E) = \{ f : \Omega \to E \mid f$ is strongly measurable with Egoroff property $\}$. $\Psi(\mu;E)$ is a vector space. The following lemma follows directly from the definition.

Lemma 1.1 Let $f \in \Psi(\mu;E)$. Then f is separably valued, and for every $S \in \Sigma^+$, and every $\varepsilon > 0$, there exists $T \subset S$, $T \in \Sigma^+$, such that $\mu(S \setminus T) < \varepsilon$ and $f(T)$ is precompact.

Definition 1.2 Let $f \in \Psi(\mu;E)$, and $S \in \Sigma^+$. The essential range of f over S is defined to be

$$er_S(f) = \{x \in E \mid \text{for every } U \in \Theta, \ \mu(\{\omega \in S \mid p_U(f(\omega) - x) < 1\}) > 0 \}.$$

Lemma 1.2 Let $f \in \Psi(\mu;E)$, and $S \in \Sigma^+$. Then (a) $er_S(f)$ is closed, (b) $er_S(f) \subset \overline{f(S)}$, and (c) if $T \subset S$, then $er_T(f) \subset er_S(f)$.

Proof : (a) Let $x \in \overline{er_S(f)}$, and $U \in \Theta$. Then there exists $V \in \Theta$ such that $V + V \subset U$, and there exists $y \in er_S(f)$ such that $y \in x + V$. Now $y \in er_S(f)$ implies that $\mu(\{\omega \in S \mid f(\omega) \in y + V\}) > 0$, and $y + V \subset x + U$; hence $\mu(\{\omega \in S \mid f(\omega) \in x + U\}) > 0$. Thus $x \in er_S(f)$.

The proof of (b) and (c) are similar. Q.E.D.

Corollary 1.3 Let $f \in \Psi(\mu;E)$. Then for every $S \in \Sigma^+$, and every $\varepsilon > 0$, there exists $T \subset S$, $T \in \Sigma^+$, such that $\mu(S \setminus T) < \varepsilon$, and $er_T(f)$ is compact.

Lemma 1.4 Let $f \in \Psi(\mu;E)$, and $S \in \Sigma^+$. Then $er_S(f) \cap f(S) \neq \emptyset$.

Proof : Let $\varepsilon > 0$. Then there exists $T \subset S$, $T \in \Sigma^+$, such that $\mu(S \setminus T) < 0$, and $f(T)$ is precompact. Without loss of generality, one can assume $f(T)$ to be separable. Thus $\overline{f(T)}$ is separable compact, and hence is metrizable by the fact that E has property (BM) . Now if $er_T(f) \cap f(T) = \emptyset$, then for every $\omega \in T$, there exists $U_\omega \in \Theta$ such that $\mu(\{\omega' \in T \mid p_{U_\omega}(f(\omega') - f(\omega)) < 1\}) = 0$. Since $f(T) \subset \bigcup_{\omega \in T} [f(\omega) + U_\omega]$, and since $f(T)$ is separable metrizable, there exists a countable subcollection $\{f(\omega_i) + U_{\omega_i}\}$ such that $f(T) \subset \bigcup_{i=1}^{\infty} [f(\omega_i) + U_{\omega_i}]$. This implies that $T \subset \bigcup_{i=1}^{\infty} \{\omega \in T \mid p_{U_{\omega_i}}(f(\omega) - f(\omega_i)) < 1\}$ and hence $\mu(T) = 0$. Thus $f(T) \cap er_T(f) \neq \emptyset$; hence $f(S) \cap er_S(f) \neq \emptyset$, by Lemma 1.2 (c).

<u>Lemma 1.5</u> Let $f, g \in \Psi(\mu;E)$, and $U_o \in \Theta$ such that $p_{U_o}(f(\omega) - g(\omega)) < 1$ a.e.(μ) on $S \in \Sigma^+$. Let $x \in er_S(f)$. Then there exists $x_o \in er_S(g)$ such that $p_{U_o}(x - x_o) < 1$.

<u>Proof</u> : Let $\varepsilon > 0$. Then there exists $T_1 \subset S$ such that $T_1 \in \Sigma^+$, $\mu(S \setminus T_1) < \varepsilon$, and $er_{T_1}(f)$ is compact. By hypothesis, $p_{U_o}(f(\omega) - g(\omega)) < 1$ a.e.(μ) on T_1. For every $\delta > 0$, let $T_\delta = \{\omega \in T_1 \mid p_{U_o}(f(\omega) - x) < \delta, p_{U_o}(f(\omega) - g(\omega)) < 1\}$. Then $\mu(T_\delta) > 0$. By Lemma 1.4 , there exists $\omega_\delta \in T_\delta$ such that $g(\omega_\delta) \in er_{T_\delta}(g)$ and $p_{U_o}(x - g(\omega_\delta)) \leq p_{U_o}(x - f(\omega_\delta)) + p_{U_o}(f(\omega_\delta) - g(\omega_\delta)) < \delta + 1$.
Thus for $n \geq 1$, let $\delta_n = 1/n$, then one can choose a sequence $\omega_n \in T_{\delta_n}$ such that $g(\omega_n) \in er_{T_{\delta_n}}(g)$ and $p_{U_o}(x - g(\omega_n)) < 1/n + 1$.

Since $er_T(g)$ is compact separable, hence metrizable by the fact that E has property (BM). Thus, the sequence $\{g(\omega_n)\} \subset er_T(g)$ has a limit point $x_o \in er_T(g)$, i.e., there exists a subsequence $\{g(\omega_{n_j})\}$ such that $\lim_j g(\omega_{n_j}) = x_o$.

Clearly, $x_o \in er_S(g)$ and in view of the inequality above, one has $p_{U_o}(x - x_o) = \lim_j p_{U_o}(x - g(\omega_{n_j})) \leq \lim_j (1/n_j + 1) = 1$. Q.E.D.

For $f \in \Psi(\mu;E)$, and $U \in \Theta$, let $q_U(f) = \int_\Omega p_U(f)d\mu$. Let $\mathcal{L}^1(\mu;E) = \{f \in \Psi(\mu;E) \mid q_U(f) < \infty$, for every $U \in \Theta\}$ denote the space of all <u>strongly integrable</u> functions, and $L^1(\mu;E) = \mathcal{L}^1(\mu;E)/\eta$ where $\eta = \{f \in \Psi(\mu;E) \mid q_U(f) = 0$, for every $U \in \Theta\}$. $L^1(\mu;E)$ is a l.c.s. topologized by the family of seminorms $\{q_U \mid U \in \Theta\}$. If E is metrizable, then so is $L^1(\mu;E)$. If E is a Banach space, then $\Psi(\mu;E)$ is just the space of all Bochner measurablefunctions and $L^1(\mu;E)$ is the Bochner integrable functions[9]. If E is a Frechet space, then $\Psi(\mu;E)$ is the space of strongly measurable functions and $L^1(\mu;E)$ is the space of strongly integrable functions [3]. In general, the structure and properties of $L^1(\mu;E)$ are not known and will not be dealt here. However, it should be pointed out that $L^1(\mu;E)$ is quite large. For, let F be a Banach space and $j : F \to E$ be a continuous injection. Then, for any $g \in L^1(\mu;F)$, $f = j \circ g \in L^1(\mu;E)$. What one should be aware of is the fact that not every $f \in L^1(\mu;E)$ is of this type. This is why one needs the lemmas of this section.

The following lemma follows directly from the definition of $\Psi(\mu;E)$.

<u>Lemma 1.6</u> Let $f \in \Psi(\mu;E)$. Then there exists an increasing sequence of measurable sets, Ω_n, in Σ^+, such that for every $n \geq 1$, $f(\Omega_n)$ is precompact and if $\Omega_o = \bigcup_{n=1}^{\infty} \Omega_n$, $\Omega'_n = \Omega_n \setminus \Omega_{n-1}$, then $\mu(\Omega_o) = 1$ and $\mu(\Omega'_n) < 2^{-n}$.

<u>Proposition 1.7</u> Let E be polar semi-reflexive (in particular, quasi-complete [19]). Then every $f \in \mathcal{L}^1(\mu;E)$ is Pettis integrable.

<u>Proof</u> : Let $f \in \mathcal{L}^1(\mu;E)$ and let $\{\Omega'_n\}_{n=1}^{\infty}$ be as in Lemma 1.6. Then $f(\Omega'_n)$ is precompact for every $n \geq 1$. Let $S \in \Sigma^+$. Since $f(S\cap\Omega'_n) \subset f(\Omega'_n)$ is precompact, hence $[f(S\cap\Omega'_n)]^o$ is a 0-neighborhood in E'_p. The inequality,

$$\left|\int_{S\cap\Omega'_n} \langle f,y\rangle d\mu\right| \leq \mu(S\cap\Omega'_n), \text{ for every } y \in [f(S\cap\Omega'_n)]^o,$$

shows that the linear form $y \to h_n(y) = \int_{S\cap\Omega'_n} \langle f,y\rangle d\mu$, for $y \in E'$ is bounded on the 0-neighborhood $[f(S\cap\Omega'_n)]^o$. Furthermore, for every $y \in [f(S\cap\Omega'_n)]^o$, $h_n(y) \leq p_{[f(S\cap\Omega'_n)]^o}(y)\ \mu(S\setminus\Omega'_n)$, since $|\langle f,y\rangle| \leq p_{[f(S\ \Omega'_n)]^o}(y)$, for every $y \in [f(S\cap\Omega'_n)]^o$. Thus $h_n \in E$, since E is polar semi-reflexive. Now for every $U \in \Theta$, and $m > n > 1$,

$$p_U\left(\sum_{i=n}^{m} h_i\right) = \sup_{y\in U^o}\left(\left|\left\langle \sum_{i=n}^{m}\int_{S\cap\Omega'_i} f\,d\mu, y\right\rangle\right|\right)$$

$$\leq \sup_{y\in U^o}\left(\sum_{i=n}^{m}\int_{S\cap\Omega'_i} |\langle f,y\rangle|\, d\mu\right)$$

$$\leq \sum_{i=n}^{m}\int_{S\cap\Omega'_i} p_U(f)d\mu$$

$$\leq \int_{\bigcup_{i=n}^{m} S\cap\Omega'_i} p_U(f)\, d\mu$$

which tends to 0 as n, m tend to ∞, since f is strongly integrable and $\mu\left(\bigcup_{i=n}^{m}(S\cap\Omega'_i)\right) \leq \sum_{i=n}^{m}\mu(S\cap\Omega'_i) \leq \sum_{i=n}^{m}\mu(\Omega'_i) \leq \sum_{i=n}^{m} 2^{-i} \to 0$ as $n, m \to \infty$.

Now every Cauchy sequence is precompact, and since E is polar semi-reflexive, hence there exists $\mu_f(S) = \lim_m \sum_{i=1}^{m} h_i$ in E, and for every $y \in E'$,

$\langle\mu_f(S),y\rangle = \int_S \langle f,y\rangle\, d\mu.$ Q.E.D.

<u>Remark 1.2</u> It is possible to prove Proposition 1.7 by appealing to the characterization theorem of Chatterji [6]. However, the above proof is retained because of its elementary nature.

Now for every $f \in \mathcal{L}^1(\mu;E)$, $\mu_f(S) = \int_S f\, d\mu$ defines a vector measure from Σ to E such that μ_f is of bounded variation, and for every $U \in \Theta$ $p_U(\mu_f(S)) \leq \int_S p_U(f)\, d\mu$ and $V(\mu_f,U)(S) = \int_S p_U(f)\, d\mu$. It follows that if $f = g$ a.e.(μ), then $\mu_f = \mu_g$.

Since the Mean Value theorem below is essential in proving the necessity part of the Radon-Nikodym theorem, its proof will be given.

<u>Theorem 1.8</u> Let $f \in L^1(\mu;E)$, and $S \in \Sigma^+$. Then $\dfrac{\mu_f(S)}{\mu(S)} \in \overline{c}[er_S(f)]$;

hence $A_S(\mu_f) \subset \overline{c}[er_S(f)]$.

<u>Proof</u>: Let $f \in L^1(\mu;E)$. Then by the Egoroff property, there exists a sequence of simple functions, ϕ_n, such that for every $U \in \Theta$, there exists $\delta > 0$ and $T \subset S$, $T \in \Sigma^+$ such that

(i) $\delta \leq \mu(S)[1 - \dfrac{6\, q_U(f)}{6\, q_U(f) + \mu(S)}]$

(ii) $V(\mu_f,U)(T) < \dfrac{\mu_f(S)}{6}$, and

(iii) There exists K such that for every $n \geq K$, $p_U(\phi_n(\omega) - f(\omega)) < 1/3$, for every $\omega \in T$.

From (i) and (ii), one has

$$p_U\left(\frac{\mu_f(S)}{\mu(S)} - \frac{\mu_f(T)}{\mu(T)}\right) < 1/3 \qquad \text{\_\_\_\_(1)}.$$

Fix an $m > K$, then by (iii),

$$p_U(\phi_m(\omega) - f(\omega)) < 1/3, \text{ for every } \omega \in T \qquad \text{\_\_\_\_(2)}.$$

Let $\{x_i\}_{i=1}^{k} \subset er_T(\phi_m)$ such that on T, $\phi_m = \sum_{i=1}^{k} x_i \chi_{T_i}$, where T_i are disjoint, $\mu(T_i) > 0$, $1 \leq i \leq k$, and $T = \bigcup_{i=1}^{k} T_i$.

By (iii) and Lemma 1.5, there exists $z_i \in er_{T_i}(f)$ such that

$p_U(z_i - x_i) < 1/3$, $1 \leq i \leq k$. Now $\sum_{i=1}^{k} \frac{\mu(T_i)}{\mu(T)} z_i \in c[er_{T_i}(f)]$, and

$$p_U\left(\frac{\phi_m(T)}{\mu(T)} - \sum_{i=1}^{k} \frac{\mu(T_i)}{\mu(T)} z_i \right) < 1/3 \qquad \text{\_\_\_\_\_(3)}.$$

From (1), (2) and (3) one obtains

$$p_U\left(\frac{\mu_f(S)}{\mu(S)} - \sum_{i=1}^{k} \frac{\mu(T_i)}{\mu(S)} z_i \right) < 1$$

which implies that $\sum_{i=1}^{k} \frac{\mu(T_i)}{\mu(T)} z_i \in \frac{\mu_f(S)}{\mu(S)} + U$.

That is, $\frac{\mu_f(S)}{\mu(S)} \in \overline{c}[er_S(f)]$, since $\overline{c}[er_T(f)] \subset \overline{c}[er_S(f)]$. Q.E.D.

__Definition 1.2__ A quasi-complete l.c.s. E is said to have the Radon-Nikodym property (RNP for short) if for every probability space (Ω,Σ,μ), and for every μ-continuous vector measure $m : \Sigma \to E$ of bounded variation, there exists an $f \in L^1(\mu;E)$ such that $m = \mu_f$.

§ 2. __Locally Convex Spaces with Property (BM)__

Let E be a l.c.s.

__Definition 2.1__ E is said to have property

(BM) if for every bounded subset $B \subset \ell_N^1\{E\}$, the space of absolutely summable sequences, there exists an absolutely convex closed bounded and metrizable subset $M \subset E$ such that $\sum_{i=1}^{\infty} p_M(x_i) < 1$, for every $(x_i) \in B$.

The property (B) of Pietsch [28] is obtained by deleting the requirement of metrizability for the set M in the above definition, and the property (CM) of [5] is obtained by replacing the words "closed bounded" by "compact". It is clear that (CM) implies (BM) which in turn implies (B). The following are some examples of l.c.s.'s having property (BM).

__Example 2.1__ Let E be a metrizable l.c.s. Then E has property (BM). For, E being metrizable, has property (B)([28],p.31).

This example covers all Frechet spaces.

Example 2.2 Let $E = s \lim_{k} E_k$ be the strict inductive limit of a sequence of l.c.s.'s E_k, where each E_k has property (BM) and each E_k is closed in E_{k+1}. Then E has property (BM). For, a subset of E is bounded iff it is a bounded subset of some E_k.

This example includes all the strict (LF)-spaces and the strict (LB)-spaces. For some particular examples of (LF)- or (LB)-spaces (strict), see [33].

In the following examples, the l.c.s.'s have the stronger property (CM). For the proof of this fact for each of these spaces, the reader is referred to [5].

Example 2.3 Let E be a metrizable Montel space. Then E has property (CM).

Example 2.4 Let E be a metrizable nuclear space. Then E has property (CM).

Example 2.5 Let E be a Montel (DF)-space. Then E has property (CM).

In the next series of examples, the dual spaces have property (CM).

Example 2.6 Let E be a separable Frechet space. Then E'_p, the precompact dual of E, has property (CM). This example is of interests, since just like the other examples, every bounded subsets of E'_p are relatively compact.

This example covers the following spaces.

Example 2.6.1 Let E be a (dF)-space of Brauner [1] such that E'_p is separable Then E has property (CM). For, E'_p is separable Frechet.

Example 2.6.2 Let E be a metrizable Montel space. Then E'_p has property (CM) since every metrizable Montel space is separable ([19],p.370).

Example 2.6.3 Let E be a metrizable Schwartz space. Then E'_p has property (CM).

Pietsch [28] used property (B) in his characterization of dual nuclear spaces. For quasi-complete dual nuclear spaces, even something stronger is true.

Example 2.7 Let E be a quasi-complete dual nuclear space. Then E has property (CM).

This example holds for instance when $E = F'_\beta$, where F is a nuclear barreled space, or when F is a nuclear (F)-space, or a complete nuclear (DF)-space, or sequential projective limits of respectively such spaces.

§ 3. The Radon-Nikodym Theorem

In this section the Radon-Nikodym theorem will be established. An important lemma will first be stated.

Let E be a l.c.s. not necessarily quasi-complete, and let $M \subset E$ be an absolutely convex closed bounded metrizable subset. Let $E_M = \bigcup_{n=1}^{\infty} n M$ and $p_M(x) = \inf\{ \lambda > 0 \mid x \in \lambda M\}$. Then (E_M, p_M) is a Banach space. Larman and Rogers ([21],p.43) showed that one can select a sequence of 0-neighborhoods

$U_i \in \Theta$ and a sequence of reals Λ_i such that $M \subset \Lambda_i U_i$ and for every $U \in \Theta$, there exists U_i such that $U_i \cap (M-M) \subset U \cap (M-M)$. Define

$$N(x) = \sum_{i=1}^{\infty} 2^{-i} \Lambda_i^{-1} \frac{p_{U_i \cap M}(x)}{1 + p_{U_i \cap M}(x)}, \text{ for every } x \in E_M .$$

In the case here, the second countability of M is not necessary. One can prove

<u>Lemma 3.1</u> (i) N defines an F-norm on E_M, (ii) N coincides with the relative topology of E on M, (iii) for every $U \in \Theta$, there exists $n \geq 1$ such that $p_U(x) \leq p_{U_n \cap M}(x)$, for every $x \in E_M$, and (iv) for every $x \in E_M$, $N(x) \leq p_M(x)$.

With this lemma and the Radon-Nikodym theorem for Frechet spaces [3], one can prove the following

<u>Theorem 3.1</u> (Radon-Nikodym Theorem) Let E be a quasi-complete l.c.s. with the property (BM), and (Ω,Σ,μ) be a complete probability space. Let $m : \Sigma \to E$ be a vector measure. Then $m = \mu_f$ for some $f \in L^1(\mu;E)$ iff (i) $m \ll \mu$, (ii) m has bounded variation, and (iii) m has locally relatively compact average range.

<u>Proof</u> : Necessity. Let $m = \mu_f$ for some $f \in L^1(\mu;E)$. Then $m \ll \mu$ and m has bounded variation. Now let $S \in \Sigma^+$, then by Corollary 1.3, for every $\varepsilon > 0$, there exists $T \subset S$, $T \in \Sigma^+$ such that $\mu(S \setminus T) < \varepsilon$ and $er_T(f)$ is compact. This implies that $\overline{c}[er_T(f)]$ is compact. Thus m has locally relatively compact average range.

Sufficiency. Let $m : \Sigma \to E$ be a vector measure satisfying (i),(ii) and (iii). Let $\Gamma = \{m(S_i)_{i=1}^{\infty} \mid S_i \in \Sigma, \text{ disjoint}\}$. Then $\Gamma \subset \ell_N^1\{E\}$ is bounded in the π - topology(For, for every $U \in \Theta$, $\sum_{i=1}^{\infty} p_U(m(S_i)) < V(m,U)(\Omega) < \infty$, for $(m(S_i)) \in \Gamma$).

Now E has property (BM), so there exists an absolutely convex closed bounded metrizable subset $M \subset E$ such that $\sum_{i=1}^{\infty} p_M(m(S_i)) < 1$, for every $(m(S_i)) \in \Gamma$. Let (E_M, p_M) be the corresponding Banach space. Then $m : \Sigma \to (E_M, p_M)$ is a μ - continuous vector measure of bounded variation (its variation is bounded by 1). However, in general, m need not have locally relatively compact average range in (E_M, p_M).

Now let N be defined as above on E_M. In general, (E_M,N) is not complete. Let $(\overline{E_M},N)$ be its completion. $m : \Sigma \to (\overline{E}_M,N)$ is clearly (from Lemma 3.1) μ - continuous of bounded variation. Furthermore, m has locally relatively compact average range in $(\overline{E}_M,N)$. For, let $S \in \Sigma^+$, and $\varepsilon > 0$. Then by hypothesis (iii), there exists $T_1 \subset S$, $T_1 \in \Sigma^+$ such that $\mu(S \setminus T) < \frac{1}{2}\varepsilon$ and $A_{T_1}(m)$ is relatively compact. Now m when considered as a measure with values in (E_M, p_M) has locally bounded average range; hence there exists $T \subset T_1$, $T \in \Sigma^+$ such that $\mu(T_1 \setminus T) < \frac{1}{2}\varepsilon$ and $A_T(m)$ is bounded in (E_M, p_M). Clearly, $\mu(S \setminus T) < \varepsilon$, and $A_T(m)$ is relatively compact in $(\overline{E}_M, N)$. For, $A_T(m) \subset A_{T_1}(m)$, $\underline{A}_T(m) \subset \lambda M$, for some $\lambda > 0$, and Lemma 3.1 (ii) together imply that $A_T(m)$ is compact in $(\overline{E}_M, N)$.

By Rieffel's Radon-Nikodym theorem for Banach space, there exists a $g \in L^1(\mu;(\overline{E}_M,N))$ such that $m = \mu_g$. Furthermore, if one let Π be the family of all finite partitions, π, on Ω, directed by inclusion, then $\lim_\pi \phi_\pi = g$ in the mean, where $\phi_\pi = \sum_{S\in\pi} \frac{m(S)}{\mu(S)} \chi_S$, and there exists a subsequence, ϕ_{π_n}, such that $\lim_n \phi_{\pi_n} = g$ a.e.(μ) ([29], Proposition 1.13) and $\mu_{\phi_{\pi_n}} \to \mu_g$. Observe that $\phi_{\pi_n} : \Omega \to E_M \subset \overline{E}_M$, since $\frac{m(S)}{\mu(S)} \in E_M$. Let $j : (E_M,N) \to E$ be the injection map, then j is continuous by Lemma 3.1 (iii). Thus $\lim_n j \circ \phi_{\pi_n} = j \circ g$ a.e.(μ).

Define

$$f(\omega) = \begin{cases} 0, \text{if } \lim_n (j \circ \phi_{\pi_n})(\omega) & \text{does not exists} \\ (j \circ g)(\omega), & \text{otherwise} \end{cases}$$

Then $f : \Omega \to E$, since E is quasi-complete. Furthermore, f is strongly measurable. From Lemma 3.1 (iii), one has

$$\int_\Omega p_U(f)\, d\mu < \alpha_U \int_\Omega N(g)\, d\mu < \infty .$$

Thus, $f \in L^1(\mu;E)$. Moreover, $m = \mu_f$ and $\mu_{(j\circ\phi_{\pi_n})} \to \mu_f$. Q.E.D.

One obtains immediately as a corollary the main result of [5].

<u>Corollary 3.3</u> If E is a quasi-complete l.c.s. with property (CM). Then in Theorem 3.2, the conditions (i) and (ii) are both necessary and sufficient. If E is a quasi-complete dual nuclear space, then the condition (i) is necessary and sufficient.

In view of the Examples 2.1, 2.2, the Radon-Nikodym theorem can be established for all Frechet spaces, all strict (LF)-spaces, and in particular, all strict (LB)-spaces. Furthermore, for the Examples 2.3 - 2.7, Corollary 3.3 shows that they all have the (RNP).

The Radon-Nikodym theorem proved here does not apply to quasi-complete l.c.s. without property (BM). For instance, if E is a quasi-complete nuclear space, or if $E = F'_\beta$, where F is an arbitrary Frechet space, then the result is not applicable. However, this does not imply necessarily that the Radon-Nikodym theorem can not be established for such spaces. Therefore, despite the fact that the class of l.c.s.'s with property (BM) is extensive, this problem remains open

Problem 3.1 Characterize the class of l.c.s.'s for which the above analogue of Rieffel's Radon-Nikodym theorem can be established.

Since the (RNP) is equivalent to the purely geometric property of dentability for Banach spaces and Frechet spaces [4,7,8,12,13,23,25,27,30,31,32], the following problem will be of interests.

Problem 3.2 Prove or disprove the equivalence of (RNP) and dentability for l.c.s.'s of Problem 3.1 .

The author wishes to thank the organizing committee, in particular Professor A.I. Tulcea and Professor D. Kölzow, for the invitation to present this talk at this Conference.

References

[1] Brauner, K. (1973) "Duals of Frechet spaces and a generalization of the Banach-Dieudonne theorem", Duke J. Math. 40, 845-853.

[2] Chi, G.Y.H. (1972) "The Radon-Nikodym theorem for vector measures with values in the duals of some nuclear barreled spaces"; Vector and Operator Valued Measures and Applications, Academic Press 1973, 85-95.

[3] ________ (1973) "The Radon-Nikodym theorem for Frechet spaces", preprint.

[4] ________ (1975) "A geometric characterization of Frechet spaces with the the Radon-Nikodym property", Proc. Amer. Math. Soc. 48, 371-380.

[5] ________ (1975) "On the Radon-Nikodym theorem and locally convex spaces with the Radon-Nikodym property", to appear in the Proc. Amer. Math. Soc.

[6] Chatterji, S.D. (1973) "Sur L'integrabilite de Pettis", preprint.

[7] Davis, W.J. and Phelps, R.R. (1975) "The Radon-Nikodym property and dentable sets in Banach spaces", To appear in the Proc. Amer. Math. Soc.

[8] Diestel, J. and Uhl, J.J. (1975) "The Radon-Nikodym theorem for Banach-spaced valued measures", to appear in the Rocky Mountain J.

[9] Dinculeanu, N. Vector Measures, Pergamon Press, New York 1967.

[10] Fernique, X. (1967) "Processus lineaires, processus generalises", Ann. Inst. Fourier Grenoble 17, 1-92.

[11] Horvath, J. Topological Vector Spaces and Distributions, Addison-Wesley Publishing Co. , New York 1966.

[12] Huff, R.E. (1974) "Dentability and the Radon-Nikodym property", Duke J. Math. 41, 111-114.

[13] ________ (1975) "Dual spaces with the Krein-Milman property have the Radon-Nikodym property", Proc. Amer. Math. Soc. 49, 104-108.

[14] Khalili, S. (1975) "Measurability of Banach spaced valued functions and Bochner integral", preprint.

[15] Khurana, S.S. (1973) "Barycenter, pinnacle points, and denting points", Trans. Amer. Math. Soc. 180, 497-503.

[16] ________ (1972) "Characterization of extreme points", J. London Math. Soc. (2) 5 , 102-104.

[17] ________ (1969) "Measures and barycenter of measures on convex sets in locally conves spaces I,II", J. Math. Anal. Appl. 27,103-115; ibid 28,222-229.

[18] Kluvanek, I. and Knowles, G., Vector Measures and Control Systems. Lecture Notes.

[19] Köthe, G., Topological Vector Spaces I, Springer-Verlag, New York 1969.

[20] Kupka, J. (1972) "Radon-Nikodym theorems for vector valued measures", Trans. Amer. Math. Soc. 169, 197-217.

[21] Larman, D.G. and Rogers, C.A. (1973) "The normability of metrizable sets", Bull. London Math. Soc. 5, 39-48.

[22] Lewis, D.R. (1971) "On the Radon-Nikodym theorem", preprint.

[23] MacGibbon, B. (1972) "A criterion for the metrizability of a compact convex set in terms of the set of extreme points", J. Functional Analysis 11, 385-392.

[24] Masani, P. (1975) "Measurability and Pettis integration in Hilbert spaces", Proc. Conf. on Measure Theory held at Oberwolfach 15-21 June, 1975.

[25] Maynard, H.B. (1972) "A geometric characterization of Banach spaces having the Radon-Nikodym property", Trans. Amer. Math. Soc. 185, 493-500.

[26] Metivier, M. (1967) "Martingales a valeurs vectorielles applications a la derivation des mesures vectorielles", Ann. Inst. Fourier, Grenoble 17, 175-208.

[27] Pietsch, A., Nuclear Locally Convex Spaces, Springer-Verlag, New York 1972.

[28] Phelps, R.R. (1974) "Dentability and extreme points in Banach spaces", J. Functional Analysis 17, 78-90.

[29] Rieffel, M.A. (1968) "The Radon-Nikodym theorem for the Bochner integral", Trans. Amer. Math. Soc. 131, 466-487.

[30] ________ (1969) "Dentable subsets of Banach spaces with applications to a Radon-Nikodym theorem", Proc. Conf. Functional Analysis, Thompson Book Co., Washongton, D.C. 1967, 71-77.

[31] Saab, E. (1974) "Dentabilite et points extremaux dans les espaces localement convexes", Seminaire Choquet, 13e annee 1973/1974 No. 13.

[32] ________ (1975) "Dentabilite, points extremaux et propriete de Radon-Nikodym", preprint.

[33] Schaefer, H.H., Topological Vector Spaces, MacMillan Co., New York 1966.

[34] Sion, M., Theory of Semi-group Valued Measures, Lecture Notes in Mathematics Mathematics 355, Springer-Verlag, New York 1973.

[35] Swartz, C. (1973) "Vector measures and nuclear spaces", Rev. Roum. Math. 18, 1261-1268.

[36] Terzioglu, T. (1969) "On Schwartz spaces", Math. Ann. 182, 236-242.

[37] Thomas, G.E.F. (1974) "The Lebesgue-Nikodym theorem for the vector valued Radon measures", Amer. Math. Soc. Memoir, 1974.

[38] Tweddle, I. (1970) "Vector valued measures", Proc. London Math. Soc. 20, 469-489.

THE RADON-NIKODÝM PROPERTY AND SPACES OF OPERATORS

J. Diestel, Kent State University
Kent, Ohio

Published in 1955, the Memoir of A. Grothendieck on "Tensor Products and Nuclear Spaces" remains today largely unknown to the functional analytic public. In this memoir Grothendieck gave a number of new directions to modern functional analysis introducing and developing homological methods of value to an area previously marked by "seat-of-the-pants" constructions. Surprisingly, some of the most wonderous results in [16] are quite easily explained given the right vantage point; it is the hope of this paper that after its reading this vantage point is a bit more easily attained. Little here is really new--we have, however, made an effort to generalize to the greatest possible extent; Banach space theorists will quickly recognize old friends among the arguments. For those new to the subject matter, it is hoped that the added degree of generality will not serve to "turn-off" interest; rather, it is hoped that newcomers will be saved the wasted time trying to improve Grothendieck on inessential matters and turn their heads to the more basic "guts" of the issues at hand.

The author takes this opportunity to thank the Institute and Professors Kölzow and Ionescu-Tulcea Bellow for their invitation to Oberwolfach and kind hospitality while at the Institute.

§1. Introduction.

The role played by the Radon-Nikodým property in the topological theory of tensor products derives largely from the beautiful theory of representation for operators on $C(\Omega)$ spaces. In the opinion of the author, the representation theory of operators on $C(\Omega)$-spaces is the most elegant aspect of the theory of vector measures and is, in itself, ample justification for studying vector-valued measures.

Though we shall be mainly concerned with classes of operators that arise naturally in the theory of tensor products, the temptation to outline the theory of representation of operators on $C(\Omega)$ is too great to withstand. Complete details of this representation theory can be found in Chapter Six of [9].

First, a few (at present unmotivated) definitions. Let $T: X \to Y$ be a continuous linear operator. T is said to be absolutely summing whenever T takes unconditionally convergent series in X into absolutely convergent series in Y; T is integral whenever T admits a factorization of the form

$$\begin{array}{ccccc} X & \xrightarrow{T} & Y & \hookrightarrow & Y^{**} \\ A \downarrow & & & \nearrow B & \\ C(\Omega) & \hookrightarrow & L_1(\mu) & & \end{array}$$

for some compact Hausdorff space Ω, some regular Borel measure μ on Ω and some operators A, B; T is nuclear whenever there exist sequences $(x_n^*) \subseteq X^*$ and $(y_n) \subseteq Y$ such that $\Sigma_n \|x_n^*\| \, \|y_n\| < \infty$ and T admits the representation $Tx = \Sigma_n x_n^*(x) y_n$ for all $x \in X$.

Now to outline the representation theory of operators on $C(\Omega)$, let $T: C(\Omega) \to X$ be a continuous linear operator. Let $B \subseteq \Omega$ be a Borel set, consider $\hat{B} \in C(\Omega)^{**}$ defined by $\hat{B}(\mu) = \mu(B)$. Define F: Borel sets in $\Omega \to X^{**}$ by $F(B) = T^{**}\hat{B}$. Then it is a routine calculation to show that for each $f \in C(\Omega)$ and each $x^* \in X^*$

$$x^*Tf = \int_\Omega f \, dx^*F.$$

Moreover, $\|T\| = \sup\{\text{variation } x^*F \text{ on } \Omega: x^* \in X^*, \|x^*\| \leq 1\}$. We call F the representing measure of T.

<u>Theorem</u>: Let T: $C(\Omega) \to X$ be bounded linear operator and F be its representing measure. Then

(1) T is weakly compact if and only if $F(B) \in X$ for each Borel set $B \subseteq \Omega$ (in which case, F is countably additive in the norm topology) ([1], [15]);

(2) T is compact if and only if the range of F is a relatively norm compact subset of X ([1]);

(3) T is absolutely summing if and only if F is an X-valued countably additive measure having <u>finite variation</u> $|F|$ ([26]);

(4) T is integral if and only if F is an X-valued countably additive measure having <u>finite variation</u> $|F|$ ([4], [33]);

--of course, (3) and (4) together yield the coincidence of the classes of absolutely summing and integral operators with $C(\Omega)$-domains--

(5) T is nuclear if and only if T is integral and F is Bochner differentiable with respect to $|F|$ ([4], [33]).

In case the $C(\Omega)$ space of the above theorem arises in the more concrete form of an $L_\infty(\mu)$ space we can sharpen to some extent the above result. Again, a few definitions are required. First, if Σ is a σ-field of sets and $F: \Sigma \to X$ is bounded and additive then F is called <u>strongly additive</u> whenever given a sequence (E_n) of pairwise disjoint members of Σ, $\Sigma_n F(E_n)$ converges (unconditionally). If F has finite variation then F is said to be <u>approximately differentiable</u> ([3]) whenever given $\epsilon > 0$ there exists an X-valued, Σ-simple function s such that the variation of $(F(\cdot) - \int_{(\cdot)} s d|F|)$ is not greater than ϵ. Of course, if F is countably additive and has finite variation it is a straightforward and somewhat tedious exercise to see that approximate differentiability and differentiability of F with respect to $|F|$ are the same.

Now suppose (Ω, Σ, μ) is a finite measure space and let $T: L_\infty(\mu) \to X$ be a continuous linear operator. Define $F: \Sigma \to X$ by $F(A) = F(X_A)$. Then it is plain that F is bounded, additive and vanishes on μ-null sets. Moreover, if $f \in L_\infty(\mu)$, then $T(f) = \int_\Omega f dF$ (integral defined in the most obvious way!); also, $\|T\| = \sup\{|x^*F|(\Omega): x^* \in X^*, \|x^*\| \leq 1\}$. Again, F is called the representing measure of T.

<u>Theorem</u>: Let $T: L_\infty(\mu) \to X$ be a bounded linear operator and $F: \Sigma \to X$ be its representing measure (remember F is generally only finitely additive!). Then

(1) T is weakly compact if and only if F is strongly additive [5]
if and only if $F(\Sigma)$ is relatively weakly compact;

(2) T is compact if and only if $F(\Sigma)$ is relatively norm-compact;

(3) T is absolutely summing if and only if F has finite variation
if and only if T is integral ([4]);

(4) T is nuclear if and only if F has finite variation and is approximately differentiable ([4]);

(5) T is weak* to weak continuous if and only if F is countably additive.

The role played by vector measure theory in the topological theory of tensor products is due largely to the characterizations of integral and nuclear operators on $C(\Omega)$ spaces in terms of their representing measures. This role will be the central topic of discussion in §2. Therein, we shall show how the Radon-Nikodým Property allows one to develop a duality theory for spaces of compact operators that closely parallels (and extends) the Dixmier-VonNeumann-Schatten theory for operators on Hilbert spaces. Some striking conclusions are drawn (in presence of Radon-Nikodým assumptions) regarding the approximation and metric approximation properties and criteria for the reflexivity of the space of operators between two Banach spaces are made transparent. Further, the possession of the Radon-Nikodým Property by the class of nuclear operators on a space X is noted for a broad class of reflexive spaces X. Finally, the Grothendieck theory is used to show how one can frequently replace a scalarly measurable function by a strongly measurable one.

In the final §3, we discuss some open problems related to the discussion of §2 and to the (Banach space) theory of spaces of operators.

§2. The Radon-Nikodým Property and Its Relation to the Topological Theory of Tensor Products.

The study of the classical Lebesgue spaces is facilitated by the density of simple functions. When studying spaces of operators, the natural analogue of simple functions is the notion of a finite rank operator. Therefore, it might be hoped that density theorems for finite rank operators can be proved and then suitably exploited to yield structural information about these spaces.

Definition. A Banach space X is said to possess the approximation property (A.P.) whenever given a compact set $K \subseteq X$ and an $\epsilon > 0$ there exists a continuous linear operator $T: X \to X$ having finite dimensional range (a so-called finite rank operator) such that for each $x \in K$, $\|Tx - x\| \leq \epsilon$; if T can always be chosen so as to satisfy $\|T\| \leq 1$, then X is said to possess the metric approximation property (M.A.P.).

Before stating a few of the more well-known equivalent formulations for A.P. and M.A.P., we recall some notions from the theory of tensor products of Banach spaces.

Let X, Y be Banach spaces.

Consider the algebraic tensor product, $X \otimes Y$, of X and Y. We will be mainly concerned with two distinct methods of endowing $X \otimes Y$ with a reasonable norm topology then completing the resultant spaces.

First, $X \otimes Y$ naturally imbeds in the space $\mathcal{L}(X^*; Y)$ of bounded linear operators from X* to Y; endow $X \otimes Y$ with the relative uniform operator norm topology and complete. This Banch space is denoted by $X \check{\otimes} Y$ and is usually referred to as the <u>injective tensor product</u> of X and Y.

Next, $X \otimes Y$ acts naturally as a set of continuous linear functionals on $X^* \check{\otimes} Y^*$. Endow $X \otimes Y$ with the relative dual norm topology and complete. This Banach space is denoted by $X \hat{\otimes} Y$ and is usually referred to as the <u>projective tensor product</u> of X and Y.

A basic defining property of the projective tensor product is the following:

<u>Universal Mapping Principle</u>: <u>If Z is any Banach space then the bilinear continuous operators $\tau: X \times Y \to Z$ are in isometric relationship to the continuous linear operators $T: X \hat{\otimes} Y \to Z$ via the correspondence $\tau \leftrightarrow T$ whenever $T(x \otimes y) = \tau(x, y)$.</u>

<u>In particular, $(X \hat{\otimes} Y)^*$ is isometrically the space $\mathcal{B}(X, Y)$ of bounded bilinear functionals on $X \times Y$.</u>

The dual of $X \check{\otimes} Y$ is a bit touchier to describe; the startlingly simple description we give is due to Grothendieck. As one should expect, the problem is that of picking out which continuous bilinear functionals on $X \times Y$ belong to $(X \check{\otimes} Y)^*$. Note that if we denote by $\Omega(X^*)$ the closed unit ball of X* <u>in its weak* topology</u>, then $X \check{\otimes} Y$ is naturally identifiable as a closed linear subspace of $C(\Omega(X^*) \times \Omega(Y^*))$. Thus if $\varphi \in (X \check{\otimes} Y)^*$ we have (by the Hahn-Banach theorem) the existence of a $\varphi' \in C(\Omega(X^*) \times \Omega(Y^*))^*$ which extends φ in a norm-preserving manner. Now, by the Riesz-Markov-Kakutani theorem, φ''s action is given by a regular Borel measure μ on $\Omega(X^*) \times \Omega(Y^*)$ where $|\mu|(\Omega(X^*) \times \Omega(Y^*)) = \|\varphi'\| = \|\varphi\|$. It follows then that

$$(\int) \quad \varphi(x, y) = \varphi'(x, y) = \int_{\Omega(X^*) \times \Omega(Y^*)} x^*(x)y^*(y)d\mu(x^*,y^*).$$

Conversely, any bilinear functional φ on $X \times Y$ of the above form defines a continuous bilinear functional on $X \check{\otimes} Y$ with norm $|\mu|(\Omega(X^*) \times \Omega(Y^*))$.

For obvious reasons such bilinear functionals are referred to as *integral bilinear forms*; the space of integral bilinear functionals is denoted by $\mathcal{B}^\wedge(X, Y)$ and is, in the dual norm of $X \check{\otimes} Y$, a Banach space.

An operator $T: X \to Y$ is integral if and only if the bilinear form $\tau: X \times Y^* \to$ scalars given by $\tau(x, y^*) = y^*Tx$ is integral. A nontrivial, though largely formal, argument shows that this notion of integrability is identical to that mentioned in the introduction. The integral norm of $T: X \to Y$ is the norm of the induced $\tau \in \mathcal{B}^\wedge(X, Y^*)$; the space of integral operators is denoted by $I(X; Y)$. Moreover, an operator $T: X \to Y$ is integral if and only if $T^*: Y^* \to X^*$ is integral.

Under natural identifications then we have

$$(X \check{\otimes} Y)^* = \mathcal{B}^\wedge(X, Y) = I(X; Y^*)$$

and

$$(X \hat{\otimes} Y)^* = \mathcal{B}(X, Y) = \mathcal{L}(X; Y^*).$$

A basic characterization of spaces with A.P. is contained in the next

Theorem. A Banach space X possesses the A.P. if and only if the natural linear injection of $X^* \hat{\otimes} X$ into $I(X; X)$ is 1-1.

Related to this is the following identification of spaces with M.A.P.

Theorem. A Banach space X possesses the M.A.P. if and only if the natural linear injection of $X^* \hat{\otimes} X$ into $I(X; X)$ is an isometry.

At this juncture, it is worth mentioning another class of operators that are somewhat easier to deal with than the integral operators: the Pietsch integral operators. The operator $T: X \to Y$ is called *Pietsch integral* whenever T admits a factorization of the form

$$\begin{array}{ccc} X & \xrightarrow{\ T\ } & Y \\ {\scriptstyle A}\downarrow & & \uparrow{\scriptstyle B} \\ C(\Omega) & \hookrightarrow & L_1(\mu) \end{array}$$

for some compact Hausdorff space Ω, some regular Borel measure μ on Ω and some bounded linear operators A, B of norm ≤ 1. The class of Pietsch integral operators from X to Y is denoted by PI(X; Y) and endowed with the norm $\|T\|_{pint} = \inf\{|\mu|(\Omega)$: T admits the above factorization$\}$ is a Banach space.

It is trivial that Pietsch integral operators are integral and $\|T\|_{int} \leq \|T\|_{pint}$; if Y is norm-one complemented in Y**, then integral operators into Y are Pietsch integral with $\|T\|_{pint} \leq \|T\|_{int}$ as well. In particular, if Y is a dual space, PI(X; Y) = I(X; Y) (isometrically).

Recall that a Banach space X has the Radon-Nikodým Property (RNP) whenever given a finite measure space (Ω, Σ, μ) and $F: \Sigma \to X$ a μ-continuous measure of bounded variation the Bochner-Radon-Nikodým derivative $dF/d\mu$ exists. It is known that weakly compactly generated duals have RNP.

The canonical inclusion of $C(\Omega)$ into $L_1(\mu)$ is Pietsch integral with Pietsch integral norm $|\mu|(\Omega)$. This simply proved fact (along with the relationship between integral and Pietsch integral maps) is the basis for most of what we do. For instance,

Proposition: If Y has RNP, then PI(X, Y) = N(X; Y) (isometrically).

Proof: The map $C(\Omega) \hookrightarrow L_1(\mu) \overset{B}{\to} Y$ is integral. Hence, the representing measure has finite variation. But Y has RNP so this measure has a Bochner Radon-Nikodým derivative with respect to its variation, it follows that $C(\Omega) \to L_1(\mu) \overset{B}{\to} Y$ is nuclear. The isometric assertion is a technical calculation which we skip.

An easy consequence of the above proposition is the

Theorem 1: Let Y be a norm-one complemented subspace of Y** and suppose Y has A.P. and R.N.P. Then Y has M.A.P. In particular, dual RNP spaces with A.P. have M.A.P.

Proof: (Throughout = means isometric). The first hypothesis on Y insures

$I(X; Y) = PI(X; Y)$ for all X. Since Y has R.N.P., $PI(X; Y) = N(X; Y)$ for all X. But now $N(X; Y)$ is easily seen to be a natural quotient (in the isometric category!) of $X^* \hat{\otimes} Y$; since Y has A.P., the quotient map of $Y^* \hat{\otimes} Y$ onto $N(Y; Y)$ is 1-1, hence, is an isometry. But this is just saying that Y has M.A.P.

It is certainly noteworthy that based upon Per Enflo's now famous example [11], T. Figiel and W. B. Johnson [14] have shown that generally A.P. need not imply M.A.P.

Probably the most striking corollary of the above Theorem is the following

Corollary 2 (Grothendieck): If X is a reflexive Banach space with A.P., then X has M.A.P.

This corollary is certainly among the most wonderous results of Grothendieck's Memoir; as we have seen, though this Corollary contains in its statement not even a hint of the measure theoretic apparatus that went into proving it, it is a measure theoretic result.

Similarly, we have the

Corollary 3 (Grothendieck): Separable dual spaces with A.P. have M.A.P.

It should be remarked here that it is open whether or not dual spaces with RNP aren't in a sense the basic spaces with RNP, i.e., unknown is the answer to the

Problem (Uhl). If X has RNP is X a subspace of a dual with RNP? In particular, if X is a separable space with RNP is X a subspace of a separable dual?

We turn now to some consequences of the above discussion concerning the duality of operators.

Theorem 4: Suppose Y^* has A.P. and RNP. Then $K(X; Y)^* = N(X^*; Y^*)$ and $N(X^*; Y^*)^* = \mathcal{L}(X^{**}; Y^{**})$. [$K(X; Y)$ = compact operators from X to Y]

Thus, if X or Y have A.P. then in order that $\mathcal{L}(X; Y)$ be reflexive it is necessary and sufficient that both X and Y be reflexive and that every operator from X to Y be compact ([18]).

<u>Proof</u>: It is a well-known consequence of Y* having A.P. that Y has A.P. and that then $X^* \check{\otimes} Y = K(X; Y)$, for all X. Thus $K(X; Y)^* = \mathcal{B}^{\wedge}(X^*; Y) = I(X^*; Y^*)$. But Y* has A.P. and RNP hence M.A.P. and $I(X^*; Y^*) = PI(X^*; Y^*) = N(X^*; Y^*) = X^{**} \hat{\otimes} Y^*$. In particular, $K(X; Y)^* = N(X^*; Y^*)$. Next, $N(X^*; Y^*)^* = (X^{**} \hat{\otimes} Y^*)^* = \mathcal{B}(X^{**}, Y^*) = \mathcal{L}(X^{**}; Y^{**})$. The first assertion is proved.

The second assertion is an easy consequence of the first: indeed, if $\mathcal{L}(X; Y)$ is reflexive, then since both X* and Y are isometric to closed subspaces of $\mathcal{L}(X; Y)$, X* and Y must be reflexive. For reflexive spaces, A.P. and M.A.P. are equivalent (Corollary 2) and a reflexive space has either of these properties if and only if its dual does (see [31], p. 198). Thus if $\mathcal{L}(X; Y)$ is reflexive $\mathcal{L}(X; Y) = K(X; Y)^{**}$ from our first assertion with the natural inclusion being what it should be, that is, $K(X; Y)$ must be $\mathcal{L}(X; Y)$. The converse is a simple consequence of the fact that if X or Y has A.P. then (since X and Y are reflexive) $K(X; Y) = X^* \check{\otimes} Y$ and the duality theory shows that $K(X; Y)^{**} = \mathcal{L}(X; Y)$. Therefore, if $\mathcal{L}(X; Y) = K(X; Y)$, reflexivity follows.

A particular consequence of Theorem 4 that is of some interest in itself is the

<u>Corollary 5</u>. <u>If X is a reflexive Banach space with A.P., then N(X; X) has RNP</u>.

<u>Proof</u>. By the duality developed in Theorem 4, $N(X; X) = K(X^*; X^*)^* = (X \check{\otimes} X^*)^*$. Moreover, if S is a separable subspace of $X \check{\otimes} X^*$ then it is easily seen that S* is separable, i.e., each separable subspace of $(X \check{\otimes} X^*)$ has separable dual. Now, a result of Uhl [35] tells us that $(X \check{\otimes} X^*)^* = N(X; X)$ has RNP.

The above result is actually a special case of a more general phenomena, namely,

Corollary 5'. If X, Y are dual spaces with RNP one of which possesses A.P., then $X \hat{\otimes} Y$ has RNP.

The proof of Corollary 5 requires (at this time) a deep result of Charles Stegall [32] and is not presented herein; the interested reader can look at [8] for a complete proof. Open at this time is the

Problem: If X, Y have RNP, then need $X \hat{\otimes} Y$?

My conjecture is yes.

Finally, we turn our attention to the question of when given a scalarly measurable function there exists a strongly measurable function equivalent to it. Our attention is restricted to bounded functions though the alert reader will note that this is not really a restriction.

Suppose we consider $f: \Omega \to X$ and suppose f is μ-essentially bounded and scalarly μ-measurable. When does there exist a strongly measurable $g: \Omega \to X$ such that g is equivalent to f in the sense that

$$x^*g = x^*f \quad \mu - \text{a.e.}$$

for each $x^* \in X^*$? We finish up this section by analyzing this problem from the point of view of integral and nuclear operators and their representation theory on $L_\infty(\mu)$ spaces. (Throughout (Ω, Σ, μ) is a finite measure space).

First, we will assume that the Banach space X satisfies "Mazur's condition": every weak* sequentially continuous linear functional on X* is weak* continuous. It should be noted that if X is an $L_1(\mu)$ space, then X satisfies Mazur's condition; this is a consequence of Kakutani's representation theorem for L-spaces and the Radon-Nikodým theorem for localizable measures. If the closed unit ball of X* is weak* sequentially compact, then X satisfies Mazur's condition; thus, anytime X imbeds in some weakly compactly generated Banach space, then X satisfies Mazur's condition ([1], [12]). Of course, separable Banach spaces satisfy Mazur's condition (this was originally shown by S. Mazur, hence, the name) as do reflexive spaces.

Now suppose $f: \Omega \to X$ is a μ-essentially bounded and scalarly μ-measurable function. Then f induces a bounded linear operator $T_f: X^* \to L_\infty(\mu)$ given by $(T_f x^*)(\cdot))$. Consider the operator

$$\overbrace{X^* \underbrace{\to L_\infty(\mu)}_{T_f} \hookrightarrow L_1(\mu)}^{T}.$$

Since X satisfies Mazur's condition, T is the adjoint of an operator

$$\overbrace{L_\infty(\mu) \hookrightarrow L_1(\mu) \underbrace{\to X}_{T_f^*}}^{S};$$

in fact, Mazur's condition and the Dominated Convergence Theorem insure that T_f is weak* continuous while $L_\infty(\mu) \hookrightarrow L_1(\mu)$ is weak* to weak continuous. Now, since $L_\infty(\mu) \hookrightarrow L_1(\mu)$ is integral and weak* to weak continuous, S is likewise. Therefore, S is represented by a countably additive X-valued vector measure F having finite variation $|F| \ll \mu$.

Whenever $dF/d|F|$ exists, i.e., whenever S is nuclear, then $g = dF/d|F|$ is equivalent to f and, of course, is strongly measurable.

A few examples of the occurrence of this phenomena:

1) Suppose <u>$f(\Omega)$ is relatively weakly compact</u>. Then X may as well be assumed to be weakly compactly generated so Mazur's condition is automatically satisfied. Further, it is straightforward to show that the average range $\{F(A)/|F|(A): A \in \Sigma\}$ is contained in the absolutely closed convex hull of $f(\Omega)$, i.e., is weakly compact. The arguments of [23] show now that $dF/d|F|$ exists.

2) <u>X is reflexive</u>.

3) <u>X is weakly compactly generated</u> and <u>X* possesses RNP</u>. In this case again X satisfies Mazur's condition and, since X* possesses RNP, $S^* = T$ is nuclear; since most everything in sight has the (metric) approximation property, S itself is nuclear.

4) <u>$X = c_0(\Gamma)$, any set Γ.</u>

5) $X = \ell_1(\Gamma)$, any set Γ. In this case, X satisfies Mazur's condition on its own and since X has RNP, S is necessarily nuclear.

6) Though it does not fall under the heading of the above results, using the methods developed above, we can show that if X^* has RNP and $f: \Omega \to X^*$ is μ-essentially bounded and $f(\cdot)(x)$ is μ-measurable for each $x \in X$ then there is a strongly measurable $g: \Omega \to X^*$ which is equivalent to f. In fact, in this case, f induces an operator T from X to $L_\infty(\mu)$ which is continuous hence weakly continuous hence weak to weak* continuous thereby the operator T is the adjoint of an operator from $L_1(\mu)$ to X^* which necessarily has a derivative g--done.

7) Finally, we note that J. Lindenstrauss and C. Stegall [21] have noted that for the dual space X of James' tree space, there is a scalarly measurable bounded function into X which is not equivalent to any strongly measurable function. This space X has the property that X^* is weakly compactly generated and, therefore, X^* has RNP. Thus some hypothesis on X like in 3) is necessary to obtain the existence of an equivalent strongly measurable function since X is not even a subspace of a weakly compactly generated space.

In developing the above approach a curious problem comes to the foreground:

Problem: Suppose X has RNP. Are weak* sequentially continuous linear functionals on X^* weak* continuous?

It should be remarked here that using the above approach and more care and cleverness, D. R. Lewis has shown that if $X = L_1(\mu)$, then each scalarly integrable X-valued function is equivalent to a strongly measurable X-valued function ([19]).

§3. Odds and Ends.

The study of spaces of operators (especially the geometry of such spaces) is inhibited by the size of such spaces. It seems that the use of Radon-Nikodým type assumptions will produce some real progress in this direction. To make for easy discussion, we start this section with a conjecture that is purposely vague yet I believe is probably quite close to being precise!

<u>Conjecture</u>? Let $\mathcal{J}$ be an operator ideal in the sense of Pietsch [28] and suppose X, Y are Banach spaces such that either X* or Y has A.P. Suppose $\mathcal{J}_{nuc}(X; Y)$ denotes the closure of $X^* \otimes Y$ in $\mathcal{J}(X; Y)$. Then $\mathcal{J}(X; Y)$ has RNP if and only if X*, Y has RNP and $\mathcal{J}(X; Y) = \mathcal{J}_{nuc}(X; Y)$.

So far the difficulty in approaching the above problem has been in understanding spaces with RNP that are <u>not</u> dual spaces. In all known operator ideals, if Y is a dual space the condition that X*, Y have RNP and $\mathcal{J}(X; Y) = \mathcal{J}_{nuc}(X; Y)$ is sufficient to conclude to $\mathcal{J}(X; Y)$ having RNP.

Without the assumption of duality only one result is presently known: <u>if X*, Y have RNP and every continuous linear operator from X to Y is compact then, $\mathcal{L}(X; Y)$ has RNP</u>. Here, <u>no</u> approximation assumptions are made ([37]).

A particular consequence of this fact is the

<u>Corollary</u>: <u>If Y has RNP, then $\ell_1(Y)$ = unconditionally converging series in Y has RNP</u>.

This gives rise to another question that is related to the above conjecture and may be useful in an eventual positive solution of the Conjecture through careful use of the results of D. R. Lewis and C. Stegall ([20]).

<u>Problem</u>: If α is an accessible $\otimes$ norm ([17]) and Y has RNP need $\ell_1 \otimes_\alpha Y$ have RNP?

For $\alpha = \lambda$, the least crossnorm, the Corollary above says yes. For $\alpha = \gamma$, the greatest crossnorm, the answer is also yes [30]. Other crossnorms have been considered by B. Faires and T. J. Morrison [13].

Finally, in all these problems, much of the interest lies in the possibility of using some of the recent progress relating the Radon-Nikodým property to the geometry of a Banach space. It would be of more than passing interest to understand the relationship between existence of many strongly exposed points in arbitrary closed bounded convex sets of X*(see [25]) with A.P. and the M.A.P. A direct derivation (internal: without recourse to integral and nuclear operators) of Theorem 1 (even in case of dual spaces) from geometric criterion would be a most interesting contribution; such a derivation might be possible using some sort of Choquet theory and the intuitive concept of measures being limits of atomic measures.

Acknowledgements: The author gratefully acknowledges the support of NSF Grant MPS08050 while preparing this paper. Also, he wishes to give special thanks to Dan Lewis and Charles Stegall for discussions in his encounters with them over the past few years; my understanding of the material contained herein comes largely from these discussions.

Bibliography

[1] D. Amir and J. Lindenstrauss, The structure of weakly compact sets in Banach spaces. Ann. of Math., 88(1968), 35-46.

[2] R. G. Bartle, N. Dunford, and J. T. Schwartz, Weak compactness and vector measures, Canad. J. Math., 7(1955), 289-305.

[3] S. Bochner, Additive set functions on groups, Ann. Math. (2) 40(1939), 769-799.

[4] J. Diestel, The Radon-Nikodým property and the coincidence of integral and nuclear operators, Revue Roum. Math. 17(1972), 1611-1620.

[5] _____, Applications of weak compactness and bases to vector measures and vectorial integration, Revue Roumaine Math., 18(1973), 211-224.

[6] J. Diestel and B. Faires, On vector measures, Trans. AMS, 198(1974), 253-271.

[7] _____, Remarks on the classical Banach operator ideals, Proc. AMS, to appear.

[8] J. Diestel and J. J. Uhl, Jr. The Radon-Nikodým theorem for Banach space valued measures, Rocky Mtn. Jour., to appear.

[9] _____, Topics in the Theory of Vector Measures, Notes presently being collected at Kent State University and the University of Illinois.

[10] J. Dixmier, Les fonctionnelles lineaires sur l'ensemble des operateurs bornés d'un espace de Hilbert, Ann. of Math., (2) 51(1950), 387-408.

[11] P. Enflo, A counterexample to the approximation problem, Acta Math.

[12] B. Faires, Grothendieck spaces and vector measures, Ph.D. dissertation, Kent State University, August, 1974.

[13] B. Faires and T. J. Morrison, (as yet unpublished).

[14] T. Figiel and W. B. Johnson, The approximation property does not imply the bounded approximation property, Proc. AMS, 41(1973), 197-200.

[15] A. Grothendieck, Sur les applications lineaires faiblement compactes d'espaces du type C(K), Canad. J. Math., 5(1953), 129-173.

[16] _____, Produits tensoriels topologiques et espaces nucléaires, Memoirs of Amer. Math. Soc., 16(k955).

[17] _____, Resumé de la théorie métrique des produits tensoriels topologiques, Bol. Soc. Matem. Sao Paolo, 8(1956), 1-79.

[18] J. R. Holub, Reflexivity of L(E, F), Proc. AMS, 39(1973), 175-177.

[19] D. R. Lewis, Weak integrability in L_1 spaces, preprint.

[20] D. R. Lewis and C. Stegall, Banach spaces whose duals are isomorphic to $\ell_1(\Gamma)$, Jour. of Funcl. Anal., 12(1973), 177-187.

[21] J. Lindenstrauss and C. Stegall, Examples of separable spaces which do not contain ℓ_1 and whose duals are nonseparable, to appear.

[22] M. Metivier, Martingales à valeurs vectorielles. Applications à la dérivations des mesures vectorielles, Ann. Inst. Fourier (Grenoble) 17(1967), 175-208.

[23] S. Moedomo and J. J. Uhl, Jr., Radon-Nikodým theorems for the Bochner and Pettis integrals, Pacific Journal Math., 38(1971), 531-536.

[24] R. Schatten, Norm Ideals of Completely Continuous Operators. Springer-Verlag, Berlin, 1960.

[25] R. R. Phelps, Dentability and extreme points in Banach spaces, J. of Functional Analysis, 16(1974), 78-90.

[26] A. Pietsch, Abbildungen von abstrakten Massen, Wiss. Zeit. Friedrich-Schiller Univ., 5(1965), 281-286.

[27] A. Persson and A. Pietsch, p-nukleare und p-integrale Abbildungen in Banachräumen, Studia Math., 33(1969), 19-62.

[28] A. Pietsch, Theorie der Operatorenideale, Friedrich-Schiller-Universitat, Jena, 1972.

[29] M. A. Rieffel, The Radon-Nikodým theorem for the Bochner integral, Transactions of Amer. Math. Soc., 131(1968), 466-487.

[30] E. Saab, Families absolument sommables et propriete de Radon-Nikodým, preprint.

[31] H. H. Schaefer, Topological Vector Spaces. MacMillan and Sons, New York, 1966.

[32] C. Stegall, The Radon-Nikodým property in conjugate Banach spaces, Trans. AMS.

[33] A. E. Tong, Nuclear mappings on C(K), Math. Ann., 194(1971), 213-224.

[34] J. J. Uhl, Jr., Orlicz spaces of finitely additive set functions, Studia Math., 29(1967), 19-58.

[35] J. J. Uhl, Jr. A note on the Radon-Nikodým property for Banach spaces, Revue Roum. Math., 17(1972), 113-115.

[36] A. Wilansky, Topics in Functional Analysis. Lecture Notes No. 45, Springer-Verlag, New York.

[37] J. Diestel and T. J. Morrison, The Radon-Nikodým Property for the Space of Operators, submitted.

THE RADON-NIKODÝM PROPERTY FOR BANACH SPACES

R.E. Huff

Pennsylvania State University
University Park, Pa. USA

§1. INTRODUCTION.

Throughout, (Ω,Σ,λ) denotes a finite measure space (i.e., Σ is a σ-algebra of subsets of Ω and λ is a non-negative measure on Σ with $0 < \lambda(\Omega) < \infty$), and X denotes a Banach space. Let $L^1(\lambda,X)$ denote the Banach space of all λ-Bochner integrable functions $f : \Omega \to X$ with norm $\|f\|_1 = \int\|f(\omega)\|d\lambda(\omega)$. If f is in $L^1(\lambda,X)$, and if $\mu : \Sigma \to X$ is defined by

$$(*) \qquad \mu(E) = \int_E f\, d\lambda, \qquad \forall E \in \Sigma,$$

then μ is countably additive (c.a.), of bounded variation (b.v.), and absolutely continuous (a.c.) with respect to λ.

The space X is said to have the Radon-Nikodým Property (RNP) provided for every finite measure space (Ω,Σ,λ) and every c.a., b.v., a.c. measure $\mu : \Sigma \to X$ there exists f in $L^1(\lambda,X)$ such that $(*)$ holds.

In 1972, J. Diestel made the conjecture that the RNP is equivalent to the Krein-Milman Property (KMP). (The space X is said to have the KMP provided every closed bounded convex subset of X has an extreme point.) This conjecture has stimulated considerable research, and the purpose here is to review some of the progress made. The discussion will center around the proof of the following theorem, the parts of which are due to several mathematicians.

MAIN THEOREM. Each of the following is equivalent to the RNP for X.

(1) $\forall$ closed, bounded set $A \subset X$, $\forall\varepsilon > 0$, $\exists x \in A$, $x \notin \overline{co}(A\backslash B_\varepsilon(x))$.

(2) $\forall$ closed, bounded set $A \subset X$, $\exists x \in A$, $\forall\varepsilon > 0$, $x \notin \overline{co}(A\backslash B_\varepsilon(x))$.

(3) $\forall$ closed, bounded set $A \subset X$, $\forall\varepsilon > 0$, $\exists x \in A$, $x \notin co(A\backslash B_\varepsilon(x))$.

(4) $\forall$ closed, bounded set $A \subset X$, $\exists x \in A$, $x \notin co(A\backslash\{x\})$.

(Here, co(A) and $\overline{co}(A)$ denote the convex hull and closed convex hull of A, respectively. $B_\varepsilon(x)$ denotes the closed ball with center x and radius ε.)

In the terminology introduced by Rieffel [16], statement (1) can be phrased "every closed, bounded set $A \subset X$ is dentable"; and statement (2) "every closed, bounded set $A \subset X$ has a denting point." Statement (4) is that every closed, bounded set $A \subset X$ has an extreme point.

By this theorem, the Diestel conjecture is now that statement (4) is still sufficient for the RNP if "convex" is added after the word "bounded".

The remainder of this paper is devoted to outlines of proofs and discussions of the various implications in the theorem.

§2. STATEMENT (1) IMPLIES THE RNP.

This particular implication is due to M.A. Rieffel, 1967 [16].

We first observe (with Rieffel [16]) that (1) is equivalent to each of the following.

(1′) $\forall$ bounded set $A \subset X$, $\forall \varepsilon > 0$, $\exists x \in A$, $x \notin \overline{co}(A \backslash B_\varepsilon(x))$.

(1″) $\forall$ closed, bounded, convex set $A \subset X$, $\forall \varepsilon > 0$, $\exists x \in A$, $x \notin \overline{co}(A \backslash B_\varepsilon(x))$.

For clearly (1′) $\Rightarrow$ (1) $\Rightarrow$ (1″), and one proves (1″) $\Rightarrow$ (1′) as follows. Let $A \subset X$ be any bounded set, $\varepsilon > 0$, and choose $k \in \overline{co}(A)$, $k \notin \overline{co}(\overline{co}(A) \backslash B_{\varepsilon/2}(k))$. By the separation theorem, choose $f \in X^*$ such that $\sup f(\overline{co}(A) \backslash B_{\varepsilon/2}(k)) = \alpha < f(k)$. Now $\{x : f(x) > \alpha\}$ must intersect A and any point x in that intersection must satisfy (1′).

Now suppose (1′) holds and $\mu : \Sigma \to X$ is c.a., b.v., and a.c. Let $\bar{\mu} : \Sigma \to R$ (= reals) denote the variation of μ, and choose $h \in L^1(\lambda, R)$, $h \geq 0$, such that $\bar{\mu}(E) = \int_E h d\lambda$, $\forall E \in \Sigma$. For $F \in \Sigma$, let

$$A(F) = \{\frac{\mu(G)}{\lambda(G)} : G \subset F,\ G \in \Sigma^+\},$$

where $\Sigma^+ = \{G \in \Sigma : \lambda(G) > 0\}$. Note that if h is bounded on F, then A(F) is a bounded set since

$$\left\|\frac{\mu(G)}{\lambda(G)}\right\| \leq \frac{\bar{\mu}(G)}{\lambda(G)} = \frac{1}{\lambda(G)} \int_G h d\lambda.$$

The heart of Rieffel's proof is contained in the following lemma.

<u>LEMMA 1</u>. $\forall \varepsilon > 0$, $\exists$ <u>countable partition</u> $\pi = (E_i)_i \subset \Sigma$ of Ω with $\operatorname{diam} A(E_i) \leq \varepsilon$, $\forall i$.

<u>PROOF</u>. We prove

(#) $\quad \forall \varepsilon > 0, \forall E \in \Sigma^+, \exists F \in \Sigma^+, F \subset E$ with $\operatorname{diam} A(F) \leq \varepsilon$.

Once (#) is proved, simply take $(E_i)_i$ to be a maximal (necessarily countable) pairwise disjoint collection in Σ^+ with daim $A(E_i) \leq \varepsilon$, $\forall i$.

To prove (#), note that $E = \bigcup_n \{\omega \in E : h(\omega) \leq n\}$, so by passing to a subset of E if necessary, we may assume h is bounded on E, and hence $A(E)$ is a bounded set. By (1′), we can choose $F_0 \subset E$, $F_0 \in \Sigma^+$ such that

$$\frac{\mu(F_0)}{\lambda(F_0)} \notin \overline{\mathrm{co}}\left[A(E)\backslash B_{\varepsilon/2}\left(\frac{\mu(F_0)}{\lambda(F_0)}\right)\right] = Q.$$

If $\operatorname{diam} A(F_0) \leq \varepsilon$ we are done; if not, $\exists B \subset F_0$, $B \in \Sigma^+$, with $\frac{\mu(B)}{\lambda(B)} \in Q$. Choose a maximal (ncessarily countable) pairwise disjoint collection $(B_n)_n \subset \Sigma^+$ with $B_n \subset F_0$ and $\frac{\mu(B_n)}{\lambda(B_n)} \in Q$, $\forall n$. Let $F = F_0\backslash(\cup B_n)$. If $\lambda(F) = 0$, then

$$\frac{\mu(F_0)}{\lambda(F_0)} = \frac{\mu(\cup B_n)}{\lambda(\cup B_i)} = \sum_n \frac{\lambda(B_n)}{\lambda(\cup B_i)} \cdot \frac{\mu(B_n)}{\lambda(B_n)}$$

would be in Q, so we must have $F \in \Sigma^+$. By the maximality of (B_n), if $G \subset F$ and $G \in \Sigma^+$, then $\frac{\mu(G)}{\lambda(G)} \notin Q$, so $\left\|\frac{\mu(G)}{\lambda(G)} - \frac{\mu(F_0)}{\lambda(F_0)}\right\| \leq \frac{\varepsilon}{2}$.

Thus $\operatorname{diam} A(F) \leq \varepsilon$, and this completes the proof of the lemma.//

Now by induction, choose a sequence of partitions $\pi_n = (E_i^n)_i$, increasing with respect to refinement, such that $\operatorname{diam} A(E_i^n) < \frac{1}{2^{n+1}}$, $\forall i,n$. Define $g_n : \Omega \to X$ by

$$g_n = \sum_i \frac{\mu(E_i^n)}{\lambda(E_i^n)} \chi_{E_i^n},$$

and note that $(g_n)_n$ converges uniformly to some function g. For any $E \in \Sigma$,

$$\|\mu(E) - \int_E g_n d\lambda\| = \|\sum_i \mu(E_i^n \cap E) - \sum_i \frac{\mu(E_i^n)}{\lambda(E_i^n)} \lambda(E_i^n \cap E)\|$$

$$\leq \sum_i \left\|\frac{\mu(E_i^n \cap E)}{\lambda(E_i^n \cap E)} - \frac{\mu(E_i^n)}{\lambda(E_i^n)}\right\| \lambda(E_i^n \cap E)$$

$$\leq \frac{1}{2^{n+1}} \lambda(E),$$

so $\mu(E) = \lim_n \int_E g_n d\lambda = \int_E g d\lambda$, $\forall E \in \Sigma$. This completes the proof that (1) implies the RNP.

<u>REMARKS</u>. (a) Suppose that $A(\Omega)$ is in fact bounded, say by 1. Let $g_0 = 0$ and $\varphi_n = g_n - g_{n-1}$ for $n \geq 1$. Then for $n \geq 1$, $\varphi_n = \sum_i x_i^n \chi_{E_i^n}$ with $\|x_i^n\| \leq \frac{1}{2^n}$. Define $\nu : \Sigma \to \ell^1 (N \times N)$ by

$$(\nu(E))_{n,i} = \frac{1}{2^n} \lambda(E \cap E_i^n),$$

and let $T : \ell^1(N \times N) \to X$ be the continuous linear map such that $T(e_{ni}) = 2^n x_i^n$, where e_{ni} is the n,i^{th} unit basis vector. Then it is easily checked that $T \circ \nu = \mu$, and we have proved the following result of D.R. Lewis and C. Stegall [10] (see also [18]).

<u>THEOREM</u>. (Lewis-Stegall). <u>If</u> X <u>has the RNP then every operator</u> $\Phi : L^1(\lambda, R) \to X$ <u>factors through</u> ℓ^1.

(b) The most important positive result about the RNP is that any separable dual space has the RNP, a result proved by N. Dunford and B.J. Pettis in 1940 [4]. It is of interest to know that a geometric proof can be given using (1).

Let $X = Y^*$ be a separable dual space, and let A be a bounded subset of X. We will show that $\forall \varepsilon > 0$, $\exists x \in A$ such that $x \notin \text{wk*-}\overline{\text{co}}(A \backslash B_\varepsilon(x))$ (where wk*-$\overline{\text{co}}$ means the weak*-closed convex hull. For more information about this strong form of (1) in dual spaces, see Namioka and Phelps [14].) As in the discussion of the equivalence of (1), (1′), and (1″) we can assume here that A is weak*-closed, bounded, and convex.

Let E denote the extreme points of A and $\overline{E}$ the wk*-closure of E, and let $\varepsilon > 0$ be given. By separability, $\overline{E}$ can be covered by countably many closed balls $(B_n)_n$, each of diameter $< \frac{\varepsilon}{2}$, and by the Baire category theorem, one of $B_n \cap \overline{E}$ has non-void relative weak*-interior. Thus, $\exists$ a weak*-open subset U of X with $U \cap \overline{E} \neq \emptyset$ (so $U \cap E \neq \emptyset$) and $\text{diam}(U \cap \overline{E}) < \frac{\varepsilon}{2}$.

Let $K_1 = \text{wk*-}\overline{\text{co}}(\overline{E}\setminus U)$ and $K_2 = \text{wk*-}\overline{\text{co}}(U \cap E)$. Let $\varphi : [0,1] \times K_1 \times K_2 \to X$ be given by $\varphi(\alpha,x,y) = \alpha x + (1-\alpha)y$. Then $A = \varphi([0,1] \times K_1 \times K_2) = \text{co}(K_1 \cup K_2)$. Next, let $C = \varphi([\frac{\varepsilon}{4\delta}, 1] \times K_1 \times K_2)$, where $\delta = \text{diam } A$. If $C = A$, it would follow that $E \subset K_1$, so $E \subset \overline{E}\setminus U$, or $E \cap U = \emptyset$, a contradiction. Thus $C \neq A$. Since C is wk*-closed and convex, it remains only to see that $A\setminus C$ has diameter $< \varepsilon$.

If $z \in A\setminus C$, then $z = \alpha x_1 + (1-\alpha)x_2$, $x_i \in K_i$, $0 \leq \alpha < \frac{\varepsilon}{4\delta}$, so

$$\|z-x_2\| = \alpha\|x_1-x_2\| \leq \frac{\varepsilon}{4\delta}\,\delta = \frac{\varepsilon}{4}.$$

Thus $\text{diam}(A\setminus C) \leq \text{diam } K_2 + \frac{\varepsilon}{4} < \varepsilon$.

The above argument is due to I. Namioka and E. Asplund ([13], [12], [14]).

§3. THE RNP IMPLIES STATEMENT (3).

By an X-valued Martingale on (Ω,Σ,λ) we shall mean a double sequence $(f_n, \Sigma_n)_n$, where each Σ_n is a sub-σ-algebra of Σ, $f_n \in L^1(\lambda|\Sigma_n,X)$, and for $n < m$, $\Sigma_n \subset \Sigma_m$ and

$$\int_E f_n\,d\lambda = \int_E f_m\,d\lambda, \qquad \forall E \in \Sigma_n.$$

For the proof that the RNP implies (3) we shall use the following lemma which is only a first step in a rather large theory concerning the relationship between the RNP and martingale convergence theory (see [1], [21]).

<u>LEMMA 2</u>. <u>If</u> X <u>has the RNP, then every uniformly bounded X-valued martingale on</u> (Ω,Σ,λ) <u>converges in</u> $L^1(\lambda,X)$.

<u>PROOF</u>. Define μ on the algebra $A = \bigcup_n \Sigma_n$ by $\mu(E) = \lim_n \int_E f_n\,d\lambda$, and observe that μ extends to a c.a., b.v., and a.c. measure on the σ-algebra $\sigma(A)$ generated by A. Let $E_n : L^1(\lambda,X) \to L^1(\lambda|\Sigma_n,X)$ and $E : L^1(\lambda,X) \to L^1(\lambda|\sigma(A),X)$ be the

conditional expectation operators (see [1]). Then $(E_n)_n$ is a sequence of continuous linear projections on $L^1(\lambda,X)$, each of norm one, which converges pointwise to the operator E (see [17] or [1] for more details). Since X has the RNP, $\exists f \in L^1(\lambda,X)$ such that $\mu(E) = \int_E f d\lambda$, $\forall E$. Then $f_n = E_n(f) \to E(f)$.//

We now show that the RNP implies (3); in fact, we show that the RNP implies the following stronger form of (3).

(3′) $\quad\forall$ <u>bounded set</u> $A \subset X$, $\forall \varepsilon > 0$, $\exists x \in A$, $x \notin \operatorname{co}(A \backslash B_\varepsilon(x))$.

For suppose (3′) fails. Then $\exists$ bounded set A and $\varepsilon > 0$ such that $x \in \operatorname{co}(A\backslash B_\varepsilon(x))$, $\forall x \in A$. For the sake of notation, assume for the moment that each point x in A can be written as $\alpha y + (1-\alpha)z$ where $y,z \in A$, $\|x-y\| > \varepsilon$, $\|x-z\| > \varepsilon$, and $0 < \alpha < 1$. By induction, choose an infinite "tree" in A:

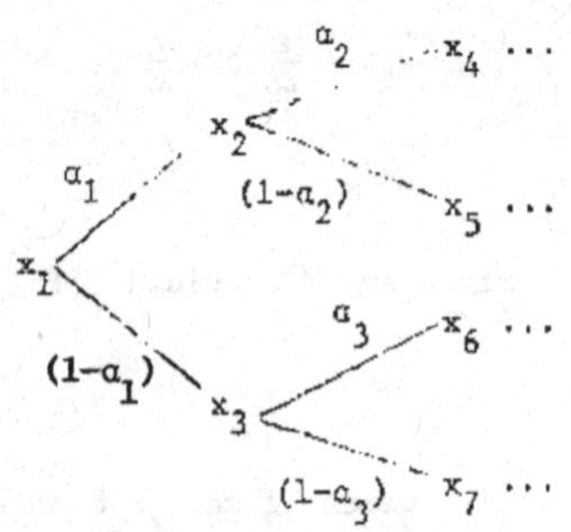

where for each n, $x_n = \alpha_n x_{2n} + (1-\alpha_n)x_{2n+1}$, $\|x_n - x_{2n}\| > \varepsilon$, $\|x_n - x_{2n+1}\| > \varepsilon$, $0 < \alpha_n < 1$.

Now choose half-open intervals I_n in $[0,1)$ such that $I_1 = [0,1)$, and for each n, $I_n = I_{2n} \cup I_{2n+1}$, $\lambda(I_{2n}) = \alpha_n \lambda(I_n)$, and $\lambda(I_{2n+1}) = (1-\alpha_n)\lambda(I_n)$, where λ is Lebesgue measure. Let Σ_k denote the σ-algebra generated by $\{I_n : 2^k \leq n < 2^{k+1}\}$, and define $f_k : [0,1) \to X$ by

$$f_k = \sum_{n=2^k}^{2^{k+1}-1} x_k \chi_{I_k} .$$

It is easily checked that $(f_k,\Sigma_k)_k$ forms a uniformly bounded martingale, and since $\|f_k(t) - f_{k+1}(t)\| > \varepsilon$ for all $t \in [0,1)$, $(f_k)_k$ is not Cauchy in $L^1(\lambda,X)$. Thus X must fail to have the RNP.

In the general case one must choose an infinite "bush" in A (i.e., each points splits' as the convex combination of a finite number of points, each at least ε-away from it), and the intervals are partitioned at each stage into a finite number of sub-intervals with lengths proportional to the convex coefficients used. Otherwise the proof is the same.

REMARKS. (c) It follows from the above proof that to check X for the RNP one needs only to consider the case when $(\Omega,\Sigma,\lambda) = ([0,1]$, Borel sets, Lebesgue measure) and $\|\mu(E)\| \leq \lambda(E)$, $\forall E \in \Sigma$.

(d) The first geometric characterization of the RNP of the type we are considering was given by H. Maynard [11] who proved that the RNP is equivalent to

(3″) $\quad \forall$ bounded set $A \subset X$, $\forall \varepsilon > 0$, $\exists x \in A$, $x \notin \sigma\text{-co}(A \setminus B_\varepsilon(x))$,

where $\sigma\text{-co}(A) = \{\sum_{i=1}^{\infty} \alpha_i x_i : x_i \in A, \alpha_i \geq 0, \sum_{i=1}^{\infty} \alpha_i = 1\}$. Observe that in Lemma 1 we could have used (3″) rather than (1′), and thus (3″) $\Rightarrow$ RNP. Maynard proved RNP $\Rightarrow$ (3″) by an argument similar to the above. (One can simply 'split' each point as an infinite convex combination of points, each at least ε-away from it, etc.). In [7] we modified Maynard's construction to show directly that the RNP implies (1′) by constructing a sequence $(f_n,\Sigma_n)_n$ which was not $L^1(\lambda,X)$-Cauchy but was sufficiently close to a martingale that it would have to be Cauchy if X had the RNP. Independently, W.J. Davis and R.R. Phelps [2] showed geometrically that (1′), (3″), and (3′) were all equivalent. That proof is given in the next section where it is also shown that (3) is sufficient for the RNP.

§4. STATEMENT (3) IMPLIES (1).

As remarked above, Davis and Phelps [2] proved that (3′) implies (1′). This is proved by the following two lemmas.

LEMMA 3. *If A and B are subsets of X, and if $A + B$ is dentable, then so are A and B.*

PROOF. Suppose $(a+b) \in A + B$ and $(a+b) \notin \overline{\text{co}}((A+B) \setminus B_\varepsilon(a+b))$. If $\|a - \sum_{i=1}^{n} \alpha_i a_i\| < \delta$,

$\|a-a_i\| > \varepsilon$, $a_i \in A$, $\alpha_i \geq 0$, $\sum_{i=1}^{n} \alpha_i = 1$, then $\|(a+b) - \sum_{i=1}^{n} \alpha_i(a_i+b)\| < \delta$, $\|(a+b) - (a_i+b)\| > \varepsilon$, and $a_i+b \in A + B$. Thus, $a \notin \overline{co}(A\setminus B_\varepsilon(a))$.//

<u>LEMMA 4</u>. (Davis and Phelps [2]). <u>Suppose</u> K <u>is a closed convex non-dentable set with non-void interior</u> $K°$. <u>Then</u> $\exists\varepsilon > 0$ <u>such that</u> $K° = co(K°\setminus B_\varepsilon(x))$ <u>for all</u> $x \in K$.

<u>PROOF</u>. Since K is non-dentable, $\exists\varepsilon > 0$ such that $x \in \overline{co}(K\setminus B_{2\varepsilon}(x))$, $\forall x \in K$. Then $K = \overline{co}(K\setminus B_\varepsilon(y))$, $\forall y \in K$. Let $x \in K$, and let $J = K\setminus B_\varepsilon(x)$. Then $K = \overline{co}(J)$ and $J° = K°\setminus B_\varepsilon(x)$.

Note that $J \subset \overline{J°}$. For, let y be any point in J. Then $y \in K$ and y is not in the closed set $B_\varepsilon(y)$. Let z be any point in $K°$. Then the half-open line segment $[z,y)$ is contained in $K°$, and for points w of $[z,y)$ sufficiently close to y, w is outside $B_\varepsilon(x)$. Hence y is the limit of points in $K°\setminus B_\varepsilon(x) = J°$.

It follows that $co\ J \subset \overline{co}(J°)$, and since the interior of a convex set (when non-empty) coincides with the interior of its closure,

$$K° = (\overline{co}\ J)° = (co\ J)° \subset (\overline{co}\ J°)° = co\ J°.//$$

Now to see that (3′) implies (1′), suppose A is a non-dentable bounded set in X, and let $B = [\overline{co}(A+B_1(0))]°$. Then B fails to satisfy the condition in (3′).

To see that (3) implies (1) is more difficult. The importance of this implication is to obtain the sufficiency of (4) for the RNP. Its proof depends on the following lemma.

<u>LEMMA 5</u>. (Huff and Morris [9]). <u>If</u> K <u>is a non-dentable closed bounded convex subset of</u> X, <u>then</u> $\exists\varepsilon > 0$ <u>such that</u> $K = \overline{co}(K\setminus(B_\varepsilon(x_1) \cup \ldots \cup B_\varepsilon(x_n)))$ <u>for every finite set</u> $\{x_1,\ldots,x_n\} \subset X$.

The proof of Lemma 5 is rather long, geometrical, and unenlighting to the discussion here. We omit it; it can be found in [9].

Using Lemma 5, a trivial modification of the proof of Lemma 2 yields the following.

LEMMA 6. Suppose K is a closed bounded convex non-dentable set with non-void interior K°. Then $\exists \varepsilon > 0$ such that $K^\circ = co(K^\circ \setminus (B_\varepsilon(x_1) \cup \ldots \cup B_\varepsilon(x_n)))$, for every finite set $\{x_1,\ldots,x_n\} \subset X$.

We can now prove that (3) implies (1). Suppose that (1) fails. Then by the lemmas there exists a bounded set K and an $\varepsilon > 0$ such that

$$K = co(K \setminus (B_\varepsilon(x_1) \cup \ldots \cup B_\varepsilon(x_n)))$$

for all $\{x_1,\ldots,x_n\} \subset K$. Choose any $x_0 \in K$ and let $F_1 = \{x_0\}$. By induction, choose a sequence of finite sets $F_1, F_2, F_3, \ldots$ such that for all n, $F_n \subset co(F_{n+1})$ and

$$x \in F_n, \quad y \in F_m, \quad n \neq m \Rightarrow \|x-y\| \geq \varepsilon.$$

Let $A = \bigcup_{n=1}^{\infty} F_n$. Then $x \in co(A \setminus B_\varepsilon(x))$, $\forall x \in A$, and it remains only to see that A is closed. But if $(x_n)_{n=1}^{\infty}$ is a Cauchy sequence in A, then it must eventually be contained in one of the finite sets F_n, and hence converges to an element of A.

REMARK. (e) It follows from the above sufficient conditions for the RNP (since e.g. the set A constructed is a countable set) that X has the RNP if every separable subspace of X does. This fact and its importance was first pointed out by J.J. Uhl, Jr. [20]. It combined with the Dunford-Pettis result on separable dual spaces gives a wide class of spaces having the RNP — so wide that the following question of Uhl is open: If X has the RNP, is every separable subspace of X isomorphic to a subspace of a separable dual space?

§5. STATEMENT (1) IMPLIES (2) - THE COMPLETION OF THE PROOF OF THE MAIN THEOREM.

Since the implications (2) ⇒ (4) ⇒ (3) are trivial, once it is shown that (1) implies (2), the proof of the main theorem will be complete.

The implications (1) ⇒ (2) is due to R.R. Phelps [15].

Let A be a closed bounded set in X, and let $K = \overline{co}(A)$. By a slice of K we mean a set of the form

$$S(f,\alpha,K) = \{x \in K : f(x) \geqq \alpha\},$$

where $f \in X^*$ and $\alpha < \sup f(K)$. For $n = 1,2,\ldots,$ let

$$U_n(K) = \{f \in X^* : \text{some slice } S(f,\alpha,K) \text{ has diameter } < \tfrac{1}{n}\}.$$

It is easy to check that $U_n(K)$ is a norm open subset of X^*. The main step of Phelps' proof is the following lemma.

LEMMA 6 (Phelps [15]). *If (1) holds, then for every n, $U_n(K)$ is norm dense in X^*.*

For the moment, suppose the lemma holds. Then by the Baire Category theorem the set

$$E(K) = \bigcap_{n=1}^{\infty} U_n = \{f \in X^* : \lim_{\alpha \to \sup f(K)-} (\operatorname{diam} S(f,\alpha,K)) = 0\}$$

is a dense G_δ set in X^*. If f is in $E(K)$, then f *strongly exposes* K; i.e., there exists a point x in K with $f(x) = \sup f(K)$ and such that

$$y_n \in K, \quad f(y_n) \to f(x) \Rightarrow y_n \to x.$$

(x is called a *strongly exposed point of* K). We have the following important Corollary.

COROLLARY (Phelps). *If (1) holds then every closed bounded convex set $K \subset X$ is the closed convex hull of its strongly exposed points.*

PROOF. Let $K_1 = \overline{co}$(strongly exposed points of K). If $K_1 \neq K$, choose $g \in X^*$ with $\sup g(K_1) = \beta < g(y)$ for some $y \in K$. Now choose $f \in E(K)$ with $\|f-g\| < \frac{1}{2}(g(y)-\beta)$. Then f strongly exposes K at some point x which cannot be in K_1, a contradiction.//

To see that (1) implies (2), note that for any f in X^*, $\sup f(A) = \sup f(K)$, so that every strongly exposed point of K lies in A. Such a point must satisfy the condition in (2).

Because of the importance of Lemma 6 we shall give a complete proof. Let f be in X^*, $\|f\| = 1$, and let $1 > \varepsilon > 0$ be given. We show $\exists g \in X^*$ and $\alpha < \sup g(K)$ such that $\|f-g\| < \varepsilon$ and $\operatorname{diam} S(g,\alpha,K) < \varepsilon$. There is no harm in translating K, so we assume K is disjoint from $f^{-1}(0)$, say $f > 0$ on K. Let $M = \sup\{\|x\| : x \in K\}$, let $\beta = 4M/\varepsilon$, and let $C = \overline{co}(K \cup \{x : x \in f^{-1}(0), \|x\| \leq \beta\})$.

CLAIM. There exists g in X^*, $\|g\| = 1$ and $\alpha < \sup g(C)$ such that diam $S(g,\alpha,C) < \varepsilon$ and $S(g,\alpha,C) \cap f^{-1}(0) = \emptyset$.

To prove the claim, let z be any point in K; then $f(z) > 0$. For every x in $C \cap f^{-1}(0)$, define $T_x : X \to X$ by

$$T_x(y) = y - 2\,\frac{f(y)}{f(z)}\,(z-x).$$

(T_x is the reflection of X thru $f^{-1}(0)$ along the line through 0 determined by (z-x).) The following are easily established:

(i) $x = \frac{1}{2}(z + T_x z)$

(ii) T_x^2 = identity; hence $T_x^{-1} = T_x$,

(iii) T_x = identity on $f^{-1}(0)$,

and

(iv) $\|T_x\| \leq N < \infty$,

where $N = 1 + (4/f(z))\sup\{\|x\| : x \in C\}$.

Let $\mathcal{K} = \{C\} \cup \{T_x C : x \in C \cap f^{-1}(0)\}$, and let K_1 denote the closed convex hull of the union of the members of $\mathcal{K}$. Then K_1 is bounded.

If $x \in C \cap f^{-1}(0)$, then by (i), x is the mid-point of a segment of K_1 of length $\|z - T_x z\| = 2\|z-x\| \geq 2f(z)$.

Since K_1 is dentable, there exists a slice $S(h,\alpha,K_1)$ of diameter d, where $d < \min\{\varepsilon/N, f(z)\}$. If $S(h,\alpha,K_1)$ were to contain a point $x \in C \cap f^{-1}(0)$, it would also contain at least one endpoint of the line segment in K_1 of which x is the mid-point, contradicting $d < f(z)$. Thus $S(h,\alpha,K_1) \cap f^{-1}(0) = \emptyset$.

Now $\sup h(K_1) = \sup h(\cup\{K' : K' \in \mathcal{K}\})$, so $\sup h(C_0) > \alpha$ for at least one $C_0 \in \mathcal{K}$. Then $S(h,\alpha,C_0) \subset S(h,\alpha,K_1)$, so $S(h,\alpha,C_0) \cap f^{-1}(0) = \emptyset$ and diam $S(h,\alpha,C_0) \leq d$.

Consider the possible choices of C_0. If $C_0 = C$, let $g = h/\|h\|$. If $C_0 = T_x C$ for some $x \in C \cap f^{-1}(0)$, then

$$T_x^{-1} S(h,\alpha,T_x C) = S(T_x \circ h,\alpha,C)$$

is a slice of C of diameter at most

$$\|T_x^{-1}\| \cdot d = \|T_x\| \cdot d \leq Nd < \varepsilon.$$

Moreover, this slice also misses $C \cap f^{-1}(0)$ since $f^{-1}(0)$ is fixed by T_x^{-1}. Let $g = T_x \circ h/\|T_x \circ h\|$. This completes the proof of the claim.

Now since $S(g,\alpha,C)$ misses $C \cap f^{-1}(0)$, and since $C = \overline{co}(K \cup (C \cap f^{-1}(0)))$, $\sup g(K) = \sup g(C)$. Since $S(g,\alpha,K) \subset S(g,\alpha,C)$, we have $\operatorname{diam} S(g,\alpha,K) < \varepsilon$. It remains to show that $\|f-g\| < \varepsilon$.

Choose any y in $S(g,\alpha,K)$. Then

$$g(y) > \sup \{g(x) : x \in f^{-1}(0),\ \|x\| \leq \beta\} = \beta\|g\|_{f^{-1}(0)},$$

so

$$\|g\|_{f^{-1}(0)} \leq \frac{1}{\beta}\, g(y).$$

Choose $\bar{g} \in X^*$ such that $\bar{g} = g$ on $f^{-1}(0)$ and $\|\bar{g}\| \leq \frac{1}{\beta}\, g(y)$. Since $g-\bar{g} = 0$ on $f^{-1}(0)$, $g-\bar{g} = \gamma f$ for some real γ. Note that

$$|1 - |\gamma|| = |\|g\| - \|g-\bar{g}\|| \leq \|\bar{g}\| \leq \frac{1}{\beta}\, g(y).$$

If $\gamma < 0$, then

$$\|f+g\| = \|(1+\gamma)f+\bar{g}\| \leq |1+\gamma| + \|\bar{g}\| \leq \frac{2}{\beta}\, g(y),$$

and since $f(y) > 0$,

$$\frac{2}{\beta}\, g(y) \geq \|f+g\| \geq (f+g)\left(\frac{y}{\|y\|}\right) > \frac{g(y)}{\|y\|} \geq \frac{g(y)}{M},$$

or $M > \beta/2$, which contradicts the choice of β. Thus $\gamma \geq 0$ and we have

$$\|f-g\| = \|(1-\gamma)f-\bar{g}\| \leq |1-\gamma| + \|\bar{g}\|$$

$$\leq \frac{2}{\beta}\, g(y) \leq \frac{2\varepsilon}{4M} \cdot M < \varepsilon.$$

This completes the proof of Lemma 6.//

REMARKS. (f) The first proof that (1) implies the KMP was given by J. Lindenstrauss. His proof appears in [15], where Phelps also proves the much stronger results given above.

See also the papers of G.A. Edgar [5][6].

(g) The conjecture of Diestel that the KMP implies the RNP remains open at this time. The strongest result known in the general case is that (4) ⇒ RNP as demonstrated above. However, it was shown in [8] that if X is a dual space then the KMP indeed does imply the RNP. The techniques of that proof are considerably different from those employed above, and the proof depends on some deep results of C. Stegall [19].

(h) The above surveys only part of the recent research on the RNP. We refer the interested reader to the papers listed below for additional results and references. In particular, see the survey paper by Diestel and Uhl [3].

REFERENCES

1. S.D. Chatterji, Martingale convergence and the Radon-Nikodým theorem in Banach spaces, Math. Scand. 22 (1968), 21-41.
2. W.J. Davis and R.R. Phelps, The Radon-Nikodým property and dentable sets in Banach spaces, Proc. Amer. Math. Soc. 45 (1973).
3. J. Diestel and J.J. Uhl, Jr., The Radon-Nikodým theorem for Banach space valued measures, Rocky Moutain J. Math.
4. N. Dunford and B.J. Pettis, Linear operations on summable functions, Trans. Amer. Math. Soc. 47 (1940), 323-392.
5. G.A. Edgar, A concompact Choquet theorem, Proc. Amer. Math. Soc. 48 (1975).
6. ________, Extremal integral representations (preprint).
7. R. Huff, Dentability and the Radon-Nikodým property, Duke Math. J. 41 (1974), 111-114.
8. ______ and P.D. Morris, Dual spaces with the Krein-Milman property have the Radon-Nikodým property, Proc. Amer. Math. Soc. 49 (1975), 104-108.
9. ______ and __________, Geometric characterizations of the Radon-Nikodým property in Banach spaces, Studia Math. (to appear).
10. D.R. Lewis and C. Stegall, Banach spaces whose duals are isomorphic to $\ell^1(\Gamma)$, J. Functional Analysis 12 (1973), 177-187.
11. H. Maynard, A geometric characterization of Banach spaces with the Radon-Nikodým property, Trans. Amer. Math. Soc. 185 (1973), 493-500.
12. I. Namioka, Neighborhoods of extreme points, Israel J. Math. 5 (1967), 145-152.
13. _________ and E. Asplund, A geometric proof of Ryll-Nardzewski's fixed-point theorem, Bull. Amer. Math. Soc. 73 (1967), 443-445.
14. _________ and R.R. Phelps, Banach spaces which are Asplund spaces (preprint).

15. R.R. Phelps, Dentability and extreme points in Banach spaces, J. Functional Analysis 16 (1974), 78-90.

16. M.A. Rieffel, Dentable subsets of Banach spaces, with applications to a Radon-Nikodym theorem, Proc. Conf. Functional Analysis, Thompson Book Co., Washington, D.C. (1967), 71-77.

17. U. Rønnow, On integral representation of vector-valued measures, Math. Scand. 21 (1967), 45-53.

18. H.P. Rosenthal, The Banach spaces $C(K)$ and $L^p(\mu)$, (preprint).

19. C. Stegall, The Radon-Nikodým property in conjugate Banach spaces, Trans. Amer. Math. Soc.

20. J.J. Uhl, Jr., A note on the Radon-Nikodým property for Banach spaces, Rev. Roum. Mat. 17 (1972), 113-115.

21. J.J. Uhl, Jr., Applications of Radon-Nikodým theorems to martingale convergence, Trans. Amer. Math. Soc. 145 (1969), 271-285.

ON THE RADON - NIKODYM DERIVATIVE OF A MEASURE TAKING VALUES IN A BANACH SPACE WITH BASIS

by Z. Lipecki and K. Musiał
Institute of Mathematics, Polish Academy of Sciences

Let (S,Σ,μ) be a positive measure space and let $(X,\|\cdot\|)$ be a Banach space.

The main result of this paper (this is a shortened version of [8]) is a criterion of the existence of the Radon-Nikodym derivative of a μ-continuous measure $\nu : \Sigma \to X$ in terms of the Radon-Nikodym derivatives of its coefficients (which are μ-continuous scalar measures on Σ) in a given (Schauder) basis in X.

In order to establish the criterion we shall need a general lemma:

LEMMA. *Suppose* $f: S \to X$ *is such that* $x^*f \in L_1(\mu)$ *for* $x^* \in M$, *where* M *is a weak** *sequentially dense subset of* X^* (*), *and* $\nu: \Sigma \to X$ *is a measure with*

$$(1) \qquad x^*\nu(E) = \int_E x^*f\,d\mu \qquad \text{for } x^* \in M \text{ and } E \in \Sigma .$$

Then f *is Pettis* μ-*integrable and*

$$\nu(E) = (P)\int_E f\,d\mu \qquad \text{for } E \in \Sigma .$$

Proof. It follows from (1) that

(*) Prof. E.Thomas called our attention to a theorem of Banach ([1], p. 126, Theorem 7), according to which, for separable X, it is sufficient to assume that M is total.

(2) $$\int_E |x^*f|\, d\mu \leq \|x^*\| \cdot \|\nu\|(E) \quad \text{for } x^* \in M \text{ and } E \in \Sigma ,$$

where $\|\nu\|$ denotes the semivariation of ν (see [2], p.293-294). Moreover, in view of the weak* density of M in X^*, (1) shows that ν is μ-continuous.

Fix $x^* \in X^*$ and choose $\{x_n^*\} \subset M$ with $x_n^*(x) \to x^*(x)$ for $x \in X$. By the Banach-Steinhaus theorem, $\sup_n \|x_n^*\| < \infty$. Hence, in virtue of (2) and the μ-continuity of ν, Corollary 2.4 of [2] shows that x_n^*f are uniformly μ-integrable. Since, moreover, $x_n^*f(s) \to x^*f(s)$ for $s \in S$, it follows from Vitali's theorem that $x^*f \in L_1(\mu)$ and $\int_E x_n^*f\, d\mu \to \int_E x^*f\, d\mu$ for $E \in \Sigma$. Hence (1) yields

$$x^*\nu(E) = \int_E x^*f\, d\mu \qquad \text{for } E \in \Sigma .$$

As $x^* \in X^*$ is arbitrary, this proves the assertion.

THEOREM 1. <u>Suppose $\{x_n\}$ is a basis in X and $\{x_n^*\}$ is the associated sequence of coefficient functionals. Let $\{f_n\} \subset L_1(\mu)$ and a measure $\nu: \Sigma \to X$ be such that</u>

(3) $$x_n^*\nu(E) = \int_E f_n\, d\mu \qquad \text{for } E \in \Sigma \text{ and } n = 1,2,\ldots$$

<u>Then the following three conditions are equivalent:</u>

(i) ν <u>has a Pettis μ-integrable Radon-Nikodym derivative.</u>

(ii) $\sum_{n=1}^{\infty} f_n(\cdot)x_n$ <u>converges strongly μ-a.e.</u>

(iii) $\sum_{n=1}^{\infty} f_n(\cdot)x_n$ <u>converges weakly in measure</u> μ.

<u>Either of them implies that</u>

(4) $$\nu(E) = (P)\int_E \sum_{n=1}^{\infty} f_n(\cdot)x_n\, d\mu \qquad \text{for } E \in \Sigma .$$

Condition (iii) was suggested to us by Prof.C.Ryll-Nardzewski.

<u>Proof</u>. To establish the implication (i) => (ii) assume that $\nu(E) = (P)\int_E f\, d\mu$ for $E \in \Sigma$, where $f: S \to X$. Then, by virtue of

(3), $x_n^* f = f_n$ μ-a.e. Hence, according to the definition of a basis, $f = \sum_{n=1}^{\infty} f_n x_n$ holds μ-a.e., which yields (ii) and (4).

Clearly, (ii) implies (iii), so it remains to derive (i) from (iii). To this end put $f = \sum_{n=1}^{\infty} f_n x_n$. We have

$$\left(\sum_{n=1}^{m} a_n x_n^*\right) \nu(E) = \int_E \left(\sum_{n=1}^{m} a_n x_n^*\right) f \, d\mu$$

whenever $E \in \Sigma$ and $a_1, \dots, a_m$ are scalars. Since the set $M = \{\sum_{n=1}^{m} a_n x_n^* : a_1, \dots, a_m$ are scalars, $m = 1, 2, \dots\}$ is weak sequentially dense in X^*, (iii) follows from the Lemma.

The following corollary is an easy consequence of the Theorem.

COROLLARY 1. <u>Let</u> $\{x_n\}$ <u>be the standard basis in</u> c_0, <u>let</u> $\{x_n^*\}$ <u>be the associated sequence of coefficient functionals, and let</u> $\{f_n\} \subset L_1(\mu)$ <u>and</u> $\nu : \Sigma \to c_0$ <u>satisfy</u> (3). <u>Then</u> ν <u>has a Pettis integrable Radon-Nikodym derivative iff</u> $f_n \to 0$ μ-<u>a.e.</u>

This corollary indicates how to construct examples of c_0-valued measures of finite variation without Radon-Nikodym derivative. Indeed, to this end it is enough to define a sequence $\{f_n\} \subset L_1(\mu)$ such that

$$\left|\int_E f_n \, d\mu\right| \leq \mu(E) \quad , \quad \int_E f_n \, d\mu \to 0 \quad \text{for } E \in \Sigma$$

and $f_n \not\to 0$ μ-a.e. (e.g. $\{\sin ns\}$ in $L_1\langle 0, 2\pi\rangle$ (see [4] p. 214; cf. also [7]). Then, putting

$$\nu(E) = \sum_{n=1}^{\infty} \left(\int_E f_n \, d\mu\right) x_n$$

for $E \in \Sigma$, we obtain a measure with the desired properties.

The next corollary is due to Dunford and Morse ([5], Theorem 5.3 and [6] p. 415).

COROLLARY 2. (Dunford-Morse). If X has a boundedly complete basis $\{x_n\}$ (for definition see [3]), then X has the Radon-Nikodym property.

Proof. Let $\nu: \Sigma \to X$ be a μ-continuous measure of σ-finite variation and let $\{x_n^*\}$ and f_n be as in the Theorem. Since $\{x_n\}$ is a basis, we have

$$\|\sum_{n=1}^{m} (x_n^*\nu(E))x_n\| \leq C\,\|\nu(E)\| \qquad \text{for } E\in\Sigma \text{ and } m=1,2,\ldots$$

for some constant C (see [3], Theorem IV,3.1i).

Hence, by virtue of (3), we have

$$\int_E \|\sum_{n=1}^{m} f_n x_n\| d\mu \leq C|\nu|(E) \qquad \text{for } E\in\Sigma \text{ and } m=1,2,\ldots$$

where $|\nu|$ denotes the variation of ν .

Consequently, $\sup\{\|\sum_{n=1}^{m} f_n x_n\| : m=1,2,\ldots\} < \infty$ except for a set of μ-measure zero. In view of the bounded completeness of the basis, the assertaion now follows from the Theorem.

References

[1] Banach S., Théorie des opérations linéaires, 2nd edition, Chelsea Publ. Co, New York.

[2] Bartle R.G., Dunford N., Schwartz J., Weak compactness and vector measures, Canad. J.Math. 7(1955), 289-305.

[3] Day M.M., Normed linear spaces, 3rd edition, Berlin-Heidelberg-New York , 1973.

[4] Diestel J., Application of weak compactness and bases to vector measures and vectorial integration, Rev.Roum.Math.Pures Appl. 18(1973), 211-224.

[5] Dunford N., Integration and linear operators, Trans.Amer. Math.Soc. 40(1936), 474-494.

[6] Dunford N., Morse A.P., Remarks ón the preceding paper of James A.Clarkson, ibidem. 40(1936), 415-420.

[7] Lewis D.R., A vector measure with no derivative, Proc.Amer. Math.Soc. 32(1972), 535-536.

[8] Lipecki Z., Musiał K., On the Radon-Nikodym derivative of a measure taking values in a Banach space with basis, Rev.Roum.Math. Pures Appl. (submitted).

RADON - NIKODYM THEOREM FOR BANACH SPACE VALUED MEASURES

by Kazimierz Musiał
Wrocław University
and
Institute of Mathematics , Polish Academy of Sciences

It is the purpose of this paper to give a new proof of Rieffel's Radon-Nikodym theorem [5] for Banach space valued measures. We formulate this theorem in the generalized form, which was proved by Moemodo and Uhl [4].

Throught the paper Σ denotes a fixed σ-algebra of subsets of a non-empty set S , X and Y are Banach spaces, $\nu: \Sigma \to X$ is an X-valued measure, μ is a finite positive measure on Σ , $\Sigma_\mu^+ = \{E \in \Sigma : \mu(E) > 0\}$ and

$$\mathcal{A}_\nu(E) = \left\{ \frac{\nu(F)}{\mu(F)} : E \supset F \in \Sigma_\mu^+ \right\}$$

is the average range of ν on $E \in \Sigma$.

We begin with a few lemmas.

LEMMA 1. [3] Let $f_n: S \to Y$, $n = 1,2,\dots$ be a sequence of strongly measurable functions on a measure space (S,Σ,μ) . If the series $\sum_{n=1}^{\infty} f_n(s)$ is divergent on a set $E \in \Sigma_\mu^+$, then there exist $E \supset F \in \Sigma_\mu^+$, $\delta > 0$, and a sequence of positive integers $n_1 < n_2 < \dots$ such that

$$\max_{n_k \leq m < n_{k+1}} \left\| \sum_{i=n_k}^{m} f_i(s) \right\| > \delta$$

for every $s \in F$ and $k = 1,2,\ldots$ ($\|\cdot\|$ is a fixed norm in Y).

LEMMA 2. *Let $f\colon S \to Y$ be a strongly measurable and Pettis integrable function. If $\varkappa(E) = \text{Pettis} - \int_E f\,d\mu$, then*

$$\{f(s) : s \in S\} \subset \overline{\text{conv}}\, \mathcal{A}_\varkappa(S) \qquad \mu - \text{a.e.}$$

Proof. For Bochner integrable functions this was proved by Rieffel ([5], Corollaries 1.6 and 1.11).

If f is arbitrary, then setting $f_n(s) = f(s)$ if $\|f(s)\| \leq n$ and $f_n(s) = \varkappa(S)/\mu(S)$ whenever $\|f(s)\| > n$, we get $\{f_n(s)\colon s \in S\} \subset \overline{\text{conv}}\, \mathcal{A}_\varkappa(S)$ μ-a.e. , which yields the required result.

LEMMA 3. *Let Y be a closed linear subspace of a separable Banach space Z , and let $\varkappa\colon \Sigma \to Y$ be a measure. Then, if $\varkappa$ is the Pettis integral on a Z-valued function f , it is also the Pettis integral of an Y-valued function (in fact it is the same function after a suitable modification on a set of μ-measure zero).*

Proof. The assertion follows from Lemma 2 and from the obvious inclusion $\mathcal{A}_\varkappa(S) \subset Y$.

Now we are able to prove the main result.

THEOREM (Rieffel, Moemodo and Uhl). *Let X be a Banach space. For any X-valued measure ν on (S,Σ) which is absolutely continuous with respect to μ , the following conditions are equivalent:*

(i) *there exists a strongly measurable function $f\colon S \to X$ such that*

$$\nu(E) = \text{Pettis} - \int_E f\,d\mu \quad , \qquad E \in \Sigma \,;$$

(ii) *$\mathcal{A}_\nu(S)$ is locally conditionally compact, that means, for every $E \in \Sigma_\mu^+$ there exists $E \supset F \in \Sigma_\mu^+$ such that $\mathcal{A}_\nu(F)$ is conditionally compact.*

Proof. i => ii. Since f is strongly measurable, there exists (by the theorem of Egoroff) for every $E \in \Sigma_\mu^+$ a set $E \supset F \in \Sigma_\mu^+$ such that $f(F) = \{f(s): s \in F\}$ is conditionally compact.

Since $\mathcal{A}_\nu(F) \subset \overline{\text{conv}}\ f(F)$ and the last set is compact by the theorem of Mazur, $\mathcal{A}_\nu(F)$ is conditionally compact.

ii => i. Using the method of exhaustion it is easy to see that there is a closed and separable subspace Y of X such that $\mathcal{A}_\nu(S) \subset Y$.

Clearly, $\mathcal{A}_\nu(S)$ is locally conditionally compact in Y as well.

Now, in view of the theorem of Banach and Mazur [1], we may assume, that Y is isometrically embedded into $C[0,1]$.

Since Y is a closed linear subspace of $C[0,1]$ and X, in order to prove (i), it is sufficient, in view of Lemma 3, to prove that ν can be represented as the Pettis integral of a $C[0,1]$-valued function.

Let $\{x_n\}$ be an orthonormal basis in $C[0,1]$ and let $\{x_n^*\}$ be the associated biothogonal sequence.

In order to prove the existence of a measurable $f: S \to C[0,1]$ such that

$$\langle x^*, \nu(E)\rangle = \int_E \langle x^*, f\rangle\, d\mu ,$$

$E \in \Sigma$ and $x^* \in C^*[0,1]$, it is sufficient, in view of the Theorem in [2], to prove the μ-a.e. convergence of the series $\sum_{n=1}^{\infty} f_n(s)x_n$, where

$$\langle x_n^*, \nu(E)\rangle = \int_E f_n\, d\mu = \nu_n(E) \quad , \quad n = 1,2,\ldots .$$

Suppose that the series is divergent on a set $E \in \Sigma_\mu^+$.

Then. in view of Lemma 1, there exist $E \supset F \in \Sigma_\mu^+$, $\delta > 0$, and a sequence of positive integers $n_1 < n_2 < \ldots$ such that

$$\max_{n_k \le m < n_{k+1}} \Big\| \sum_{i=n_k}^{m} f_i(s)\, x_n \Big\| > \delta ,$$

for every $s \in F$ and $k = 1,2,\ldots$ (the norm is taken in $C[0,1]$).

Without loss of generality we may assume, that $\mathcal{A}_\nu(F)$ is conditionally compact in $C[0,1]$. In particular, there exists N such that for every $n > N$ and every $H \in \{A \in \Sigma: A \subset F\}$ the inequality

$$(*) \qquad \left\| \sum_{k=n}^{\infty} \nu_k(H)\, x_k \right\| < \frac{1}{2}\, \delta\mu(H)$$

holds.

Let k and m be such that $m \geq n_k > N$ and $G = \{s \in F:\ \|\sum_{i=n_k} f_i(s)x_i\| > \delta\} \in \Sigma_\mu^+$. Moreover, let $\varkappa$ be a measure defined on $\Sigma_0 = \{A \in \Sigma : A \subset G\}$ by the equality

$$\varkappa(A) = \sum_{i=n_k}^{m} \nu_i(A)\, x_i$$

Now, $(*)$ yields

$$\|\varkappa(A)\| \leq \delta\, \mu(A) \quad , \qquad A \in \Sigma_0$$

and hence

$$|\varkappa|(A) \leq \delta\, \mu(A) \quad , \qquad A \in \Sigma_0 \, ,$$

where $|\varkappa|$ is the variation of $\varkappa$.

On the other hand, it follows from the definition of $\varkappa$, that

$$\varkappa(A) = \int_A \left(\sum_{i=n_k}^{m} f_i(s)\, x_i \right) \mu(ds) \, , \quad A \in \Sigma_0$$

and hence

$$|\varkappa|(A) = \int_A \left\| \sum_{i=n_k}^{m} f_i(s) x_i \right\| \mu(ds) > \delta\mu(A) \, ,$$

whenever $A \in \Sigma_0$.

This contradiction proves the assertion.

<u>References</u>

[1] Banach S., Théorie des opérations linéaires, Warszawa, Monografie Matematyczne, 1932 .

[2] Lipecki Z., Musiał K., On the Radon-Nikodym derivative of a measure taking values in a Banach space with basis (this volume of Lecture Notes in Math.).

[3] Musiał K., Ryll-Nardzewski C., Woyczyński W.A., Convergence presque sûre des séries aléatoires vectorielles à multiplicateurs bornés, C.R.Acad. Sc. Paris, t.279, Série A, 225-228 (1974).

[4] Moemodo S., Uhl J.J.Jr., Radon-Nikodym theorems for the Bochner nad Pettis integrals, Pacific J. Math., 38 (1971), 531-536.

[5] Rieffel M.A., The Radon-Nikodym theorem for the Bochner integral, Trans.Amer.Math.Soc. 131, (1968), 466-487.

A NOTE ON ADAPTED CONES

Jürgen Bliedtner

Let X be a locally compact Hausdorff space, C(X) the vector space of real continuous functions on X, and $C_c(X)$ the subspace of C(X) consisting of all functions which have a compact support. Furthermore, let $P \subset C^+(X)$ be a convex cone such that for every $x \in X$, there is a $p \in P$ such that $p(x) > 0$. We say that a positive linear form $T: P \to \mathbb{R}$ is represented by a positive Radon measure μ on X if

(i) every $p \in P$ is μ-integrable and

(ii) $T(p) = \mu(p)$ for all $p \in P$.

By a theorem of Choquet [1], p.283, every positive linear form on P is represented by a positive Radon measure if P is an adapted cone, i.e. for every $p \in P$ there is a $q \in P$ such that for every $\varepsilon > 0$ there exists a compact set $K \subset X$ such that $p \leq \varepsilon q$ on $X \setminus K$ ("p is dominated by q at infinity").

If one is only interested that a certain given positive linear form on P is represented by a positive Radon measure(e.g.existence of balayaged measures in potential theory), the following proposition gives a sufficient condition.

PROPOSITION: Let $T: P \to \mathbb{R}$ be a positive linear form satisfying the following condition:

(*) For every $p \in P$ and for every $\varepsilon > 0$ there exist a $q \in P$ and a compact set $K \subset X$ such that $T(q) \leq \varepsilon$ and $p \leq q$ on $X \setminus K$.

Then T is represented by a positive Radon measure μ on X.

Proof: By a version of the Hahn-Banach theorem [1], p.269, T has an extension to a positive linear form $\hat{T}$ on $C_P(X) := \{f \in C(X) : \exists\, p \in P,\ |f| \leq p\}$. Since $C_c(X) \subset C_P(X)$ the restriction of $\hat{T}$ to $C_c(X)$ defines a positive Radon measure μ on X. For any $p \in P$ we have

$$\mu(p) = \sup\{\mu(f) : f \in C_c(X),\ 0 \leq f \leq p\} \leq \hat{T}(p) = T(p),$$

hence p is μ-integrable with $\mu(p) \leq T(p)$.

To obtain the converse inequality let $\varepsilon > 0$ be given. By condition $(*)$ there exist $q \in P$ and a compact set $K \subset X$ such that $T(q) \leq \varepsilon$ and $p \leq q$ on $X \setminus K$. Choose $f \in C_c(X)$ such that $0 \leq f \leq 1$ and $f = 1$ on K. Then $pf \in C_c(X)$ and $p(1-f) \leq q$ on X, hence the assertion follows from

$$T(p) = \hat{T}(p) = \hat{T}(pf) + \hat{T}(p(1-f)) \leq \mu(pf) + T(q) \leq \mu(p) + \varepsilon . \qquad \lrcorner$$

The adaptedness of P and the condition $(*)$ are related as follows:

1. Suppose P is an adapted cone. Then every positive linear form T on P satisfies the condition $(*)$.

<u>Proof:</u> Let $p \in P$, and choose a $q \in P$ dominating p at infinity. We may assume that $T(q) > 0$. If $\varepsilon > 0$ is given there exists a compact set $K \subset X$ such that $p \leq \frac{\varepsilon}{T(q)} \cdot q$ on $X \setminus K$, hence $q_1 := \frac{\varepsilon}{T(q)} \cdot q \in P$ satisfies $T(q_1) = \varepsilon$ and $p \leq q_1$ on $X \setminus K$. $\lrcorner$

2. If X is countable at infinity and if P is inf-stable and every positive linear form on P satisfies the condition $(*)$ then the convex cone $P_\sigma = \{ \sum p_n \in C(X) : (p_n) \subset P \}$ is an adapted cone.

<u>Proof:</u> By [2], p.34, it suffices to show that for every $p \in P$, $\varepsilon > 0$ and $K \subset X$ compact there exist $K' \subset X$ compact and $q \in P$ such that

$$q \leq \varepsilon \text{ on } K \text{ and } p \leq q \text{ on } X \setminus K'.$$

Since for every $x \in X$, evaluation at x defines a positive linear form on P there exist a $q_x \in P$ and a compact set $K_x \subset X$ such that

$$q_x(x) < \varepsilon \text{ and } p \leq q_x \text{ on } X \setminus K_x .$$

Since K is compact there are finitely many points $x_1, \dots, x_n \in K$ such that $q := \inf(q_{x_1}, \dots, q_{x_n}) \in P$ satisfies $q \leq \varepsilon$ on K and $p \leq q$ on $X \setminus K'$ where $K' := K_{x_1} \cup \dots \cup K_{x_n}$. $\lrcorner$

<u>References:</u>

[1] CHOQUET, G.: Lectures on Analysis II. New York: Benjamin 1969.

[2] SIBONY, D.: Cônes de fonctions et potentiels. Lecture Notes, McGill University, Montreal 1968.

Integraldarstellungen linearer Funktionale

Benno Fuchssteiner

Gesamthochschule Paderborn

Das Problem dieser Arbeit ist die Charakterisierung derjenigen konvexen Kegel F(X) von reellwertigen Funktionen auf einer Menge X, welche die Eigenschaft haben, daß jede monotone lineare Abbildung $F(X) \to R$ eine dominierende Integraldarstellung hat. Zur Charakterisierung dieser Kegel erhalten wir eine einfache Bedingung, die hinreichend und notwendig ist. Kegel, welche diese Bedingung erfüllen, werden wir zukünftig Dini-Kegel nennen. Diese Dini-Kegel sind nicht zu verwechseln mit den von Portenier in [11] eingeführten Dini-Räumen.

Der angegebene Hauptsatz (Abschnitt I) verallgemeinert in durchsichtiger Weise die verschiedensten Integraldarstellungssätze, so zum Beispiel den Satz von Choquet und den Satz von Riesz. Der Beweis des Hauptsatzes fußt auf drei Teilergebnissen, die im zweiten Abschnitt bereitgestellt werden. Am wichtigsten von diesen Teilergebnissen scheint mir der (abzählbare) Zerlegungssatz zu sein, der auch in anderem Zusammenhang von einiger Bedeutung ist. Im dritten Abschnitt werden einige wenige Folgerungen und Anwendungen des erhaltenen Hauptsatzes aufgezeigt. Im letzten Kapitel geben wir ein weiteres Ergebnis (ohne Beweis) und einen kurzen Überblick über in diesem Zusammenhang wichtig erscheinende offene Probleme.

Die Beweise dieses Aufsatzes sind, soweit sie in vorhandenen oder noch erscheinenden Arbeiten enthalten sind, bewußt kurz gehalten.

I. Der Hauptsatz

Es sei X eine beliebige Menge, und $F = F(X)$ sei ein konvexer Kegel beschränkter reeller Funktionen auf X. Außerdem setzen wir voraus, daß F die konstanten Funktionen enthält. Eine Abbildung $\mu : F \to R$ heißt wie üblich *linear*, wenn sie additiv und positiv-homogen ist. μ wird *monoton* genannt, wenn für $f, g \in F$ aus $f \geq g$ immer $\mu(f) \geq \mu(g)$ folgt. Man beachte, daß im Falle eines Vektorraumes F, ein lineares μ genau dann monoton ist wenn $f \geq 0 \longrightarrow \mu(f) \geq 0 \quad \forall f \in F$

Im allgemeinen Fall reicht dies jedoch nicht aus. Da aber F die Konstanten enthält, läßt sich ein monotones lineares μ immer zu einem positiven linearen Funktional auf dem Vektorraum der reellen beschränkten Funktionen auf X fortsetzen (siehe etwa [3, Cor. 1.3]).

Mit Σ_F bezeichnen wir die von F erzeugte σ-Algebra in X. Ein positives Σ_F- Maß τ heißt *Darstellungsmaß* von μ wenn:

$$(*) \qquad \mu(f) \leq \int_X f \, d\tau \qquad \text{für alle } f \in F.$$

Gilt sogar Gleichheit bei (*), so heißt τ *echtes* Darstellungsmaß. Für jedes Darstellungsmaß τ des linearen μ gilt $\tau(X) < \infty$, da F die Konstanten enthält. Dies sieht man sofort mit:

$$-\infty < \mu(-1_X) \leq -\int_X 1_X \, d\tau = -\tau(X) .$$

Darstellungsmaße sind in vielen Fällen automatisch echt. So zum Beispiel wenn F ein Vektorraum ist. Oder wenn μ ein maximales monotones lineares Funktional ist. Das soll heißen, für jedes monotone lineare ν , welches μ *dominiert* ($\nu(f) \geq \mu(f) \quad \forall f \in F$) gilt schon $\mu = \nu$.

Es gibt in der Tat genügend viele dieser maximalen Funktionale. Denn da jedes μ-dominierende monotone lineare ν seinerseits von dem sublinearen Funktional

$$f \to \mu(1_X) \sup_{x\in X} f(x)$$

dominiert wird, folgt aus dem Zornschen Lemma, daß jedes monotone lineare μ von einem maximalen monotonen Funktional dominiert wird.

Wir kommen nun zur Formulierung unseres Hauptergebnisses:

Hauptsatz ([6, Main theorem]) *Folgendes ist äquivalent:*

(i) *Jedes monotone lineare* $\mu : F \to R$ *besitzt ein Darstellungsmaß.*

(ii) F *ist ein* Dini-Kegel, *das heißt: Für jede punktweise fallende Folge* (f_n) *in* F *gilt*

$$\inf_{n\in N} \sup_{x\in X} f_n(x) = \sup_{x\in X} \inf_{n\in N} f_n(x).$$

II Die wesentlichen Hilfsmittel

Sei F wie im letzten Kapitel. Ein monotones lineares $\mu : F \to R$ nennen wir *Zustand* wenn $\mu(1_X) = 1$. Wir sagen, daß ein Zustand μ die *abzählbare Zerlegungseigenschaft* hat, wenn für jede Überdeckung $\{ X_n | n \in N \}$ von X durch Teilmengen Zahlen $\lambda_n \geq 0$ mit $\sum_{n\in N} \lambda_n = 1$ existieren, so daß

$$\mu(f) \leq \sum_{n\in N} \lambda_n \sup_{x\in X_n} f(x) \qquad \forall f \in F.$$

Zerlegungssatz: *Es ist äquivalent:*

(i) *Jeder Zustand von* F *hat die abzählbare Zerlegungseigenschaft.*

(ii) F *ist ein Dini-Kegel.*

Beweis: (i) $\longrightarrow$ (ii): Da der Zustandsraum St(F) unter der Topologie punktweiser Konvergenz auf F kompakt ist, gibt es für jede punktweise fallende Folge (f_n) in F einen Zustand μ mit

$$(*):\qquad \alpha = \inf_{n\in N} \sup_{x\in X} f_n(x) = \inf_{n\in N} \mu(f_n).$$

Ist nun

$\sup_{x\in X} \inf_{n\in N} f_n(x) = \beta < \alpha$, so wäre durch $X_n = \{ x\in X \mid f_n(x) \leq \frac{\alpha+\beta}{2} \}$, $n\in N$

eine Oberdeckung von X gegeben, und die Zerlegung von μ bezüglich dieser Oberdeckung stünde im Widerspruch zu (*).

(ii) $\longrightarrow$ (i): Es sei $\{ X_n \mid n \in N \}$ irgendeine Oberdeckung von X und μ sei ein beliebiger Zustand. Wir betrachten die kompakten Teilmengen

$Y_n = \{ z \in St(F) \mid z(f) \leq \sup_{x\in X_n} f(x) \quad \forall f \in F \}$ von St(F), den σ-kompakten

Raum $Z = \bigcup \{ Y_n \mid n \in N \}$ und $F(Z) = \{ \hat{f} \mid f \in F \}$, wobei $\hat{f}(z) = z(f) \; \forall z \in Z$.

Da F Dini-Kegel ist, ist auch $F(Z)$ Dini-Kegel. μ läßt sich als Zustand auf $F(Z)$ auffassen, da es offensichtlich eine bijektive ordnungserhaltende lineare Ab= bildung von F nach $F(Z)$ gibt, die 1_X in 1_Z überführt. Nach [5, Satz 1] gibt es ein Wahrscheinlichkeitsmaß τ auf Z mit $\mu(\hat{f}) \leq \int_Z \hat{f}\, d\tau \quad \forall \hat{f} \in F(Z)$.

Daraus erhalten wir sofort die gesuchte Zerlegung von μ indem wir setzen

$\lambda_n = \tau(Y_n \setminus \bigcup \{ Y_k \mid k < n \})$. ▮

Die Bedeutung des Zerlegungssatzes liegt darin, daß man bei Zuständen von Dini-Kegeln immer abzählbare Zerlegungen finden kann. Betrachtet man nur endliche Oberdeckungen von X, so ist für alle Kegel die Existenz der entsprechenden endlichen Zerlegungen eine Konsequenz des Satzes von Hahn-Banach (siehe [9] oder [3]). Es sollte noch erwähnt werden, daß M. Neumann kürzlich einen etwas anderen Beweis des Zerlegungssatzes angegeben hat [10].

<u>Satz 3:</u> *Es ist äquivalent:*

(i) F *ist ein Dini-Kegel*

(ii) *Der Kegel* $VF = \{ \max(f_1,\dots,f_n) \mid n \in N, f_1,\dots,f_n \in F \}$ *ist ein Dini-Kegel.*

Beweis: (ii) ⟶ (i) ist trivial.

(i) ⟶ (ii): Sei (g_n) eine beliebige fallende Folge in VF. Wir setzen

$\alpha = \inf_{n\in N} \sup_{x\in X} g_n(x)$, $\beta = \sup_{x\in X} \inf_{n\in N} g_n(x)$. Offensichtlich gilt $\beta \leq \alpha$.

Es muß also nur noch $\beta \geq \alpha$ gezeigt werden. Wir nehmen an, daß $\sup_{x\in X} g_n(x) \leq \alpha + \frac{1}{n}$ (wenn dies nicht der Fall ist, gehen wir zu einer Teilfolge über, und verschaffen uns durch eventuelle Wiederholung von Folgegliedern eine Folge mit der angenommenen Eigenschaft). Weiter nehmen wir einen maximalen Filter Φ auf X, der die Mengen $X_n = \{ x \in X \mid g_n(x) \geq \alpha - \frac{1}{n} \}$ enthält. Da die einzelnen g_n von der Form $g_n = \max(f_n^1, f_n^2, \dots, f_n^{k_n})$ mit $f_n^i \in F$ sind, gibt es wegen der Maximalität von Φ ganze Zahlen $\rho_n \leq k_n$, so daß die $Y_n = \{ x \in X \mid f_n^{\rho_n}(x) \geq \alpha - \frac{1}{n} \}$ Elemente von Φ sind. Wir untersuchen nun die durch $h_m = \sum_{n\in N} \frac{1}{n} (f_n^{\rho_n} - \alpha - \frac{1}{n})$ definierte fallende Folge in F und wählen Elemente y_m aus den nichtleeren Mengen $\bigcap \{ Y_i \mid i \leq m \}$.

Da F Dini-Kegel ist, folgt die Existenz eines $x_o \in X$ mit:

$$\sum_{n\in N} \frac{1}{n} (f_n^{\rho_n}(x_o) - \alpha - \frac{1}{n}) \geq \inf_{n\in N} \sup_{x\in X} h_n(x) - 1 \geq$$

$$\geq \inf_{n\in N} h_n(y_n) - 1 \geq -1 - \sum_{n\in N} \frac{2}{n^2} > -\infty .$$

Kombiniert man dies mit $\inf_{n\in N} f_n^{\rho_n}(x_o) \leq \inf_{n\in N} g_n(x_o) \leq \beta$,

so erhält man für alle $\delta > o$, daß $(\beta + \delta - \alpha) \sum_{n\in N} \frac{1}{n} > -\infty$.

Da die harmonische Reihe divergiert, bedeutet dies $\beta \geq \alpha$. ▮

Der dritte Pfeiler für den Beweis unseres Hauptsatzes besteht aus dem folgenden:

Satz 4: *Es sei* E *ein Vektorverband (bezüglich punktweiser Maxima und Minima) von beschränkten reellen Funktionen auf* X *, der die konstanten Funktionen enthält, und es sei* μ *ein Zustand auf* E*. Dann ist äquivalent:*

(i) μ *hat ein* Σ_E *- Darstellungsmaß auf* X.

(ii) μ *hat die abzählbare Zerlegungseigenschaft.*

Beweis: Wir beweisen hier nur (ii)$\longrightarrow$(i), da wir die andere Richtung für den Beweis des Hauptsatzes nicht benötigen. Wir führen den Beweis als Anwendung des Satzes von P. Daniell und M.H. Stone (vergleiche etwa [2, S.160]). Es genügt deshalb zu zeigen, daß für jede punktweise absteigende Folge (f_n) in E mit $f = \inf_{n\in N} (f_n) \in E$ die Beziehung $\mu(f) = \inf_{n\in N} \mu(f_n)$ gültig ist.

Seien deshalb $\delta > o$ beliebig und $X_n = \{ x \mid f_n(x) \leq f(x) + \delta \}$. Dann ist $\{ X_n \mid n \in N \}$ eine Überdeckung von X , und es folgt aus der Zerlegungseigenschaft zusammen mit dem Hahn-Banach-Satz (z.B. [3, Theorem 3]) die Existenz von Zuständen μ_n und von Zahlen $\lambda_n \geq o$ mit $\sum_{n\in N} \lambda_n = 1$

und $\mu = \sum_{n\epsilon N} \lambda_n \mu_n$, so daß $\mu_n(g) \leq \sup_{x\epsilon X_n} g(x) \quad \forall\, g \epsilon E$.

Aus dieser Darstellung erhält man sehr einfach, daß $\mu(f) + \delta \geq \inf_{n\epsilon N} \mu(f_n)$.

Da $\delta > o$ beliebig war, und da sich $\mu(f) \leq \inf_{n\epsilon N} \mu(f_n)$ als Folge der

Monotonie ergibt, haben wir die gesuchte Gleichung bewiesen. ▮

Die Zusammenfügung dieser drei Bausteine ergibt den:

Beweis des Hauptsatzes: (i)⟶(ii) ist eine unmittelbare Folge des Lebesgueschen Satzes über Monotone Konvergenz (siehe etwa [5, erster Teil des Beweises von Satz 1]).

(ii)⟶(i): Es genügt zu zeigen, daß ein beliebiger Zustand ν ein Darstellungsmaß hat. Mit dem Satz von Hahn-Banach (etwa [3, Cor. 1.1]) verschaffen wir uns einen Zustand δ auf dem Dini-Kegel (Folge von Satz 3) VF, so daß $\delta(f) \geq \nu(f) \quad \forall\, f \epsilon F$. Mit dem Lemma von Zorn sichern wir uns die Existenz eines maximalen Zustandes $\bar{\mu}$ auf VF, der δ dominiert. $\bar{\mu}$ hat die abzählbare Zerlegungseigenschaft (Zerlegungssatz) und läßt sich *eindeutig* zu einem Zustand μ auf dem Vektorverband $E = VF - VF$ fortsetzen. Ist nun $\{ X_n \mid n \epsilon N \}$ irgendeine Überdeckung von X, dann läßt sich wegen der Zerlegungseigenschaft und des Satzes von Hahn-Banach $\bar{\mu}$ schreiben als

$$\bar{\mu} = \sum_{n\epsilon N} \lambda_n \bar{\mu}_n \text{ mit } \lambda_n \geq o, \quad \sum_{n\epsilon N} \lambda_n = 1 \text{ und } \bar{\mu}_n(f) \leq \sup_{x\epsilon X_n} f(x) \ \forall\, f \epsilon VF,$$

wobei die einzelnen $\bar{\mu}_n$ ebenfalls maximal sein müssen. Andererseits läßt sich aber $\bar{\mu}_n$ dominiert fortsetzen auf E zu einem μ_n mit

$$\mu_n(f) < \sup_{x\epsilon X_n} f(x) \quad \forall\, f \epsilon E.$$

Da aber $\bar{\mu}_n$ maximal ist, sind die Zustände $\bar{\mu}_n$ und μ_n auf VF gleich. Damit ist μ_n die eindeutige Fortsetzung von $\bar{\mu}_n$, und es gilt $\mu = \sum \lambda_n \mu_n$.

Also hat μ die abzählbare Zerlegungseigenschaft und nach Satz 4 ein Darstellungsmaß. ∎

III Beispiele und Anwendungen

1. *Ausdehnung des Meßraumes* (X, Σ_F) *im topologischen Fall.*

Seien X ein Hausdorffraum und F ein Dini-Kegel bestehend aus oberhalbstetigen Funktionen auf X .

Satz 5: *Zu jedem monotonen linearen* $\mu : F \to R$ *gibt es ein positives Maß* τ *auf der kleinsten* σ*-Algebra in* X , *die von* F *und den kompakten Teilmengen von* X *erzeugt wird, so daß*

$$\mu(f) \leq \int_X f \, d\tau \quad \forall f \in F .$$

Beweis: Sei $UC^+_\infty(X)$ die Menge der nichtnegativen oberhalbstetigen Funktionen auf X , die im Unendlichen verschwinden. Unschwer verifiziert man [6] , daß $\Phi = F + UC^+_\infty(X)$ ein Dini-Kegel ist. Nach Hahn-Banach kann μ monoton linear auf Φ fortgesetzt werden, und diese Fortsetzung hat gemäß unseres Hauptsatzes ein Σ_Φ - Darstellungsmaß τ . Da nun die charakteristischen Funktionen der kompakten Mengen in Φ liegen, sind sie Σ_Φ - meßbar. ∎

2. *Der Satz von Choquet - Bishop - de Leeuw.*

Seien Z kompakte konvexe Teilmenge eines lokalkonvexen Vektorraumes, Kon(Z) die Menge der stetigen konvexen Funktionen auf Z und ∂Z, die Extrempunkte von Z.

<u>Satz 6:</u> *Zu jedem monotonen linearen* $\mu : \mathrm{Kon}(Z) \to R$ *existiert ein positives Maß* τ *bezüglich der von* $\mathrm{Kon}(Z)$ *erzeugten* σ*-Algebra in* ∂Z*, so daß*

$$\mu(f) \leq \int_{\partial Z} f \, d\tau \quad \forall \; f \in \mathrm{Kon}(Z).$$

<u>Beweis:</u> Sei $F = \mathrm{Kon}(Z)_{/\partial Z}$. Auf F definieren wir durch

$$\delta(f) = \sup\{ \mu(g) \mid g \in \mathrm{Kon}(Z),\ g_{/\partial Z} = f \},\ p(f) = \mu(1_Z) \sup_{x \in \partial Z} f(x)$$

ein superlineares δ und ein sublineares p. Wegen des Maximumprinzips [1, S.46] gilt $p(f) \geq \delta(f) \ \forall \ f \in F$. Nach Hahn-Banach [3. Theorem 1] gibt es ein monotones lineares $\bar{\mu}$ mit $p(f) \geq \bar{\mu}(f) \geq \delta(f) \ \forall \ f \in F$. Sei nun (g_n) eine beliebige auf ∂Z punktweise fallende Folge in Kon(Z).

Dann ist $Y = \{ x \in Z \mid \inf_{n \in N} g_n(x) = \inf_{n \in N} \sup_{z \in Z} g_n(z) \}$ eine kompakte Seite von Z, die nach dem Satz von Dini nichtleer ist. Also enthält Y nach Krein-Milman einen Extrempunkt, und F muß ein Dini-Kegel sein. Damit hat $\bar{\mu}$ ein Darstellungsmaß τ. Dieses ist wegen $\bar{\mu}(g_{/\partial Z}) \geq \mu(g)$ auch Darstellungsmaß für μ. ∎

3. *Gewichtete Maße.*

Seien $\omega \geq 0$ eine Funktion auf X und F ein konvexer Kegel reeller Funktionen auf X, so daß alle Elemente von $\omega F = \{ \omega f \mid f \in F \}$ beschränkt sind. Eine einfache Anwendung des Hahn-Banach-Satzes liefert nun zusammen mit dem Hauptsatz:

<u>Satz 7</u> [6, Theorem 2] *Es ist äquivalent:*

(i) *Für jedes lineare* $\mu : F \to R$ *mit* $\mu(f) \leq \sup_{x \in X} \omega(x) f(x) \ \forall \ f \in F$ *existiert ein positives* $\Sigma_{\omega F}$ *- Maß auf* X *mit:*

$$\mu(f) \leq \int_X f\,\omega\, d\tau \qquad f \in F$$

(ii) $\omega F + R$ *ist Dini-Kegel*

4. *Pseudokompakte Räume*

Seien X ein vollständig-regulärer Raum und βX seine Stone - Čech - Kompaktifizierung. Wir erinnern daran, daß X pseudokompakt genannt wird, wenn jedes $f \in C(X)$ (stetige Funktionen auf X) sein Maximum auf X annimmt.

Satz 8: X *ist genau dann pseudokompakt, wenn* βX *die einzige* F_σ *- Teilmenge von* βX *ist, die* X *enthält.*

Beweis: Sei X pseudokompakt, und sei $Y \supset X$ eine σ-kompakte Teilmenge von βX. Da für $C(\beta X) = C(X)$ der Satz von Dini gilt, ist $C(X)$ ein Dini-Kegel. Also ist auch $C(\beta X)_{/Y}$ ein Dini-Kegel. Nach dem Hauptsatz hat dann jedes $z \in \beta X$ ein Darstellungsmaß τ_z auf Y, welches wegen der σ- Kompaktheit von Y ein Borelmaß auf βX sein muß. Da aber das Diracmaß δ_z das einzige z- darstellende Borelmaß ist, folgt $z \in Y$. Mithin $Y \supset \beta X$. Die andere Richtung des Beweises ist eine leichte Übung. ∎

IV Ein weiteres Ergebnis und Probleme

Verzichtet man auf die Positivität der Darstellungsmaße, so wird man schwächere Bedingungen an die Kegel erwarten, als sie die Dini-Kegel erfüllen. In [7] wurde dieses Problem behandelt und das folgende Ergebnis bewiesen.

Satz 9: *Sei F ein konvexer Kegel beschränkter reeller Funktionen auf X (nicht notwendig die Konstanten enthaltend). Dann ist äquivalent:*

(i) *Für jedes lineare* $\mu : F \to R$ *mit* $|\mu(f)| \leq \sup_{x \in X} |f(x)| \quad \forall f \in F$

existiert ein signiertes Σ_F *- Maß* τ *von Totalvariation* ≤ 1, *so daß*

$\mu(f) \leq \int_X f \, d\tau \quad \forall f \in F.$

(ii) *Für jede Folge* (α_n, f_n) *in* $R \times F$, *so daß die Folgen* $(\alpha_n + f_n)$ *und* $(\alpha_n - f_n)$ *punktweise fallen, gilt:*

$$\sup_{x \in X} \inf_{n \in N} (\alpha_n + |f_n(x)|) = \inf_{n \in N} \sup_{x \in X} (\alpha_n + |f_n(x)|).$$

Zum Schluß dieser Arbeit sollen noch einige offene Probleme angeführt werden. Dabei sind meiner Meinung nach die ersten zwei Probleme schwierig zu lösen.

Problem 1: Wenn F kein Dini-Kegel ist, so charakterisiere man diejenigen Zustände, die trotzdem Darstellungsmaße auf X besitzen.

Problem 2: Wenn F nicht die im Satz 9 (ii) geforderte Eigenschaft besitzt, so charakterisiere man diejenigen linearen μ, für welche Darstellungsmaße entsprechend Satz 9 (i) existieren.

Problem 3: Man charakterisiere diejenigen Kegel, für welche die nach Satz 9 existierenden Darstellungsmaße eindeutig sind.

Problem 4: Sei X topologischer Raum. Unter welchen Zusatzforderungen kann man die nach Satz 9 existierenden Maße auf eine σ-Algebra ausdehnen, welche alle kompakten Teilmengen von X enthält.

Literatur

1. E.M. Alfsen, Compact convex sets and boundary integrals (Springer Verlag) Berlin-Heidelberg-New York (1971)

2. H. Bauer, Wahrscheinlichkeitstheorie und Grundzüge der Maßtheorie (De Gruyter Verlag) Berlin (1968)

3. B. Fuchssteiner, Sandwich theorems and Lattice semigroups, J. Functional Analysis 16, 1-14 (1974)

4. B. Fuchssteiner, Lattices and Choquet's theorem , J. Functional Analysis 17, 377-387 (1974)

5. B. Fuchssteiner, Maße auf σ-kompakten Räumen, Math. Z. 142, 185-190 (1975)

6. B. Fuchssteiner, When does the Riesz representation theorem hold? preprint (1975)

7. B. Fuchssteiner, Signed representing measures, preprint (1975)

8. I. Glicksberg, The representation of functionals by integrals, Duke Math. J. 19, 253-261 (1952)

9. H. König, Sublineare Funktionale, Arch.Math. 23, 500-508 (1972)

10. M. Neumann, Varianten zum Konvergenzsatz von Simons und Anwendungen in der Choquettheory, preprint (1975) (erscheint in Arch.Math.)

11. C. Portenier, Caractérisation de certains espaces de Riesz, Séminaire Choquet: Initiation à l'analyse, 10e année, n° 6, 21p (1970/71).

MEASURES WITH VALUES IN PARTIALLY ORDERED SPACES: REGULARITY AND σ-ADDITIVITY

J. D. Maitland Wright
Mathematics Department,
University of Reading, England

PREAMBLE. The first part of this article is entirely expository and contains no proofs; the results given in the second part are believed to be new and proofs are provided.

PART I

Let us recall that a partially ordered vector space V is said to be monotone σ-complete if, whenever (a_n) $(n = 1, 2, \ldots)$ is an upper bounded, monotone increasing sequence in V, then it has a least upper bound $\bigvee_1^\infty a_n \in V$. If, for each upper bounded, upward directed family (a_λ) in V, there exists a least upper bound $\bigvee a_\lambda \in V$, then V is said to be monotone complete. In all that follows we shall suppose V to be monotone σ-complete and, for certain results, shall require that it satisfy the stronger condition of monotone completeness.

For simplicity, let us confine our attention to (finite) V-valued measures on σ-fields and compact spaces, ignoring generalizations to locally finite measures on σ-rings and locally compact spaces.

Definition: Let $(X, \mathcal{B})$ be a measurable space (i.e. $\mathcal{B}$ is a σ-field of subsets of X). A (finite) V-valued measure on $(X, \mathcal{B})$ is a map $m : \mathcal{B} \to V^+$ such that

(i) $m E \geq 0, \forall E \in \mathcal{B}$;

(ii) $m E + m F = m(E \cup F)$ when $E \cap F = \emptyset$ and $E \in \mathcal{B}$, $F \in \mathcal{B}$;

(iii) If (E_n) $(n = 1, 2, \ldots)$ is a monotone increasing sequence in $\mathcal{B}$ then

$$m \bigcup_1^\infty E_n = \bigvee_1^\infty m E_n .$$

(Observe that $(m E_n)$ $(n = 1, 2, \ldots)$ is an order bounded sequence, since mX is an upper bound.)

Given such a V-valued measure m, it is easy to construct the corresponding integral and to prove that an "order" version of the Monotone Convergence Theorem holds.

Call a topology $\mathcal{J}$ for V, σ-compatible if, whenever (a_n) $(n = 1, 2, \ldots)$ is an upper-bounded monotone increasing sequence in V, then (a_n) $(n = 1, 2, \ldots)$ converges in the $\mathcal{J}$-topology to $\bigvee_1^\infty a_n$. The basic reason that results and methods for V-valued measures differ from those for measures in topological vector spaces is the fact, pointed out by E. E. Floyd, that there need not exist any Hausdorff vector topology for V which is σ-compatible [4].

EXAMPLE. Let $B^{\infty}[0, 1]$ be the space of bounded (real) Borel functions on [0,1] and let $\mathcal{M}$ be the order ideal

$$\left\{f \in B^{\infty}[0, 1] : \text{the set } \{x \in [0, 1] : f(x) \neq 0\} \text{ is meagre}\right\}.$$

The quotient space $B^{\infty}[0, 1]_{/\mathcal{M}}$ is easily shown to be a monotone complete vector lattice. It follows from the work of E. E. Floyd that there is no T_1-vector topology for this space which is σ-compatible.

As already remarked, it is easy to construct an integral corresponding to a V-valued measure. A more delicate question is - does an analogue of the Riesz representation theorem hold?

The natural way to approach this question is to attempt to generalize the usual Daniell-Bourbaki extension method. Indeed, it follows from (different) results of McShane, Kantorovich and Matthes that, provided sufficiently strong additional conditions are imposed on V, then this can be done. But, in general, the Daniell-Bourbaki extension process fails and different methods are needed.

Theorem 1 [22]. Let X be a compact Hausdorff space; $\mathcal{B}$ the σ-field of Borel subsets of X; $\mathcal{B}_0$ the σ-field of Baire subsets of X. Let $\phi : C(X) \to V$ be a positive linear operator. Then there exists a unique V-valued Baire measure m, such that

$$\phi(f) = \int_X f \, dm \quad \text{for all} \quad f \in C(X) .$$

Further, if V is monotone complete then there exists a unique "quasi-regular" V-valued Borel measure q, such that

$$\phi(f) = \int_X f \, dq \quad \text{for all} \quad f \in C(X) .$$

Note: q is quasi-regular if, for each compact set K

$$qK = \bigwedge\{m\,O : K \subset O \;\&\; O \text{ open}\} .$$

The essential reason for the failure of the Daniell method for general V is that, when it works, it produces a regular Borel measure q (regular Baire measure m). But there exist quasi-regular Borel measures which are not regular.

e.g. Let V be the Dedekind complete vector lattice $B^{\infty}[0, 1]_{/\mathcal{M}}$. For each Borel set $E \subset [0, 1]$, let qE be $\chi_E + \mathcal{M}$. Then it is easy to see that q is quasi-regular. But q is not regular. For, if Q is the set of rationals in [0, 1] then, since Q is meagre, $qQ = 0$. For any open $O \supset Q$, $\overline{O} \supset \overline{Q} = [0, 1]$, so $qO = q\overline{O} = q[0, 1]$.

Hence $\bigwedge\{qO : O \supset Q \text{ and } O \text{ open}\} = q[0, 1] \neq qQ = 0$.

It is not hard to prove theorem 1, by using an appropriate device, but the only proofs known to me make implicit use of the Axiom of Choice by appealing to representation theorems. But it must be possible to find a genuinely constructive proof.

Problem. Give a constructive proof of Theorem 1.

Let X be a non-empty set; $\mathfrak{F}$ a field of subsets of X; and $\mathfrak{F}^{\infty}$ the σ-field generated by $\mathfrak{F}$.

Definition: A V-valued pre-measure on $(X, \mathfrak{F})$ is a function $m_o : \mathfrak{F} \to V$ such that

(i) $m_o F \geq 0$, $\forall F \in \mathfrak{F}$;

(ii) $m_o F_1 + m_o F_2 = m_o(F_1 \cup F_2) + m_o(F_1 \cap F_2)$; $\forall F_1 \forall F_2 \in \mathfrak{F}$;

(iii) If (F_n) $(n = 1, 2, \ldots)$ is a monotone decreasing sequence in $\mathfrak{F}$ with $\bigwedge_1^{\infty} F_n = \emptyset$ then $\bigwedge_{n=1}^{\infty} m_o F_n = 0$.

The classical Hopf-Kolmogorov extension theorem tells us that, when $V = \mathbb{R}$, each $\mathbb{R}$-valued pre-measure on $(X, \mathfrak{F})$ can be extended to a (σ-additive) measure on $(X, \mathfrak{F}^{\infty})$. It is natural to ask, if the Hopf-Kolmogorov extension theorem can be generalized to arbitrary V-valued pre-measures. The answer is - no. But it will work for some V.

Definition: A monotone σ-complete partially ordered vector space V is said to have the measure extension property if, for each $(X, \mathfrak{F})$, every V-valued pre-measure on $(X, \mathfrak{F})$ can be extended to a V-valued (σ-additive) measure on $(X, \mathfrak{F}^{\infty})$.

Definition: We say that V has the Baire regularity property if, and only if, each V-valued Baire measure on each compact Hausdorff space is regular.

These two properties are intimately related. In fact:

Theorem 2. (See [20,25].) V has the measure extension property $\Leftrightarrow$ V has the Baire regularity property.

Let us now specialize, by supposing that V is a vector lattice (= Riesz space).

Definition: A vector lattice V is said to be weakly σ-distributive if, whenever (F_n) $(n = 1, 2, \ldots)$ is a sequence of non-empty, countable, downward directed, subsets of V where, for each n, $\bigwedge F_n = 0$, then

$$\bigwedge \left\{ \bigvee_{n=1}^{\infty} \alpha(n) : \alpha \in \prod_1^{\infty} F_n \right\} = 0 .$$

The importance of this condition is that (i) it is an intrinsic condition and (ii), it follows by results of K. Matthes [16] that, whenever V is a weakly σ-distributive (monotone σ-complete) vector lattice then V has the measure extention property. (The most recent and most elegant proof of this result is due to Fremlin [5].)

Using different methods, I showed that V is weakly σ-distributive if, and only if, V has the Baire regularity property. This, together with Theorem 2, gives a converse to Matthes' result. So we have:

Theorem 3. [20] Let V be a Dedekind σ-complete vector lattice (Riesz space). Then

V has the measure extension property

<=> V has the Baire regularity property

<=> V is weakly σ-distributive.

Definition: When V is monotone complete, we say that V has the Borel regularity property if, and only if, each V-valued, quasi-regular Borel measure on each compact Hausdorff space is regular.

Definition: A Dedekind complete vector lattice V is said to be weakly (σ, ∞)-distributive if, whenever (F_n) $(n = 1, 2, \ldots)$ is a sequence of non-empty, downward directed, subsets of V where, for each n, $\wedge F_n = 0$, then

$$\wedge \left\{ \bigvee_{n=1}^{\infty} \alpha(n) : \alpha \in \prod_{1}^{\infty} F_n \right\} = 0 .$$

It follows from the work of Traczyk [18] that weak (σ, ∞)-distributivity is a strictly stronger condition than weak σ-distributivity.

Theorem 4. [21] A Dedekind complete vector lattice V has the Borel regularity property if, and only if, V is weakly (σ, ∞)-distributive.

PART II

Regularity and countable additivity

A classical theorem of A. D. Alexandroff, see Page 138 of [3], states that when $\mathcal{R}$ is a ring of subsets of a compact Hausdorff space X and $\mu : \mathcal{R} \to \mathbb{R}^+$ is a bounded regular measure then μ is σ-additive on $\mathcal{R}$. It turns out to be quite straightforward to obtain the equivalent theorem for finitely additive V-valued measures. But, since there exist countably additive, quasi-regular V-valued Borel measures which are not regular, the assumption of regularity seems excessively strong.

We shall see that, when V is a Dedekind complete vector lattice, if V is weakly (σ, ∞)-distributive then each finitely additive, quasi-regular V-valued Borel measure is countably additive. This result is, in a precise sense, best possible. For, if V is not weakly (σ, ∞)-distributive it can be proved that there always exists a finitely additive, V-valued quasi-regular Borel measure which is not σ-additive.

Henceforth $\mathcal{R}$ will be a (non-trivial) ring of subsets of a compact Hausdorff space X. Further, $m : \mathcal{R} \to V$ will be a finitely additive V-valued measure, that is,

(i) $m\emptyset = 0$ and $mA \geq 0$ for $A \in B$;

(ii) $m(A \cup B) + m(A \cap B) = mA + mB$ for all A and all B in $\mathcal{R}$;

The finitely additive measure m is said to be σ-additive on $\mathcal{R}$ if, whenever $\{A_n\}$ $(n = 1, 2, \ldots)$ is a sequence of pairwise disjoint sets in $\mathcal{R}$ and $\bigcup_1^\infty A_n \in \mathcal{R}$ then

$$m\left(\bigcup_1^\infty A_n\right) = \bigvee_{k=1}^{\infty} \sum_{r=1}^{k} m A_r .$$

A finitely additive measure m on $(X, \mathcal{R})$ is defined to be <u>regular</u> if, whenever $E \in \mathcal{R}$ there can be found a downward filtering family $\mathcal{G} \subset \mathcal{R}$ and an upward filtering family $\mathcal{H} \subset \mathcal{R}$ such that the following conditions are satisfied.

(i) $\bigwedge\{mG : G \in \mathcal{G}\} = mE = \bigvee\{mH : H \in \mathcal{H}\}$

(ii) For each $G \in \mathcal{G}$, the open interior of G contains E and, for each $H \in \mathcal{H}$, the closure of H is contained in E .

The measure m is defined to be <u>Baire regular</u> if, whenever $E \in \mathcal{R}$ there can be found a downward filtering family $\mathcal{G} \subset \mathcal{R}$ and an upward filtering family $\mathcal{H} \subset \mathcal{R}$ such that the following conditions are satisfied.

(i)' $\bigwedge\{mG : G \in \mathcal{G}\} = mE = \bigvee\{mH : H \in \mathcal{H}\}$

(ii)' For each $G \in \mathcal{G}$ and each $H \in \mathcal{H}$, there can be found an open Baire set U and a closed Baire set F such that

$$H \subset F \subset E \subset U \subset G .$$

Clearly, if m is Baire regular then m is regular.

<u>Theorem 5</u>. <u>Let</u> $\mathcal{R}$ <u>be a ring of subsets of a compact Hausdorff space</u> X . <u>Let</u> m <u>be a bounded, finitely additive</u> V-<u>valued measure on</u> $(X, \mathcal{R})$. <u>When</u> m <u>is regular then</u> m <u>is</u> σ-<u>additive on</u> $\mathcal{R}$.

Since m is a bounded measure, there is no loss of generality in supposing V equipped with an order-unit e . Hence V may be identified with $A(S)$, the space of affine continuous functions on the state space of V . Then, by applying [26] and adapting the proof on page 138 [3], this theorem is readily established.

<u>Corollary 6</u>. <u>Let</u> $\mathcal{R}$ <u>be the</u> σ-<u>algebra of Baire subsets of</u> X . <u>Let</u> V <u>be a weakly</u> σ-<u>distributive vector lattice. Then a finitely additive</u> W-<u>valued measure</u> m <u>on</u> $(X, \mathcal{R})$ <u>is</u> σ-<u>additive if, and only if,</u> m <u>is Baire regular</u>.

By Theorem T [20], whenever V is weakly σ-distributive and m is a σ-additive Baire measure on X then m is Baire regular. Conversely, if m is Baire regular then it is regular and so, by the above theorem, is σ-additive.

<u>Lemma 7</u>. <u>Let</u> V <u>be a boundedly complete vector lattice. Let</u> m <u>be a</u> V-<u>valued</u>

quasi-regular Borel measure on a compact Hausdorff space X. Then there exists a unique σ-additive quasi-regular Borel measure q on X such that $q\,U = m\,U$ for each open set U.

Let $m\,X = e$ so that m takes its value in the order-ideal $V[e]$. Since e is an order-unit for $V[e]$ we may equip $V[e]$ with the corresponding order-unit norm so that $V[e]$ then becomes a Banach space.

Call a function $s : X \to R$ very simple if there exists a finite collection of open sets $\{U_1, V_1, U_2, V_2, \ldots, U_n, V_n\}$ and real numbers $\{\alpha_1, \alpha_2, \ldots, \alpha_n\}$ such that

$$s = \sum_{i=1}^{n} \alpha_i \chi_{(U_i \setminus V_i)} .$$

We define $\phi(s)$, for such a very simple function, to be

$$\sum_{i=1}^{n} \alpha_i \big(m\,U_i - m(U_i \cap V_i)\big) .$$

Let $f : X \to R$ be the uniform limit of a sequence of very simple functions (s_n) $(n = 1, 2, \ldots)$. Then $(\phi(s_n))$ $(n = 1, 2, \ldots)$ is a Cauchy sequence in $V[e]$, with respect to the order-unit norm and hence converges in norm. Further it can be shown that if (s'_n) $(n = 1, 2, \ldots)$ is another sequence of very simple functions converging uniformly to f then $\lim \phi(s'_n) = \lim \phi(s_n)$. Accordingly we define $\phi(f)$ to be $\lim \phi(s_n)$. Let $\mathscr{S}$ be the space of all uniform limits of sequences of very simple functions on X. Then $\phi : \mathscr{S} \to V$ is a positive linear operator. It is not difficult to show that $C(X) \subset \mathscr{S}$.

By Theorem 1, Part I, there exists a unique quasi-regular σ-additive V-valued Borel measure q on X such that $\phi(f) = \int_X f\,dq$ whenever $f \in C(X)$.

Let U be an open subset of X and let F be any closed subset of U. Then there exists a continuous function $f : X \to \mathbb{R}$ such that $\chi_F \le f \le \chi_U$. Thus

$$m\,F = \phi(\chi_F) \le \phi(f) \le \phi(\chi_U) = m\,U .$$

So, by the quasi-regularity of m,

$$m\,U = \bigvee\{\phi(f) : f \in C(X) \text{ and } 0 \le f \le \chi_U\} .$$

Similarly,

$$q\,U = \bigvee\{\phi(f) : f \in C(X) \text{ and } 0 \le f \le \chi_U\} .$$

Thus $$q\,U = m\,U \text{ for each open set } U .$$

Suppose q' is any σ-additive quasi-regular Borel measure on X such that $q'\,U = m\,U$. Then $\phi(f) = \int_X f\,dq'$ whenever f is a very simple function and so, by an easy approximation argument, whenever $f \in C(X)$. It now follows from the uniqueness part of Theorem 1, Part I that $q = q'$.

Theorem 8. Let W be a weakly (σ, ∞)-distributive, boundedly complete vector lattice. Let m be a W-valued quasi-regular Borel measure on a compact Hausdorff space X. Then m is σ-additive and regular.

The proof of this theorem is independent of Theorem 5. By Lemma 7, there exists a unique σ-additive, quasi-regular, W-valued Borel measure q which coincides with m on the open subsets of X. Because W is weakly (σ, ∞)-distributive we have, by Corollary 3.4 [21], that q is regular. Hence, if E is any Borel subset of X we have

$$\bigvee\{q F : F \subset E \text{ and } F \text{ closed}\} = q E = \bigwedge\{q U : E \subset U \text{ and } U \text{ is open}\} .$$

So

$$\bigvee\{m F : F \subset E \text{ and } F \text{ closed}\} = q E = \bigwedge\{m U : E \subset U \text{ and } U \text{ is open}\} .$$

But, whenever $F \subset E \subset U$, where F is closed and U is open, then $m F \le m E \le m U$ and so $m E = q E$. Since $m = q$, m is σ-additive and regular.

Corollary 9. Let μ be a quasi-regular (positive) real valued Borel measure on X. Then μ is σ-additive and regular.

The following theorem shows that Theorem 8 is 'best possible'.

Theorem 10. Let W be a boundedly complete vector lattice which is not weakly (σ, ∞)-distributive. Then there exists a compact Hausdorff space X and a quasi-regular W-valued Borel measure m on X such that m is not σ-additive.

Because W is not weakly (σ, ∞)-distributive, it follows from Corollary 3.4 [21] that there exists a compact Hausdorff space X and a σ-additive quasi-regular Borel measure q on X such that q is not regular.

Let $B^\infty(X)$ be the space of all bounded real valued Borel functions on X. Let L be the subspace of $B^\infty(X)$ consisting of all functions of the form $f^* - g^*$, where f^* and g^* are bounded and lower semicontinuous.

Let $p : B^\infty(X) \to W$ be defined by

$$p(h) = \bigwedge\{\textstyle\int_X f^* dq : f^* \ge h \text{ where } f^* \text{ is lower semicontinuous}\} .$$

Then p is a W-valued sublinear functional. Let $\psi : L \to W$ be defined by

$$\psi(h) = \textstyle\int_X h dq .$$

Then ψ is a W-valued linear operator which is dominated by p.

So, (see the proof of Theorem 4.5 [23]) when h_* is a bounded upper semicontinuous function on X then

$$\psi(h_*) = \bigwedge\{\psi(f) : f \in C(X)\} = p(h_*) .$$

Thus, whenever f^* and g^* are bounded and lower semicontinuous, we have

$$\psi(f^*-g^*) \le p(f^*-g^*) \le p(f^*) + p(-g^*) = \psi(f^*) + \psi(-g^*) = \psi(f^*-g^*) .$$

So $\psi(h) = p(h)$ for all $h \in L$.

But p is not linear on the whole of $B^\infty(X)$ For, if it were, then for each Borel set E we would have

$$-p(-\chi_E) \leq q\,E \leq p(\chi_E) = -p(-\chi_E) .$$

This would imply

$$q\,E = \bigwedge\{q\,O : E \subset O \text{ and } O \text{ open}\} ,$$

so that q would be regular, which is false.

Because W is a boundedly complete vector lattice a generalization of the Hahn-Banach Theorem is valid for p-dominated W-valued linear operators, see Proposition 2.1 of Chapter 2 [17] (we do not need the more powerful and delicate results of [19]). Also, exactly as for real valued functionals, since p is not linear on $B^\infty(X)$ but coincides with ψ on L, there will exist more than one p-dominated linear extension of ψ to $B^\infty(X)$. So let $\Phi : B^\infty(X) \to W$ be a p-dominated extension where $\Phi(f) \neq \int_X f dq$ for some $f \in B^\infty(X)$.

For $h \leq 0$, we have $\Phi(h) \leq p(h) \leq 0$ so that Φ is a positive linear operator. Let $m\,E = \Phi(\chi_E)$ for each Borel set E. Then $m \neq q$ but, as $mA = q\,A$ whenever A is an open set or is a closed set, m is quasi-regular. By Lemma 7, q is the unique σ-additive quasi-regular Borel measure which coincides with m on the open subsets of X. Hence m is not σ-additive.

Corollary 11. *Let W be a boundedly complete vector lattice and let q be a σ-additive quasi-regular W-valued Borel measure on a compact Hausdorff space X. If q is not regular then there exists a W-valued quasi-regular Borel measure on X which is not σ-additive.*

We recall [7] that whenever F is a closed Baire subset of a compact Hausdorff space X then there exists a decreasing sequence (O_n) $(n = 1, 2, \ldots)$ of open Baire subsets of X such that $F = \bigcap_{n=1}^{\infty} O_n$. The following theorem can be proved in essentially the same way as Theorem 8.

Theorem 12. *Let V be a boundedly σ-complete vector lattice. Let m be a V-valued measure on the Baire subsets of a compact Hausdorff space X such that, whenever F is a closed Baire set with $F = \bigcap_{n=1}^{\infty} O_n$, where (O_n) $(n = 1, 2, \ldots)$ is a decreasing sequence of open Baire sets,*

$$m\,F = \bigwedge_1^\infty m\,O_n .$$

Then, if V is weakly σ-distributive, m is σ-additive.

Corollary 13. *Let μ be a (positive) real valued Baire measure on a compact*

Hausdorff space X. Then μ is σ-additive if, and only if, whenever (O_n) $(n = 1, 2, \ldots)$ is a decreasing sequence of open Baire sets whose intersection $\bigcap_1^\infty O_n$ is closed,

$$\mu \bigcap_1^\infty O_n = \lim_n \mu O_n .$$

Corollary 14. Let μ be a (positive) finitely additive real valued Borel measure on a compact Hausdorff space X. Then μ is regular and σ-additive if, and only if, whenever F is a closed subset of X then

$$\mu F = \inf\{\mu O : F \subset O \text{ and } O \text{ is open}\} .$$

There are a number of topics, involving measures with values in partially ordered spaces, which have not been touched on here. For example, Radon-Nikodym theorems, which are considered in [9, 10, 6, 24]. In [2], product measures with values in non-commutative partially ordered algebras are applied to quantum theory and, in [25], the measure theoretic aspects of [2] are further extended.

References

1. E. Alfsen, Compact convex sets and boundary integrals (Springer, 1971).

2. E. B. Davies and J. T. Lewis, 'An operational approach to quantum probability', Commun. Math. Phys. 17 (1970), 239-260.

3. N. Dunford and J. T. Schwartz, Linear operators I (Interscience, 1967).

4. E. E. Floyd, 'Boolean algebras with pathological order properties', Pacific J. Math. 5 (1955), 687-689.

5. D. H. Fremlin, 'A direct proof of the Matthes-Wright integral extension theorem', J. London Math. Soc. 11 (1975), 276-284.

6. W. Hackenbroch, 'Zum Radon-Nikodymschen Satz für positive Vektormaße', Math. Ann. 206 (1973), 63-65.

7. P. R. Halmos, Measure theory (Van Nostrand, 1950).

8. L. V. Kantorovich, B. Z. Vulich and A. G. Pinsker, Functional analysis in partially ordered spaces (Gostekhizdat, 1950) (Russian).

9. D. Maharam, 'The representation of abstract integrals', Trans. Amer. Math. Soc. 75 (1953), 154-184.

10. ———, 'On kernel representations of linear operators', Trans. Amer. Math. Soc. 79 (1955), 229-255.

11. E. J. McShane, 'Order-preserving maps and integration processes', Annals, Math. Studies, 31 (1953)

12. K. Matthes, 'Über eine Schar von Regularitätsbedingungen', Math. Nachr. 22 (1960), 93-128.

13. K. Matthes, 'Über eine Schar von Regularitätsbedingungen II', Math. Nachr. 23 (1961), 149-159.

14. ———, 'Über die Ausdehnung von $\aleph$-Homomorphismen Boolescher Algebren', Z. Math. Logik u. Grundl. Math. 6 (1960), 97-105.

15. ———, 'Über die Ausdehnung von $\aleph$-Homomorphismen Boolescher Algebren, II' Z. Math. Logik u. Grundl. Math. 7 (1961), 16-19.

16. ———, 'Über die Ausdehnung positiver linearer Abbildungen', Math. Nachr. 23 (1961), 223-257.

17. A. L. Peressini, Ordered topological vector spaces (Harper and Row, 1967).

18. T. Traczyk, 'Minimal extensions of weakly distributive Boolean algebras', Coll. Math. 11 (1963), 17-24.

19. G. F. Vincent-Smith, 'The Hahn-Banach theorem for modules', Proc. London Math. Soc. 17 (1967), 72-90.

20. J. D. M. Wright, 'The measure extension problem for vector lattices', Annales Inst. Fourier (Grenoble) 21 (1971), Fasc. 4, 65-85.

21. ———, 'An algebraic characterization of vector lattices with the Borel regularity property', J. London Math. Soc. 7 (1973), 277-285.

22. ———, 'Measures with values in a partially ordered vector space', Proc. London Math. Soc. 25 (1972), 675-688.

23. ———, 'Stone-algebra-valued measures and integrals', Proc. London Math. Soc. 19 (1969), 107-122.

24. ———, 'A Decomposition Theorem', Math. Z. 115 (1970), 387-391.

25. ———, 'Products of positive vector measures', Quart. J. Math. Oxford 24 (1973), 189-206.

26. ———, 'On approximating concave functions by convex functions', Bull. London Math. Soc. 5 (1973), 221-222.

MEASURES WITH VALUES IN NON-LOCALLY CONVEX SPACES

Klaus Bichteler
The University of Texas at Austin

Measures with values in non-locally compact topological vector spaces have been studied by various authors; a thorough treatment can be found in THOMAS [12], who considers continuous linear maps

$$m: C(K) \to F ,$$

where K is compact and F an arbitrary topological vector space. Such m is called a Radon measure, or *extendable*, if there is an extension to a larger class of functions on which the dominated convergence theorem is satisfied. Applications of this theory seem to be sparse as yet.

There is an instance of great interest of a vector measure with values in a non-locally convex space, the stochastic integral. Let $(\Omega, (\mathcal{F}_t : t \in \mathbb{R}_+), P)$ be a probability space with an increasing and right-continuous family of σ-algebras. On the set $X = \Omega \times \mathbb{R}_+$ let $\mathcal{C}$ denote the ring of sets

$$\mathcal{C} = \bigcup_{i=1}^{n} (S_i, T_i]$$

that are finite disjoint unions of increasing elementary stochastic intervals $(S_i, T_i]$, where S_i, T_i are bounded stopping times taking only finitely many values. Every adapted process Z in L^p $(0 \leq p \leq \infty)$ defines an additive set function

$$m_Z = dZ : \mathcal{C} \to L^p \quad \text{by} \quad m_Z(C) = \sum_{i=1}^{n} (Z_{T_i} - Z_{S_i}) .$$

This can be extended to the vector lattice $\mathcal{R}$ of step functions over $\mathcal{C}$ by linearity, and the question arises when m_Z is extendable. One is also interested in stochastic measures with values in the Lorentz spaces L^{pq} and in Orlicz spaces. For instance, if Z is a right continuous martingale in L^1 then m_Z is extendable as a measure into L^p $(o \leq p < 1)$, but not, in general, into L^1. Concerning stochastic differential equations, it is sufficient from the engineers point of view to consider existence of solutions, convergence etc. in L^o.

Note that the underlying set X has no topological structure, so THOMAS' theory is not applicable directly. I shall now outline how it can be generalized so as to apply to this and other examples, which I will indicate, but for lack of time only in the most superficial way.

The data one is led to consider are the following. X is a set, and $\mathcal{R}$ is a vector lattice of bounded real-valued functions on X that separates points and is closed under the STONE operation $\phi \to \phi \wedge 1$. In all applications $\mathcal{R}$ is also *dominated*;

i.e., for every $\phi \in R$ there is a $\psi \in R$ with $\phi = 0$ on $[|\psi| < 1]$. We assume this henceforth. Given further is a linear map

$$m: R \to F \ ,$$

where F is a Hausdorff topological vector space with a gauge D of translation-invariant pseudometrics d. If F is locally convex we assume D to consist of seminorms.

The problem is to determine under which conditions on m there is an extension satisfying the dominated convergence theorem

$$\int \cdot \, dm : L \to \tilde{F} \ ,$$

from a vector space L containing R into the completion of F. Necessary for the extendability of m are evidently these three conditions:

(E_1) The map m is continuous on R.

(The subspaces $R[\psi] = \{\phi \in R : \phi = 0 \text{ on } [\psi < 1]\}$ exhaust R as $\psi \in R_+$ increases and continuity is understood in the sense of the inductive limit topology of the sup-normed spaces $R[\psi]$, the SCHWARTZ topology of R.)

(E_2) If the sequence (ϕ_n) of R_+ decreases pointwise to zero then $m(\phi_n)$ converges to zero in F.

(E_3) For every disjoint sequence (ϕ_n) in R_+ whose sum is majorized by an element $\psi \in R_+$, $\lim m(\phi_n) = 0$ in F.

Given (E_1), (E_3) is automatically satisfied when F is a *C-space*; i.e. every sequence (ξ_n) in F whose finite partial sums form a bounded subset of F converges to zero. Merely take $\xi_n = m(\phi_n)$. This provides the most useful criterion for (E_3) to hold. Concerning stochastic integrals, (E_3) hardly ever poses a problem since the spaces L^p $(0 \leq p < \infty)$, most of the Lorentz spaces, and the Orlicz spaces defined by a YOUNG function satisfying the Δ_2-condition are C-spaces. They are, in fact *Σ-complete* in the sense that sequences (ξ_n) as above are summable (see below). (E_1) is the difficult condition in this case.

The conditions (E_{1-3}) are also sufficient for the extendability of m. The easiest way to see this is the DANIELL construction. For every $d \in D$ and function $h \geq 0$ that is the supremum of a sequence in R ($h \in R^S_\uparrow$) set

(1) $$G_d(h) = \sup \{d(m(\phi)) : |\phi| \leq h\} \ ,$$

and then set

(2) $$G_d(f) = \inf \{G_d(h) : |f| \leq h \in R^S_\uparrow\}$$

for arbitrary numerical functions f on X. The following properties of G_d are easy consequences of ($E_{1,2}$):

(M) $$|f| \leq |g| \text{ implies } 0 \leq G_d(f) \leq G_d(g) \leq \infty$$

(CSA) $G_d(\sum_{n=1}^{\infty} f_n) \leq \sum_{n=1}^{\infty} G_d(f_n)$

(F) $G_d(\lambda\phi) \to 0$ as $\lambda \to 0$, for all $\phi \in R$.

We call a numerical function f finite for the collection G of function metrics G_d, $d \in D$, if $G_d(\lambda f) \to 0$ as $\lambda \to 0$, for all $d \in D$. Their collection is $F(G)$. If G is countable then $F(G)$ is complete and so is the adherence $L^1(R,G) = L^1$ of R in $F(G)$, the (R,G)-*integrable* functions. It is natural to speak of convergence in *G-mean* on $F(G)$ and L^1. The measure m is continuous in G-mean on R and so has an extension by continuity, $\int \cdot dm: L^1 \to F$.

The dominated convergence theorem follows in a number of steps from this obvious consequence of (E_3):

(WUG) A disjoint sequence of R_+ whose finite partial sums form an order bounded set in R_+ converges to zero in G-mean.

If F is a C-space then (WUG) can be replaced, evidently, by the following stronger statement:

(UG) A disjoint sequence of R_+ whose finite partial sums form a G-mean bounded set converges to zero in G-mean.

Study of [weak] upper gauges. I have found it convenient to call collections G of function metrics G satisfying (M), (CSA), (F), and (WUG) *weak upper gauges* and *upper gauges* if they satisfy (UG), and to study their theory independently from their provenience (here a vector measure). This because they appear in different contexts, e.g. the theory of topological vector lattices (see below) and in interpolation theory.

On the way to a proof that the dominated convergence theorem holds in $L^1(R,G)$ when G is a weak upper gauge, let us show sketchily that (WUG) and (UG) remain true with the word 'disjoint' eliminated. We do this for (WUG) only - note that in the case of a vector measure into a C-space the stronger version of (UG) is automatic (given (E_1)). The first step is to reduce the siuation to the case when the underlying set X is locally compact, so we have the help of the theorems of Urysohn, Tietze, and Stone-Weierstrass:

We equip X with the initial uniformity for the bounded functions $\phi \in R$ and note that it is precompact. It is called the *R-uniformity* on X. Deleting the points of its completion at which the extensions by uniform continuity of the functions of R all vanish, we are left with a locally compact space $\hat{X}$, the *spectrum* of R, which contains X densely. Every function $\phi \in R$ has a unique extension $\hat{\phi} \in C^{oo}(\hat{X})$ (vanishing off a compact set since R is dominated), and by the Stone-Weierstrass theorem the vector lattice $\hat{R}$ is dense in $C^{oo}(\hat{X})$ in its SCHWARTZ topology. We define $g(\hat{\phi}) = G(\phi)$ for $G \in G$, $\phi \in R$, and extend g to all numerical functions on $\hat{X}$ by the procedure (1),(2), with d∘m replaced by g and allowing for h any lower semicontinuous function on $\hat{X}$. We arrive at function metrics $\hat{G}$ on $\hat{X}$, that are weak upper gauges for $C^{oo}(\hat{X})$. It suffices evidently to show that their collection $\hat{G}$ satisfies (WUG) with the word 'disjoint' deleted.

Let $U \subset \hat{X}$ be a relatively compact open set and $\hat{G} \in \hat{G}$. Then

$$\inf \{G(U - K): K \text{ compact} \subset U\} = 0 \ ^{1)}.$$

Indeed, if this is not so one constructs in an obvious way a disjoint sequence $(\hat{\phi}_n)$ in $\hat{R}$ underneath U and such that $\hat{G}(\hat{\phi}_n)$ is bounded away from zero, contradicting (WUG). Using Urysohns lemma and the density of $\hat{R}$ in $C^{oo}(\hat{X})$, one sees that every relatively compact open set is $\hat{G}$-integrable. A bounded lower semicontinuous function of compact support is the uniform limit of linear combinations of such sets [1] and therefore is integrable as well. Let then $(\hat{\phi}_n)$ be a sequence in $\hat{R}_+$ whose sum h is majorized by an element $\hat{\psi}$ of $\hat{R}_+$. From the above, h is integrable. Given an $\varepsilon > 0$, one can find a compact subset $K \subset [\psi > 0]$ with $G([\psi > 0] - K) < \varepsilon$ such that $h|K$ is uniformly continuous. From Dini's theorem, $\sum\hat{\phi}_n$ converges uniformly on K to h, and the triangle inequality gives $\hat{G}(\hat{\phi}_n) \to 0$. So the stronger versions of (WUG) and (UG) hold for $(\hat{R},\hat{G})$ and (R,G).

By approximation one shows now that they hold for (L^1,G), whence the Monotone Convergence Theorem: $L^1 \ni f_n \uparrow f$ implies $f_n \to f$ in G-mean provided (a) f is majorized by an integrable function or (b) G is an upper gauge and $f \in F(G)$. Statement (b) characterizes the upper gauges. The Dominated Convergence Theorem follows routinely (Cf. BOURBAKI [4]).

Measurability. A family F of (R,G)-integrable sets is termed (R,G)-*dense* if for every $G \in G$, integrable set A, and $\varepsilon > 0$, there exists a set $K \subset A$ in F with $G(A - K) < \varepsilon$. A function f with values in a uniform space is called (R,G)-*measurable* if the family $U(f)$ of integrable sets on which f is uniformly continuous (in the R-uniformity on X) is dense.

If $f_n: X \to F_n$ $(n=1,2,\ldots)$ are measurable then $\bigcap U(f_n)$ is clearly also dense and so $\Pi f_n: X \to \Pi F_n$ is measurable. Let $\Phi : \Pi F_n \to F$ be a map and assume either it is uniformly continuous or it is continuous and the F_n are complete. Then $\Phi(f_1,f_2,\ldots)$ is measurable. This is clear in the first case, and in the second case follows from the observation that $\Pi f_n(K)$ is precompact, hence relatively compact, for $K \in \bigcap U(f_n)$, and that Φ is uniformly continuous on this set. Applying this to functions Φ: $R\times R \to R$ such as $(a,b) \to a+b$, $(a,b) \to a\wedge b$, one sees that linear and order combinations of measurable functions are measurable. One has the Localization Principle: If f is measurable on all sets of a dense family [2] then f is measurable. Egoroff's theorem for sequences of measurable functions into a metric space is proved literally as in BOURBAKI [4]. The measurable sets therefore form a tribe $T = T(R,G)$, which can be shown to consist exactly of those sets that intersect integrable

[1] We identify a set with its characteristic function.
[2] I.e., $U(f)$ is dense in every set of that family.

sets in integrable sets. One has the Integrability Criterion: A measurable function f is integrable provided (a) $|f|$ is majorized by an integrable function or (b) G is an upper gauge and f vanishes outside a σ-finite set and belongs to $F(G)$. The second statement also characterizes the upper gauges among the weak upper gauges.

One has a Generalized Version of PETTIS' Measurability Criterion, which I have not seen before in ordinary measure theory: Let f be a function on X with values in a uniform space Y. Then f is measurable provided (a) the family $S(f)$ of integrable sets L such that $f(L)$ is contained in a Suslin subspace of Y is dense and (b_1) there is a collection H of continuous functions h on Y separating the points and such that $h \circ f$ is measurable or (b_2) $f^{-1}(B) \in T$ for all open (or, equivalently, Borel) subsets B of Y.

The equivalence of (b_1) with (b_2) is known [13]. Here is a sketch of proof of the Criterion. We may assume that Y is Suslin, by the localization principle. We define $G'(g) = G(g \circ f)$ for $g: Y \to \overline{R}$ and notice that G' is a weak upper gauge for the Q-vector lattice H generated by the functions $(-1) \vee h \wedge 1$, $h \in H$, which can be assumed to be countable. We do this for every $G \in G$. It can be shown by a highly technical argument (but very simply if the $G \in G$ are positive-homogeneous) that there is a weak upper gauge $G'^{\#}$ that coincides with G' on H and satisfies $f_n \uparrow f \Rightarrow G'^{\#}(f_n) \uparrow G'^{\#}(f)$. On the sets of Y, $G'^{\#}$ is a capacity. A maximal family K of mutually disjoint $G'^{\#}$-non-negligible compact (hence H-compact and $G'^{\#}$-integrable) subsets of Y is countable, $K = \{K_1, K_2, \dots\}$, and exhausts Y $G'^{\#}$-a.e. by the theorem of Choquet. The sets $L_n = f^{-1}(K_n)$ are integrable, either from (b_1) directly or since K_n is the infimum of a sequence in H. By the localization principle it suffices to show that f is measurable on each of them (a function measurable on two measurable sets is measurable on their union). To this end we find, given $\varepsilon > 0$, a set $L' \subset L_n$ in $\bigcap\{U(h \circ f) : h \in H\}$ with $G(L_n - L') < \varepsilon$ and remark that the given and the H-uniformity coincide on K_n.

Here are two amusing consequences. First, the measurability of a function into a Suslin space Y does not change if the uniformity on Y is replaced by another one having the same Borel sets. Secondly, a Generalization of Egoroff's Theorem: Let (f_n) be a sequence of measurable functions in a (uniform) Suslin space Y, converging a.e. to a function f. Then f is measurable. Indeed, the pseudometrics of the uniformity provide enough continuous functions h to separate the points, and for each of them $h \circ f = \lim h \circ f_n$ is measurable by the scalar version of Egoroff's theorem.

Control Measures. For simplicity assume that $G = \{G\}$ is a singleton. When $L^1(R,G)$ is locally convex then there exists, for every σ-finite set $A \subset X$, a positive and finite measure μ_A on the integrable subsets of A which has precisely the same negligible sets as G. This is an easy consequence of the Hahn-Banach theorem in conjunction with an exhaustion argument. If G is positive-homogeneous μ_A can be taken to be dominated by $G \cdot \mu_A$ is called a *control measure* for G on A. When L^1 is not locally convex, A splits essentially uniquely into a set A' on which G is controlled by a measure and a set A'' on which no measure is absolutely continuous with respect to G. The question whether $A'' = \emptyset$ is equivalent with an old open problem posed by D. MAHARAM-STONE [ ].

The upper gauges $G_o^A(f) = G(A(|f| \wedge 1))$, $A \subset X$ integrable, define the topology of *convergence in G-measure* on the G-measurable a.e. finite functions $L^{oo}(R,G)$, which coincides with the topology of convergence in μ_X-measure if X is σ-finite and G controllable. From Khintchine's inequality one concludes that in this case L^{oo} is Σ-complete [6]. Also, the topology of convergence in G-measure on integrable sets always has the Orlicz property on $L^1(R,G)$ (i.e., any sequence all of whose subsequences are summable in this topology to an element of $L^1(R,G)$ are summable in G-mean). This follows from a gliding hump argument. Putting these two bits together one sees that $L^1(R,G)$ is Σ-complete provided that (R,G) is a locally controllable upper gauge. Checking the Orlicz and Lorentz spaces for Σ-completeness, which is needed for stochastic integration, is done easiest using this result.

Suppose X is σ-finite and G positive homogeneous and let $\mu \leq G$ be a measure controlling G on X. The Radon-Nikodym theorem yields a representation of the dual of L^1: it consist of all measurable functions h with

$$G^\mu(h) := \sup \{\int fh d\mu : f \in L^1, G(f) \leq 1\} < \infty .$$

One may analyze the function metric G^μ; if it is an upper gauge, L^1 is reflexive and vice versa, etc.

Application to Topological Vector Lattices. It has been shown by MEYER-NIEBERG [9] that a Banach lattice in which disjoint order-bounded sequences converge to zero, and these Banach lattices have been the object of much study [1 ,7 ,10]. This study can be much simplified if one observes that such a Banach lattice is a space $L^1(R,G)$. To see this, represent it as a space L of continuous numerical functions on a locally compact space X and let R be the bounded functions with compact support in L. The norm on R satisfies (WUG), and one concludes easily that $L = L^1(R,G)$. The dual has again a nice representation by functions, which facilitates the duality theory of such Banach lattices. Furthermore, all this generalizes to arbitrary topological vector lattices in which disjoint order bounded sequences (or disjoint sequences with topologically bounded partial sums - case of upper gauges) tend to zero.

Application to linear maps on $R \otimes E$. Let E be a Banach space with norm and $R \otimes E$ the vector space of functions

$$\psi(x) = \sum_{i=1}^{n} \phi_i(x)\xi_i \qquad (x \in X,\ \xi_i \in E,\ \phi_i \in R) .$$

It is clear how the SCHWARTZ topology on $R \otimes E$ is to be defined. Consider a linear map

$$U : R \otimes E \to F .$$

In order that U have an extension satisfying the dominated convergence theorem, the conditions (E_{1-3}) are again necessary and sufficient. They read in this case

(E_1') U is continuous ;

(E_2') If $R_+ \ni \phi_n \downarrow 0$ and $\xi \in E$ then $U(\phi_n \xi) \to 0$ in F ;

(E_3') If (ψ_n) is a disjoint sequence in $R \otimes E$ with $\sum ||\psi_n|| < \phi \in R_+$ then $U(\psi_n) \to 0$ in F.

Given (E_1'), (E_3') is automatic when F is a C-space. The extension procedure is the same as above where E was R. One uses (1) and (2) to define weak upper gauges G_d - replacing m by U and $|\cdot|$ by $||\cdot||$ throughout; one defines $L^1_E(R,G)$ as the G-mean closure of $R\otimes E$ in $F_E(G) = \{f: X \to E \; ; \; ||f|| \in F(G)\}$ and extends U by continuity.

Let me note here that, given (E_1') and (E_3'), (E_2') can be weakened to the requirement that $U(\phi_n \xi) \to 0$ for some Hausdorff topology on F weaker than the given one: The BAUER-STONE transform $\hat{U} : \hat{\phi} \to U(\phi)$ is extendable since (E_2') holds for it by Dini's theorem; and if $d(U(\phi_n \xi)) > a > 0$ for some $d \in D$ and all n, then $\inf \hat{\phi}_n = \hat{k}$ has $\int \hat{k}\xi d\hat{U} = \int \hat{k}\xi d\hat{U}_\sigma = \int (\hat{k}|X)\xi dU_\sigma \neq 0$, which is impossible since $\hat{k}|X = 0$ (here U_σ denotes the extendable map U viewed as a map into F_σ).

In the case of stochastic integrals we take σ to be convergence in measure. A classical argument on each trajectory of the process Z shows that (given continuity (E_1) and Σ-completeness of the range space) m_Z is extendable iff Z has right-continuous paths.

One has the theorem of DIEUDONNE of Vitali-Hahn-Saks type for sequences (U_n) of extendable maps from $R\otimes E$ to F : If $U_\infty(h) = \lim U_n(h)$ exists for all functions $h: X \to E$ that are infinite sums of disjoint functions in $R\otimes E$ and whose modulus is majorized by a function in R, then this limit exists for all bounded measurable with R-majorized modulus, and U_∞ is extendable [3]. If F is locally convex and $E = R$ or C it suffices to check convergence on sets h [1]. Also, $\{U_1, U_2, .., U_\infty\}$ is *uniformly extendable* in the sense that there exist weak upper gauges H_d majorizing all the $d \circ U_n$, for each $d \in D$. For the proof one considers the corresponding linear map $V : R\otimes E \to c_F$ into the space of convergent sequences in F. It is extendable for the pointwise topology on c_F, which, roughly, is the topology of convergence in measure, and therefore has the Orlicz property, for the topology of convergence in mean of the weak upper gauge $(c^o, ||\cdot||)$.

Application to Summability. The linear map $U: R\otimes E \to F$ gives rise to a an operator valued measure $m: R \to L(E,F)$. Let us retain (E_1'), but instead of $(E_{2,3}')$ make the much weaker assumption that m is extendable for the strong operator topology on $L(E,F)$. We get, for each $d \in D$ and $\xi \in E$, a weak upper gauge $G_{d,\xi}$ such that $d(m(\phi)\xi) \leq G_{d,\xi}(\phi)$ $(\phi \in R)$. Let us assume henceforth in addition that m is *controllable* in the sense that there exists, for each $d \in D$, a weak upper gauge H_d

such that

(C) $$G_{d,\xi} \ll H_d \qquad \text{for all } \xi \in E .$$

This is trivially satisfied when m is extendable for the uniform operator topology or when E is separable. Set $H = \{H_d : d \in D\}$.

One can extend U and m by elementary methods to R^*_E and R^* , respectively, the closures of $R \otimes E$ and R under pointwise limits of R-dominated sequences.

For any function $f \in R^* \otimes E$ or a uniform R-dominated limit thereof,

$$\mu_f(T) = \int fTdU \in \tilde{F} \qquad (T \in T(R,H))$$

defines an extendable measure $\mu_f : T(R,H) \to \tilde{F}$, the *indefinite integral* of f . The integration theories of PETTIS [11], BARTLE [2], and DOBRAKOV [5] adress themselves to the question for which more general class of functions f μ_f can be defined reasonably, the criterion of reasonableness being that the Vitali-Hahn-Saks theorem on the interchange of pointwise limit and indefinite integrals hold. The question can be answered in a way generalizing and unifying the three theories mentioned above, using weak upper gauges through condition (C). Note here that $\int fhdm$ exists in $\tilde{F}$ for every function $h \in R^*$ vanishing off a set K in $U(f)$. For on the precompact set K fh is the uniform dominated limit of functions in $R \otimes E$. The main points of the summability theories mentioned above are contained in the following theorem:

Theorem. Let $f: X \to E$ be an (R,H)-measurable function. Equivalent are:

1) There is an extendable measure $\mu_f: L^\infty(R,H) \to \tilde{F}$ such that $\int Kd\mu_f = \int Kfdm$ for $K \in U(f)$ [1].

2) $\int \psi_n fdU \to 0$ for every sequence (ψ_n) in R^* whose carriers lie in mutually disjoint sets of $U(f)$.

3) The sequence $\int \psi_n fdU$ above is subsequence summable in $\tilde{F}$ (When F is locally convex the ψ_n can be taken to be sets[1]).

4) There is net (f_α) in $R^* \otimes E$ converging in measure on the sets in $U(f)$ to f and such that $\{\mu_{f_\alpha}\}$ is uniformly extendable.

5) There is a net (f_α) of functions satisfying either of 1)-4), converging in measure on the sets in $U(f)$ to f , and such that $\{\mu_{f_\alpha}\}$ is uniformly extendable. Then $\mu_{f_\alpha} \to \mu_f$ on $L^\infty(R,H)$.

6) f is the limit of a net in $R^* \otimes E$ under the pseudometrics

$$g \to M_d(g) := \sup \{d(\int hd\mu_{Kf}): h \in R^*, |h| \le 1, K \in U(g)\} \qquad (d \in D).$$

5) is the desired theorem on interchange of limit and indefinite integral. The proof is a direct consequence of DIEUDONNE's theorem above, the f_α of 4) being functions Kf , $K \in U(f)$. The functions f described are called m-*summable*, their collection is $M^1_E(m)$. It is easy to see that M^1_E is complete metrizable with $R \otimes E$ dense, under the *natural topology* of convergence both in measure and under the pseudometrics M_d , provided F is metrizable.

The Pettis integral emerges as the special case when $E = F$ and $m(\phi)$ is a multiple of the identity operator for each $\phi \in R$. (C) is satisfied with $H_d = m^*$. If $E = F$ is not Banach, one uses the theorem on each of the seminormed spaces $(E,d) = (F,d)$, $d \in D$, separately.

The integrals of BARTLE and DOBRAKOV emerge when one notices that their summable functions take values in separable subspaces, whence condition (C) in this case.

References

[1] ANDO,T. Banachverbaende und positive Projektionen. Math. Zeit.109(1969)121-130.

[2] BARTLE, R.G. A general bilinear vector integral. Studia Math. 15(1956)337-352.

[3] BICHTELER,K. On sequences of measures. Bull.Amer.Math.Soc. 80/5(1974)839-844.

[4] BOURBAKI,N. Integration. Hermann, Paris, 1965.

[5] DOBRAKOV,I. On integration in Banach spaces I&II. Czech.Math.J. 20(95)(1970).

[6] KWAPIEN,S. Complement au theoreme de Sazonov-Minlos. C.R.Ac.Sci.Paris 267(1968).

[7] LUXEMBURG,W. & ZAANEN,A. Notes on Banach function spaces. Ind.Math 25-27(1963-5).

[8] MAHARAM,D. An algebraic characterization of measure algebras. Ann.Math 48(1947).

[9] MEYER-NIEBERG,P. Charakterisierung einiger topologischer und ordnungstheoretischer Eigenschaften von Banachverbaenden mit Hilfe disjunkter Folgen. Arch. der Math. 24(1973)640-647.

[10] OGASAWARA, T. Vector Lattices. Tokyo, 1948. In Japanese.

[11] PETTIS,J. On integration in vector spaces. Trans.Amer.Math.Soc. 44(1938)277-304.

[12] THOMAS,E. On Radon maps with values in arbitrary topological vector spaces, and their integral extensions. Dept.Math.Yale U.,1972.

[13] ____ The Lebesgue-Nikodym theorem for vector valued Radon measures. Memoir AMS.

SUMMEN HALBADDITIVER INTEGRALNORMEN VOM LEBESGUE-TYP

Dieter Hoffmann und Hans Weber

Universität Konstanz

Einleitung

In [3] betrachten W.M.BOGDANOWICZ und J.N.WELCH zwei klassische (d.h.: $[0,\infty)$-wertige) Maße ν_1 und ν_2, beweisen naheliegende Beziehungen zwischen den zu $\nu := \nu_1 + \nu_2$ und den zu den ν_i gebildeten integrationstheoretischen Größen und erhalten als Anwendung eine wesentliche Beweisvereinfachung eines Satzes über Spektralintegrale (vgl. [2],Theorem 2). Allgemeiner erhalten die genannten Autoren in [4] entsprechende Beziehungen für unendliche Summen klassischer Maße. N.DINCULEANU zeigt in [5] Ähnliches auch für vektorwertige Maße endlicher Totalvariation auf lokalkompakten Räumen.

Die Tatsache, daß sich alle Resultate aus [3], die dort einige Mühe bereiten, in einfachster Weise aus der sofort zu verifizierenden Integralnormbeziehung $\| \|^{(\nu)} = \| \|^{(\nu_1)} + \| \|^{(\nu_2)}$ ablesen lassen, war ein Ausgangspunkt unserer Überlegungen. Wir machen in der vorliegenden Note in einem sehr allgemeinen Rahmen entsprechende Aussagen über Summen starker, auf einem Elementarbereich halbadditiver endlicher Integralnormen vom LEBESGUEschen Typ und die zugehörige "integrablen" Funktionen.

Dabei sind diese Integralnormen aus drei Gründen von vorrangiger Bedeutung: 1. Für eine Integralnorm sind die Eigenschaften "stark" und "halbadditiv" *charakteristisch für Vollständigkeit und Gültigkeit von Konvergenzsätzen* (LEBESGUE, LEVI, FATOU) (vgl. dazu [8]). 2. Für operatorwertige Maße, bei denen die zugehörige LEBESGUEsche Integralnorm auf den einfachen Funktionen endlich und halbadditiv ist, erhält man die wesentlichen für klassische Maße vertrauten Ergebnisse (vgl. [6] und auch Kapitel 1.). 3. Starke Integralnormen, die auf dem Ausgangsbereich endlich und *halbadditiv* aber *nicht additiv* sind, treten zum Beispiel bei Spektralintegralen auf (vgl. dazu [7]).

Die Arbeit gliedert sich in 5 Abschnitte. In 0. stellen wir Bezeichnungen und Vorbemerkungen voran, die im Wesentlichen Ergebnisse von F.W.SCHÄFKE ([8]) referieren. Im 1. Kapitel, das wir nur für Korollare und Ergänzungen benötigen, zeigen wir für (auch gruppenwertige) meßbare Funktionen insbesondere die Abgeschlossenheit gegenüber fast-überall-Konvergenz und die Beschreibung integrierbarer Funktionen durch meßbare endlicher Integralnorm. Dabei bringen die Beweise selbst für bekannte Spezialfälle (vgl. [6] und z.B. [1]) noch wesentliche Vereinfachungen gegenüber den uns bekannten Beweisen. In 2. zeigen wir, daß für eine Menge halbadditiver Integralnormen $\|\ \|_t$ $(t \in T)$ auch $\sum_{t\in T} \|\ \|_t$ halbadditiv ist und daß sich eine Integralnormbeziehung $\|\ \| = \sum_{t\in T} \|\ \|_t$ bei LEBESGUEscher Erweiterung von dem Ausgangsbereich auf alle Funktionen überträgt. Daraus folgen dann Beziehungen über gruppenwertige integrierbare Funktionen und Integrale in 3.. Kapitel 4. gibt noch Spezialfälle und Ergänzungen. Für den in [4] behandelten Fall geben unsere Überlegungen mit dann offensichtlichen Vereinfachungen einen einfachen und durchsichtigen Beweis der dortigen Ergebnisse.

In den Bezeichnungen schließen wir an [7], [8] und [9] an.

0. Vorbemerkungen

Während der gesamten Arbeit seien

$\mathfrak{R}$ eine nicht-leere Menge und $\mathfrak{L}'$ ein SV-System in
$\mathfrak{P}(\mathfrak{R}) := \{\psi \mid \psi : \mathfrak{R} \longrightarrow [0,\infty]\}$.

(d.h.: $\emptyset \neq \mathfrak{L}' \subset \mathfrak{P}(\mathfrak{R})$ und $\mathfrak{L}'$ abgeschlossen bezüglich $+$, Δ, $\wedge$ (inf) (und damit bezüglich $\vee$ (sup)), wobei $+$, Δ, $\wedge$ und $\vee$ auf $\mathfrak{P}(\mathfrak{R})$ punktweise erklärt sind und für $(\alpha,\beta) \in [0,\infty)^2$ $\alpha\Delta\beta := |\alpha-\beta|$, $\alpha\Delta\infty := \infty\Delta\alpha := \infty$ und $\infty\Delta\infty := 0$ bedeutet.)

Ist $0 \in \mathfrak{M} \subset \mathfrak{P}(\mathfrak{R})$, dann heißt eine Abbildung $\|\ \| : \mathfrak{M} \longrightarrow [0,\infty]$ *Integralnorm*, wenn $\|0\| = 0$ und $\phi \le \phi_1 + \phi_2 \curvearrowright \|\phi\| \le \|\phi_1\| + \|\phi_2\|$, *starke Integralnorm*, wenn $\|0\| = 0$ und $\phi \le \sum_{\nu=1}^{\infty} \phi_\nu \curvearrowright \|\phi\| \le \sum_{\nu=1}^{\infty} \|\phi_\nu\|$ und *halbadditiv*, wenn $\sup_n \|\sum_{\nu=1}^{n} \phi_\nu\| < \infty \curvearrowright \|\phi_\kappa\| \longrightarrow 0$ gilt.

Für eine Integralnorm $\|\ \| : \mathfrak{P}(\mathfrak{R}) \longrightarrow [0,\infty]$ ist $\{\{(\phi,\psi) \in \mathfrak{P}(\mathfrak{R})^2 : \|\phi\,\Delta\psi\| < \varepsilon\} : \varepsilon \in (0,\infty)\}$ eine symmetrische Nachbarschaftsbasis einer Uniformität - der $\|\ \|$-Uniformität - auf $\mathfrak{P}(\mathfrak{R})$, bezüglich der die Operationen $+$, Δ, $\wedge$, $\vee$ gleichmäßig stetig sind. Eine Menge $\mathfrak{N} \subset \mathfrak{R}$ heißt $\|\ \|$-*Nullmenge*, wenn $\|\chi_{\mathfrak{N}}\| = 0$ ist; wir benutzen diesbezüglich wie üblich die Abkürzung $\|\ \|$- *f.ü.* (fast-überall).

In 0.1 bis 0.3 (vgl. dazu [8]) stellen wir für das Folgende wichtige Hilfsmittel zusammen:

<u>Hilfssatz 0.1</u> *Ist* $\|\ \|_0 : \mathfrak{L}' \longrightarrow [0,\infty]$ *eine Integralnorm, dann definiert* $\|\psi\|_3 := \inf\{\sup_n \|\phi_n\|_0 : (\phi_\kappa) \in \mathfrak{L}'^{\mathbb{N}} \wedge \phi_1 \le \phi_2 \le \cdots \le \sup_n \phi_n \ge \psi\}$ *für* $\psi \in \mathfrak{P}(\mathfrak{R})$ *- mit* $\inf \emptyset := \infty$ *- eine starke Integralnorm* $\|\ \|_3 : (=(\|\ \|_0)_3) : \mathfrak{P}(\mathfrak{R}) \longrightarrow [0,\infty]$ *mit* $\|\phi\|_3 \le \|\phi\|_0$ *für* $\phi \in \mathfrak{L}'$.

$\| \ \|_3$ *ist Fortsetzung von* $\| \ \|_o$, *falls zusätzlich gilt*

(*) $\forall\ (\phi_\kappa) \in \mathfrak{L}'^{\mathbb{N}}\ \forall\ \phi \in \mathfrak{L}'\quad \phi_1 \le \phi_2 \le \cdots \le \sup_n \phi_n \ge \phi \curvearrowright \sup_n \|\phi_n\|_o \ge \|\phi\|_o$.

Für den Rest des Abschnitts sei $\| \ \|: \mathfrak{P}(\mathfrak{R}) \longrightarrow [0,\infty]$ eine Integralnorm. $\mathfrak{Z}' := \overline{\mathfrak{L}'}^{\| \ \|}$ bezeichne den topologischen Abschluß von $\mathfrak{L}'$; mit $\mathfrak{L}'$ ist auch $\mathfrak{Z}'$ ein SV-System.

<u>Satz 0.2</u> *Vor.:* $\| \ \|: \mathfrak{P}(\mathfrak{R}) \longrightarrow [0,\infty]$ *sei eine starke Integralnorm; die Einschränkung* $\| \ \|_{/\mathfrak{L}'}$ *von* $\| \ \|$ *auf* $\mathfrak{L}'$ *sei halbadditiv.*

Beh.: Sind $(\psi_\kappa) \in \mathfrak{Z}'^{\mathbb{N}}$, $\psi \in \mathfrak{P}(\mathfrak{R})$, $\sup_n \|\sup_{i=1}^{n} \psi_i\| < \infty$ *und* $\psi_n(x) \longrightarrow \psi(x)$ $\| \ \|$*-f.ü., so gelten* $\|\psi_n \,\Delta\, \psi\| \longrightarrow 0$, $\psi \in \mathfrak{Z}'$ *und* $\|\psi_n\| \longrightarrow \|\psi\|$.

Unter der <u>Annahme</u>

Ⓐ | *Es seien* $(\mathfrak{G}_1, +, |\ |)$ *normierte Gruppe und* $\mathfrak{L}$ *Untergruppe von* $\mathfrak{F} := \{f \mid f: \mathfrak{R} \longrightarrow \mathfrak{G}_1\}$ *mit* $|\mathfrak{L}| \subset \mathfrak{L}'$

(,wobei für $f \in \mathfrak{F}$ und $x \in \mathfrak{R}$ $|f|(x) := |f(x)|$ gesetzt ist.)

hat man mit $\| \ \|^* := \| \ \| \circ |\ |$ für die Untergruppe $\mathfrak{Z} := \overline{\mathfrak{L}}^{\| \ \|^*}$ $|\mathfrak{Z}| \subset \mathfrak{Z}'$ und

<u>Satz 0.3</u> *Vor.:* $\| \ \|: \mathfrak{P}(\mathfrak{R}) \longrightarrow [0,\infty]$ *sei starke Integralnorm und* $\| \ \|_{/\mathfrak{L}'}$ *halbadditiv.*

Beh.: Sind $f \in \mathfrak{F}$, $(f_\kappa) \in \mathfrak{Z}^{\mathbb{N}}$ *und* $\phi \in \mathfrak{P}(\mathfrak{R})$ *mit* $\|\phi\| < \infty$, $f_n(x) \longrightarrow f(x)$ $\| \ \|$*-f.ü. und* $|f_n(x)| \le \phi(x)$ $\| \ \|$*-f.ü., so gelten* $f \in \mathfrak{Z}$ *und* $\|f_n - f\|^* \longrightarrow 0$.

<u>Hilfssatz 0.4</u> *Es seien* $\| \ \|_o: \mathfrak{L}' \longrightarrow [0,\infty]$ *eine Integralnorm (dazu* $\| \ \|_3$ *gemäß 0.1 gebildet),* $\| \ \|: \mathfrak{P}(\mathfrak{R}) \longrightarrow [0,\infty]$ *eine starke Integralnorm und* $\| \ \|_{/\mathfrak{L}'}$ *halbadditiv. Ist* π *eine* $\| \ \|_3$*-Nullmenge, dann existiert eine* $\| \ \|_3$*-Nullmenge* $\pi^* \supset \pi$, *so daß für alle* $\phi \in \mathfrak{Z}' := \overline{\mathfrak{L}'}^{\| \ \|}$ *mit* $\|\phi\| < \infty$ *gilt* $\chi_{\pi^*}\phi \in \mathfrak{Z}'$.

Beweis: Zu $i \in \mathbb{N}$ existiert $(\phi_{\kappa,i}) \in \mathfrak{L}'^{\mathbb{N}}$ mit $\phi_{1,i} \le \phi_{2,i} \le \cdots \le \sup_n \phi_{n,i} =: \psi_i \ge \chi_\pi$ und $(\|\psi_i\|_3 \le)$ $\sup_n \|\phi_{n,i}\|_o \le \frac{1}{i}$. Für $\psi := \inf_i \psi_i$ und $\pi^* := \{x \in \mathfrak{R} \mid \psi(x) \ne 0\}$ gelten also $\pi \subset \pi^*$, π^* $\| \ \|_3$-Nullmenge und für ein $\phi \in \mathfrak{Z}'$ mit $\|\phi\| < \infty$ $\chi_{\pi^*}\phi = \sup_n((n\psi) \wedge \phi) \in \mathfrak{Z}'$ auf Grund von 0.2 .

<u>1. Meßbare Abbildungen</u>

Für eine Integralnorm $\| \ \|: \mathfrak{P}(\mathfrak{R}) \longrightarrow [0,\infty]$ bezeichnen wir

$\mathfrak{M}' := \{\psi \in \mathcal{P}(\mathbb{R}) : \exists\, (\phi_\kappa) \in \mathcal{E}'^{\mathbb{N}}\ \ \phi_n(x) \longrightarrow \psi(x)\ \ \|\ \|\text{-f.ü.}\}$ *(meßbare* $[0,\infty]$*-wertige Abbildungen)* und - falls Ⓐ gegeben - $\mathfrak{M} := \{f \in \mathcal{F}: \exists (f_\kappa) \in \mathcal{E}^{\mathbb{N}}\ \ f_n(x) \longrightarrow f(x)\ \ \|\ \|\text{-f.ü.}\}$ *(meßbare* $\mathcal{G}_1$*-wertige Abbildungen)*. Offenbar gilt dann $|\mathfrak{M}| \subset \mathfrak{M}'$. Bezeichnet man für $\psi \in \mathcal{P}(\mathbb{R})$ $\|\psi\|_\sigma := 0$, falls $(\phi_\kappa) \in \mathcal{E}'^{\mathbb{N}}$ mit $\psi \le \sup_n \phi_n$ existiert, und $\|\psi\|_\sigma := \infty$, sonst, dann ist $\|\ \|_\sigma : \mathcal{P}(\mathbb{R}) \longrightarrow [0,\infty]$ starke Integralnorm.

Damit machen wir in diesem Abschnitt die Annahme

Ⓑ $\left|\begin{array}{l} \|\ \| : \mathcal{P}(\mathbb{R}) \longrightarrow [0,\infty] \ \ \textit{starke Integralnorm}, \ \|\psi\| = 0 \curvearrowright \|\psi\|_\sigma = 0, \\ \|\ \|_{/\mathcal{E}'} \ \ \textit{endlich und halbadditiv.} \end{array}\right.$

und notieren dazu den

<u>Satz 1.1</u> *a)* $\mathfrak{M}' = \{\psi \in \mathcal{P}(\mathbb{R}) : \|\psi\|_\sigma = 0 \ \wedge\ \forall \phi \in \mathcal{E}'\ \ \phi \wedge \psi \in \mathcal{J}'\}$

b) Für $(\psi_\kappa) \in \mathfrak{M}'^{\mathbb{N}}$ *und* $\psi \in \mathcal{P}(\mathbb{R})$ *mit* $\psi_n(x) \longrightarrow \psi(x)\ \|\ \|$*-f.ü. gilt* $\psi \in \mathfrak{M}'$.

c) $\mathcal{J}' = \{\psi \in \mathfrak{M}' : \|\psi\| < \infty\}$

(Der Beweis ergibt sich als vereinfachte Version des Beweises zu 1.3)

Zur Behandlung von gruppenwertigen meßbaren Abbildungen machen wir die <u>Zusatzannahmen</u> Ⓐ und (vgl. dazu [9])

Ⓒ $\left|\begin{array}{lll} \multicolumn{3}{l}{\textit{Es seien } \circledwedge \textit{ eine Abbildung von } \mathcal{P}(\mathbb{R}) \times \mathcal{F} \textit{ in } \mathcal{F} \textit{ und } k \textit{ eine natürliche}} \\ \textit{Zahl mit} & (C1) & \mathcal{E}' \circledwedge \mathcal{E} \subset \mathcal{E} \\ & (C2) & 0 \circledwedge f = \psi \circledwedge 0 = 0 \\ & (C3) & |\psi_1 \circledwedge f_1 - \psi_2 \circledwedge f_2| \le k\cdot(|f_1 - f_2| + \psi_1 \Delta \psi_2) \\ & (C4) & f = |f| \circledwedge f \end{array}\right.$

(Ist $\mathcal{G}_1$ normierter Vektorraum, dann ist in wichtigen Spezialfällen (z.B. $\mathcal{E}$ = Menge der stetigen (mit kompakten Träger) oder der bezüglich eines Semirings einfachen Funktionen) Ⓒ mit $\mathcal{E}' := |\mathcal{E}|$ und $k = 2$ erfüllt durch

$$(\psi \circledwedge f)(x) := \left\{\begin{array}{l} f(x), \text{ falls } |f(x)| \le \psi(x) \\ \frac{\psi(x)}{|f(x)|} f(x), \text{ sonst} \end{array}\right\}.)$$

Offenbar gilt die

<u>Bemerkung 1.2</u> *a)* $\mathcal{J}' \circledwedge \mathcal{J} \subset \mathcal{J}$ *b)* $|\psi \circledwedge f| \le k\cdot(\psi \wedge |f|)$

Wir zeigen

<u>Satz 1.3</u> *a)* $\mathfrak{M} = \{f \in \mathcal{F}: \|f\|^*_\sigma = 0 \ \wedge\ \forall \phi \in \mathcal{E}'\ \ \phi \circledwedge f \in \mathcal{J}\}$

b) Für $(f_\kappa) \in \mathfrak{M}^{\mathbb{N}}$ *und* $f \in \mathcal{F}$ *mit* $f_n(x) \longrightarrow f(x)\ \|\ \|$*-f.ü. gilt* $f \in \mathfrak{M}$.

c) $\mathcal{J} = \{f \in \mathfrak{M}: \|f\|^* < \infty\}$

Beweis: Wir bezeichnen die rechte Seite in a) mit $\mathfrak{M}_1$ und zeigen:

1) Für $(f_\kappa) \in \mathfrak{M}_1^{\mathbb{N}}$ und $f \in \mathcal{F}$ mit $f_n(x) \longrightarrow f(x)\ \|\ \|$-f.ü. gilt $f \in \mathfrak{M}_1$.

Beweis: Ist $\mathfrak{N}$ die zugehörige Ausnahmemenge, dann gilt $|f| \leq \sum_{n=1}^{\infty} (|f_n| + \chi_{\mathfrak{N}})$, also $\|f\|_\sigma^* \leq \sum_{n=1}^{\infty} (\|f_n\|_\sigma^* + \|\chi_{\mathfrak{N}}\|_\sigma) = 0$. Für $\phi \in \mathfrak{C}'$ hat man $\phi \circledA f_n \in \mathfrak{F}$, $|\phi \circledA f_n| \leq k\cdot\phi \in \mathfrak{C}'$ (nach 1.2), $(\phi \circledA f_n)(x) \longrightarrow (\phi \circledA f)(x)$ $(x \in \mathfrak{R}\setminus\mathfrak{N})$ (nach (C3)), also nach 0.3 $\phi \circledA f \in \mathfrak{F}$.
2) Da $\mathfrak{C}$ Teilmenge von $\mathfrak{M}_1$ ist, gilt nach 1): $\mathfrak{M} \subset \mathfrak{M}_1$
3) Für $f \in \mathfrak{M}_1$ gilt $f \in \mathfrak{M}$ und : $\|f\|^* < \infty \curvearrowright f \in \mathfrak{F}$
Beweis: Für $(\phi_\kappa) \in \mathfrak{C}'^{\mathbb{N}}$ mit $\phi_1 \leq \phi_2 \leq \cdots \leq \sup_n \phi_n \geq |f|$ hat man $g_n := \phi_n \circledA f \in \mathfrak{F}$, $|g_n - f| = |\phi_n \circledA f - |f| \circledA f| \leq k\cdot(\phi_n \,\Delta\, |f|)$, also $g_n(x) \longrightarrow f(x)$ $(x \in \mathfrak{R})$, und $|g_n| \leq k\cdot|f|$. Ist $\|f\|^* < \infty$, dann liefert 0.3 $f \in \mathfrak{F}$. In jedem Fall existiert $(h_\kappa) \in \mathfrak{C}^{\mathbb{N}}$ mit $\sum_{n=1}^{\infty} \|g_n - h_n\|^* < \infty$; dann gilt $g_n(x) - h_n(x) \longrightarrow 0$ $\|\ \|$-f.ü., also $h_n(x) \longrightarrow f(x)$ $\|\ \|$-f.ü. und somit $f \in \mathfrak{M}$. 4) $\mathfrak{F} \subset \{f \in \mathfrak{M}: \|f\|^* < \infty\}$ (vgl. z.B. [8],2.4.5) schließt dann den Beweis von 1.3 ab.

2. Summen von Integralnormen

In diesem Abschnitt seien

(D) *T eine nicht-leere Menge und für jedes* $t \in T$ $\|\ \|_{o,t}: \mathfrak{C}' \longrightarrow [0,\infty]$ *eine halbadditive Integralnorm mit* (*) *(siehe* 0.1*) und* $\sum_{t \in T} \|\phi\|_{o,t} < \infty$ *für* $\phi \in \mathfrak{C}'$.

Wir benutzen die folgenden Bezeichnungen:

Für $S \subset T$ seien $\|\ \|_{o,S} := \sum_{t\in S} \|\ \|_{o,t}$, $\|\ \|_S := (\|\ \|_{o,S})_3$ (gemäß 0.1 gebildet), $\|\ \|_\Sigma := \sum_{t\in T} \|\ \|_t$, $\mathfrak{F}'_S := \overline{\mathfrak{C}'_S}^{\|\ \|_S}$ und $\mathfrak{M}'_S$ die bezüglich $\|\ \|_S$ meßbaren $[0,\infty]$-wertigen Abbildungen; (hierbei notieren wir natürlich $\|\ \|_t := \|\ \|_{\{t\}}$, $\mathfrak{F}'_t := \mathfrak{F}'_{\{t\}}$ (und später entsprechend).)

Dabei ist offenbar nach 0.1 $\|\ \|_S$ *eine starke Integralnorm und Fortsetzung von* $\|\ \|_{o,S}$; ferner ist $\|\ \|_\Sigma$ *eine starke Integralnorm und Fortsetzung von* $\|\ \|_{o,T}$.

Als erster Schritt zum Beweis von 2.3 und 2.4 dient

Lemma 2.1 *Vor.:* S *sei eine endliche Teilmenge von* T.

Beh.: a) Für $T_o \subset T$ *ist* $\|\ \|_T = \|\ \|_{T_o} + \|\ \|_{T\setminus T_o}$, *also auch* $\|\ \|_S = \sum_{t\in S} \|\ \|_t$.

b) $\|\ \|_{o,S}$ *ist halbadditiv*.

c) Ist $\psi \in \bigcap_{t\in S} \mathfrak{F}'_t$ *und* $\psi \leq \phi \in \mathfrak{C}'$, *dann ist* $\psi \in \mathfrak{F}'_S$.

Der *Beweis* von a) und b) ist trivial. Beweis von c): Sei $S = \{t_1, \cdots, t_m\}$. Nach 1.1 (bzw. [8],2.4.5) gibt es für $\mu \in \{1, \cdots, m\}$

Folgen $(\phi_{\kappa,\mu}) \in \mathfrak{L}'^{\mathbb{N}}$ und $\| \|_{t_\mu}$-Nullmengen Π_μ mit $\phi_{n,\mu}(x) \longrightarrow \psi(x)$ für $x \in \mathcal{R} \setminus \Pi_\mu$ und $n \longrightarrow \infty$. Nach b) und 0.4 kann ohne Einschränkung $\chi_{\Pi_\mu}\phi \in \mathfrak{J}'_S$ angenommen werden. Daher ist

$$\psi_n := \sum_{\mu=1}^{m} \chi_{\bigcap_{\rho=1}^{\mu-1} \Pi_\rho \setminus \Pi_\mu} \phi_{n,\mu} \wedge \phi = \sum_{\mu=1}^{m} (\phi \Delta \chi_{\Pi_\mu}\phi) \wedge \bigwedge_{\rho=1}^{\mu-1} \chi_{\Pi_\rho}\phi \wedge \phi_{n,\mu} \in \mathfrak{J}'_S ,$$

$\psi_n \leq \phi$ und $\psi_n(x) \longrightarrow \psi(x)$ für $x \in \mathcal{R} \setminus \bigcap_{\mu=1}^{m} \Pi_\mu$, also nach a) $\| \|_S$-f.ü. . Hieraus folgt nach 0.2 $\psi \in \mathfrak{J}'_S$.

Lemma 2.2 *a)* $\| \|_\Sigma \leq \| \|_T$

b) Für $\psi \in \mathfrak{J}'_T$ *ist* $\|\psi\|_T = \|\psi\|_\Sigma < \infty$.

Beweis: Nach 2.1.a ist für $\psi \in \mathcal{P}(\mathcal{R})$

$\|\psi\|_\Sigma := \sup\{ \sum_{t \in S} \|\psi\|_t : S \text{ endlich} \subset T\} = \sup\{\|\psi\|_S : S \text{ endlich} \subset T\} \leq \|\psi\|_T$.

b) Ist $(\phi_\kappa) \in \mathfrak{L}'^{\mathbb{N}}$ mit $\|\psi \Delta \phi_n\|_T \leq \frac{1}{n}$ $(n \in \mathbb{N})$, so ist nach a) auch $\|\psi \Delta \phi_n\|_\Sigma \leq \frac{1}{n}$, also $\|\psi\|_T = \lim\|\phi_n\|_T = \lim\|\phi_n\|_\Sigma = \|\psi\|_\Sigma \leq \|\psi \Delta \phi_1\|_\Sigma + \|\phi_1\|_\Sigma \leq 1 + \|\phi_1\|_\Sigma < \infty$.

Proposition 2.3 $\| \|_{o,T}$ *ist halbadditiv*

($\| \| := \| \|_T$ erfüllt also die Annahme Ⓑ .)

Beweis: Sei $(\phi_\kappa) \in \mathfrak{L}'^{\mathbb{N}}$ und $M := \sup_n \| \sum_{\nu=1}^{n} \phi_\nu\|_{o,T} < \infty$. Für jede endliche Teilmenge S von T ist $\sup_n \| \sum_{\nu=1}^{n} \phi_\nu\|_{o,S} \leq M$, also nach 0.2 und 2.1.b $\| \sum_{\nu=1}^{\infty} \phi_\nu\|_S \leq M$ und daher $\| \sum_{\nu=1}^{\infty} \phi_\nu\|_\Sigma \leq M$. Folglich gibt es zu gegebenem $\varepsilon > 0$ eine endliche Teilmenge T_o von T mit $\sum_{t \in T \setminus T_o} \| \sum_{\nu=1}^{\infty} \phi_\nu\|_t < \varepsilon$; wegen der Halbadditivität von $\| \|_{o,T_o}$ existiert ferner ein $N \in \mathbb{N}$ mit $\|\phi_n\|_{o,T_o} < \varepsilon$ für $n \geq N$. Insgesamt hat man daher für $n \geq N$ $\|\phi_n\|_T = \|\phi_n\|_{T \setminus T_o} + \|\phi_n\|_{T_o} < 2\varepsilon$.

Satz 2.4 $\mathfrak{J}'_T = \{\psi \in \bigcap_{t \in T} \mathfrak{J}'_t : \|\psi\|_\Sigma < \infty\}$.

Beweis: Offenbar ist nur die Inklusion "$\supset$" zu beweisen. Wir zeigen zunächst (*) $\psi \in \bigcap_{t \in T} \mathfrak{J}'_t$, $\psi \leq \phi \in \mathfrak{L}' \curvearrowright \psi \in \mathfrak{J}'_T$.

Sei dazu $\varepsilon > 0$ und T_o eine endliche Teilmenge von T mit $\|\phi\|_{T \setminus T_o} < \varepsilon$. Nach 2.1.c ist $\psi \in \mathfrak{J}'_{T_o}$; also gibt es ein $\phi_o \in \mathfrak{L}'$ mit $\|\psi \Delta \phi_o\|_{T_o} < \varepsilon$ und $\phi_o \leq \phi$. Damit hat man $\|\psi \Delta \phi_o\|_T = \|\psi \Delta \phi_o\|_{T_o} + \|\psi \Delta \phi_o\|_{T \setminus T_o} \leq \|\psi \Delta \phi_o\|_{T_o} + \|\phi\|_{T \setminus T_o} < 2\varepsilon$.— Ist nun $\psi \in \bigcap_{t \in T} \mathfrak{J}'_t$ mit $\|\psi\|_\Sigma < \infty$, dann existiert - wegen $\|\phi\|_t < \infty$ - $(\phi_\kappa) \in \mathfrak{L}'^{\mathbb{N}}$ mit $\phi_1 \leq \phi_2 \leq \cdots \leq \sup_n \phi_n \geq \psi$, Nach (*) ist $\psi_n := \psi \wedge \phi_n \in \mathfrak{J}'_T$. Da

$\psi_n(x) \longrightarrow \psi(x) \quad (x \in \mathcal{R})$ und - nach 2.2.b - $\|\psi_n\|_T = \|\psi_n\|_\sum \le \|\psi\|_\sum < \infty$, folgt nach 0.2 und 2.3 $\psi \in \mathfrak{J}_T'$.

Mit 1.1.a erhält man aus 2.4 sofort das

Korollar 2.5 $\mathfrak{M}_T' = \bigcap_{t \in T} \mathfrak{M}_t'$

Lemma 2.6 *Ist* $\psi \in \mathfrak{P}(\mathcal{R})$ *und für jedes* $t \in T$ $\|\psi\|_t < \infty$, *dann gibt es ein* $\phi \in \bigcap_{t\in T} \mathfrak{J}_t'$ *mit* $\phi \ge \psi$ *und* $\|\phi\|_\sum = \|\psi\|_\sum$.

Beweis: Sei $(\phi_{\kappa,o}) \in \mathfrak{E}'^{\mathbb{N}}$ mit $\psi \le \sup_n \phi_{n,o} =: \phi_o$. Da $\|\phi_{n,o}\|_{o,T} < \infty$, gibt es eine Folge $(t_r) \in T^{\mathbb{N}}$, so daß für alle $t \in T \setminus \{t_r : r \in \mathbb{N}\}$ und $n \in \mathbb{N}$ $\|\phi_{n,o}\|_t = 0$, also auch $\|\phi_o\|_t = 0$. Wir zeigen zunächst:

(*) Zu $\varepsilon > 0$ gibt es ein $\phi_\varepsilon \in \bigcap_{t\in T} \mathfrak{J}_t'$ mit $\phi_\varepsilon \ge \psi$ und $\|\phi_\varepsilon\|_t \le \|\psi\|_t + \varepsilon$ für $t \in T$. Zum Beweis wählt man für $r \in \mathbb{N}$ monoton wachsende Folgen $(\phi_{\kappa,r}) \in \mathfrak{E}'^{\mathbb{N}}$ mit $\sup_n \phi_{n,r} \ge \psi$, $\|\phi_{n,r}\|_{t_r} \le \|\psi\|_{t_r} + \varepsilon$ und $\phi_{n,r+1} \le \phi_{n,r} \le \phi_{n,o}$. Für $i \ge r$ ist $\|\phi_{n,i}\|_{t_r} \le \|\phi_{n,r}\|_{t_r} \le \|\psi\|_{t_r} + \varepsilon$, also nach 0.2 $\psi_i := \sup_n \phi_{n,i} \in \mathfrak{J}_{t_r}'$ und $\|\psi_i\|_{t_r} \le \|\psi\|_{t_r} + \varepsilon$. Hieraus folgt - wieder mit 0.2 - $\phi_\varepsilon := \inf_i \psi_i = \inf_{i\ge r} \psi_i \in \mathfrak{J}_{t_r}'$. Da ferner $\|\phi_\varepsilon\|_{t_r} \le \|\psi\|_{t_r} + \varepsilon$, $\phi_\varepsilon \ge \psi$ und für $t \in T \setminus \{t_\rho : \rho \in \mathbb{N}\}$ $\|\phi_\varepsilon\|_t \le \|\phi_o\|_t = 0$, ist (*) bewiesen. — Wählt man (gemäß (*)) $\phi_k \in \bigcap_{t\in T} \mathfrak{J}_t'$ mit $\phi_k \ge \psi$ und $\|\phi_k\|_t \le \|\psi\|_t + \frac{1}{k}$ $(t \in T)$, dann leistet nach 0.2 $\phi := \inf_k \phi_k$ das Gewünschte.

Satz 2.7 $\| \ \|_T = \| \ \|_\sum$

Beweis: Nach 2.2.a ist nur $\|\psi\|_T \le \|\psi\|_\sum$ für $\psi \in \mathfrak{P}(\mathcal{R})$ mit $\|\psi\|_\sum < \infty$ zu zeigen. Zu ψ sei ϕ gemäß 2.6 gewählt. Dann gehört nach 2.4 ϕ zu $\mathfrak{J}_T'$, und nach 2.2.b gilt $\|\psi\|_T \le \|\phi\|_T = \|\phi\|_\sum = \|\psi\|_\sum$.

3. Integrable Funktionen und Integrale

In diesem Abschnitt gelten die Annahmen: Ⓐ, Ⓒ und Ⓓ.
Neben den Bezeichnungen des Abschnitts 2. benutzen wir noch für $S \subset T$:

$\mathfrak{J}_S := \overline{\mathfrak{E}}^{\| \ \|_S^*}$ und $\mathfrak{M}_S := \{f \in \mathfrak{F} : \exists\, (f_\kappa) \in \mathfrak{E}^{\mathbb{N}} \ f_n(x) \longrightarrow f(x) \ \| \ \|_S\text{-f.ü.}\}$.

Satz 3.1 $\mathfrak{J}_T = \{f \in \bigcap_{t\in T} \mathfrak{J}_t : \|f\|_\sum^* < \infty\}$

Der *Beweis* verläuft analog zu dem von 2.1.c und 2.4: Es ist nur die Inklusion "$\supset$" zu beweisen. Dazu zeigen wir zunächst:

(*) $S = \{t_1, \cdots, t_m\} \subset T$, $f \in \bigcap_{t\in S} \mathfrak{J}_t \curvearrowright f \in \mathfrak{J}_S$

Nach [8],2.4.5 (bzw. 1.3) gibt es für $\mu \in \{1, \cdots, m\}$ Folgen $(f_{\kappa,\mu})$

$\in \mathfrak{E}^{\mathbb{N}}$ und $\| \|_{t\mu}$-Nullmengen $\mathfrak{N}_\mu$ mit $f_{n,\mu}(x) \longrightarrow f(x)$ für $n \longrightarrow \infty$ und $x \in \mathfrak{R} \setminus \mathfrak{N}_\mu$. Nach 2.4 ist $|f| \in \mathfrak{F}'_S$, und nach 0.4 kann ohne Einschränkung $\chi_{\mathfrak{N}_\mu}|f| \in \mathfrak{F}'_S$ angenommen werden. Daher ergibt sich aus 1.2 mit $f_\mu := \chi_{\bigcap_{\rho=1}^{\mu-1} \mathfrak{N}_\rho \setminus \mathfrak{N}_\mu} f$, $g_n := \sum_{\mu=1}^{m} |f_\mu| \circledwedge f_{n,\mu} \in \mathfrak{F}_S$,

$|g_n| \le \sum_{\mu=1}^{m} k\, |f_\mu| \le k\, |f|$, $|g_n - f| \le |f - \sum_{\mu=1}^{m} f_\mu| + \sum_{\mu=1}^{m} ||f_\mu| \circledwedge f_\mu - |f_\mu| \circledwedge f_{n,\mu}| \le |f - \sum_{\mu=1}^{m} f_\mu| + \sum_{\mu=1}^{m} (k\cdot|f_\mu - f_{n,\mu}|) \wedge (2k\cdot|f_\mu|)$, also $g_n(x) \longrightarrow f(x)$ für $x \in \mathfrak{R} \setminus \bigcap_{\mu=1}^{m} \mathfrak{N}_\mu$ und somit nach 0.3 $f \in \mathfrak{F}_S$.

Sei nun $f \in \bigcap_{t\in T} \mathfrak{F}_t$ mit $\|f\|^*_\Sigma < \infty$; zu $\varepsilon > 0$ sei T_o eine endliche Teilmenge von T, so daß $\|f\|^*_{T\setminus T_o} < \varepsilon$. Nach (*) gibt es ein $g \in \mathfrak{E}$ mit $\|f - g\|^*_{T_o} < \varepsilon$. Nach 2.4 und 1.2 ist $h := |f| \circledwedge g \in \mathfrak{F}_T$; wegen $|f - h| = ||f| \circledwedge f - |f| \circledwedge g| \le k\,|f - g|$ und $|f - h| \le |f| + ||f| \circledwedge g| \le (k + 1)\,|f|$ hat man ferner $\|f - h\|^*_T = \|f - h\|^*_{T\setminus T_o} + \|f - h\|^*_{T_o} \le (k + 1)\|f\|^*_{T\setminus T_o} + k\,\|f - g\|^*_{T_o} \le (2k + 1)\varepsilon$. Dies zeigt $f \in \mathfrak{F}_T$.

Mit 1.3.a erhält man aus 3.1 sofort das

<u>Korollar 3.2</u> $\mathfrak{M}_T = \bigcap_{t\in T} \mathfrak{M}_t$

Für den nächsten Satz nehmen wir noch an

(E) $(\mathfrak{G}_2, +, |\ |)$ *sei eine vollständige normierte abelsche Gruppe,* $i_t : \mathfrak{E} \longrightarrow \mathfrak{G}_2$ *Homomorphismus mit* $|i_t(f)| \le \| |f| \|_{o,t}$ $(t \in T)$.

Dann ist für $f \in \mathfrak{E}$ $\sum_{t\in T} i_t(f)$ absolut konvergent, und *durch* $i_T(f) := \sum_{t\in T} i_t(f)$ *ist ein Homomorphismus* $i_T : \mathfrak{E} \longrightarrow \mathfrak{G}_2$ *definiert mit* $|i_T(f)| \le \|f\|^*_{o,T}$ $(f \in \mathfrak{E})$.

$\bar{\imath}_T$ (bzw. $\bar{\imath}_t$) bezeichne die $\| \|^*_T$-stetige (bzw. $\| \|^*_t$-stetige) Fortsetzung von i_T (bzw. i_t) auf $\mathfrak{F}_T$ (bzw. $\mathfrak{F}_t$).

<u>Satz 3.3</u> *Für* $f \in \mathfrak{F}_T$ *ist* $\sum_{t\in T} \bar{\imath}_t(f)$ *absolut konvergent und*

$$\sum_{t\in T} \bar{\imath}_t(f) = \bar{\imath}_T(f).$$

Beweis: Es ist $\sum_{t\in T} |\bar{\imath}_t(f)| \le \sum_{t\in T} \|f\|^*_t = \|f\|^*_T < \infty$. Ist $\varepsilon > 0$ und $g \in \mathfrak{E}$ mit $\|g - f\|^*_T < \varepsilon$, dann hat man $|\bar{\imath}_T(f) - \sum_{t\in T} \bar{\imath}_t(f)| \le |\bar{\imath}_T(f) - \bar{\imath}_T(g)| + |\sum_{t\in T} \bar{\imath}_t(g) - \sum_{t\in T} \bar{\imath}_t(f)| \le \|f - g\|^*_T + \|f - g\|^*_\Sigma < 2\varepsilon$.

4. Spezialfälle, Ergänzungen

Wir nehmen jetzt gleich wesentlich spezieller an

(F)
> $\mathfrak{R}$ *und* T *seien nicht-leere Mengen,*
> $\mathfrak{Y}_1$ *und* $\mathfrak{Y}_2$ *(B)-Räume (über* $\mathbb{R}$ *oder* $\mathbb{C}$*),*
> $\mathfrak{I}$ *ein Semiring über* $\mathfrak{R}$ (d.h.: $\mathfrak{I}$ ist nicht-leere Teilmenge der Potenzmenge von $\mathfrak{R}$, für $A,B \in \mathfrak{I}$ gilt $A \cap B \in \mathfrak{I}$ und $\exists\, n \in \mathbb{N}\ \exists\ (A_1, \cdots, A_n) \in \mathfrak{I}^n \quad A \setminus B = \biguplus_{\nu=1}^{n} A_\nu$.) *und*
> $\forall\, t \in T \quad \mu_t\colon \mathfrak{I} \longrightarrow \mathfrak{L}$ *schwaches Maß, wobei* $\mathfrak{L}$ *die Menge der linearen Abbildungen von* $\mathfrak{Y}_1$ *in* $\mathfrak{Y}_2$ *bezeichnet* (d.h.: Für jedes $b \in \mathfrak{Y}_1$ ist $\mu_t(\cdot)b$ $\mathfrak{Y}_2$-wertiges Maß .)

Wir bezeichnen mit $\mathfrak{E}$ den Unterraum der $\mathfrak{I}$-einfachen Abbildungen von $\mathfrak{F} := \{f \mid f\colon \mathfrak{R} \longrightarrow \mathfrak{Y}_1\}$, mit i_t die durch $i_t(\chi_A b) = \mu_t(A)b$ $(A \in \mathfrak{I},\ b \in \mathfrak{Y}_1)$ eindeutig bestimmte lineare Abbildung von $\mathfrak{E}$ in $\mathfrak{Y}_2$ ([7],5.2.1) und damit für $\phi \in \mathfrak{E}' := |\mathfrak{E}|$

$\|\phi\|_{o,t} := \sup\{|i_t(f)| : f \in \mathfrak{E} \wedge |f| \le \phi\}$ $(t \in T)$. Nach [7],5.4.1 und [7],5.6.1 ist $\|\ \|_{o,t}\colon \mathfrak{E}' \longrightarrow [0,\infty]$ *Integralnorm mit* (*) (siehe 0.1) *und* $\|\alpha\phi\|_{o,t} = \alpha\,\|\phi\|_{o,t}$ $(\alpha \in (0,\infty),\ \phi \in \mathfrak{E}')$.

Wir verlangen noch

> $\forall\, t \in T$ $\|\ \|_{o,t}$ *halbadditiv und* $\forall\, \phi \in \mathfrak{E}' \quad \sum_{t \in T} \|\phi\|_{o,t} < \infty$.

Damit sind alle Annahmen aus 2. und 3. erfüllt, und es gilt:

Satz 4.1 *Durch* $\mu(A)b := \sum_{t\in T} \mu_t(A)b$ $(A \in \mathfrak{I},\ b \in \mathfrak{Y}_1)$ *wird ein schwaches Maß* $\mu\colon \mathfrak{I} \longrightarrow \mathfrak{L}$ *definiert. Für die zu* μ *(entsprechend wie oben) gebildeten Größen* i , $\|\ \|_o$, $\|\ \| := (\|\ \|_o)_3$, $\mathfrak{J}'$, $\mathfrak{M}'$, $\mathfrak{J}$, $\mathfrak{M}$ *und* τ *gilt*

a) $i = i_T$ *und* $\|\ \| \le \|\ \|_T$ *und daher:*

b) $\{\psi \in \bigcap_{t\in T} \mathfrak{J}'_t : \|\psi\|_\Sigma < \infty\} = \mathfrak{J}'_T \subset \mathfrak{J}'$

c) $\bigcap_{t\in T} \mathfrak{M}'_t = \mathfrak{M}'_T \subset \mathfrak{M}'$

d) $\{f \in \bigcap_{t\in T} \mathfrak{J}_t : \|f\|^*_\Sigma < \infty\} = \mathfrak{J}_T \subset \mathfrak{J}$

e) Für $f \in \mathfrak{J}_T$ *gilt* $\sum_{t\in T} \tau_t(f) = \tau_T(f) = \tau(f)$.

f) $\bigcap_{t\in T} \mathfrak{M}_t = \mathfrak{M}_T \subset \mathfrak{M}$

Beweis: Für $A \in \mathfrak{I}$ und $b \in \mathfrak{Y}_1$ gilt $\mu(A)b := \sum_{t\in T} \mu_t(A)b =$

$\sum_{t\in T} i_t(\chi_A b) = i_T(\chi_A b)$ (*) , also nach 3. und [7],5.2.1
$\mu : \mathfrak{I} \longrightarrow \mathfrak{X}$ Inhalt, $i = i_T$ und für $f \in \mathfrak{E}$ $|i(f)| = |i_T(f)| \leq$
$\||f|\|_{o,T}$; daher gilt $\|\ \|_o \leq \|\ \|_{o,T}$ und somit $\|\ \| \leq \|\ \|_T$. (*), 2.3 und 0.3 liefern dann: μ ist schwaches Maß (vgl. auch [7],4.3.1). Die restlichen Behauptungen sind nun mit den Ergebnissen aus 2. und 3. trivial.

Wir setzen noch für $1 \leq p < \infty$

$\|\psi\|_p := (\|\psi^p\|)^{\frac{1}{p}}$ $(\psi \in \mathfrak{P}(\mathfrak{R}))$, $\|\ \|_p^* := \|\ \|_p \circ |\ |$ und $\mathcal{L}_p := \overline{\mathfrak{E}}^{\|\ \|_p^*}$; außerdem sei $\|\psi\|_\infty := \inf\{K \in [0,\infty]: \psi(x) \leq K \quad \|\ \|\text{-f.ü.}\}$ $(\psi \in \mathfrak{P}(\mathfrak{R}))$, $\|\ \|_\infty^* := \|\ \|_\infty \circ |\ |$ und $\mathcal{L}_\infty := \{f \in \mathfrak{M}: \|f\|_\infty^* < \infty\}$. Entsprechend seien $\|\ \|_{r,T}$, $\|\ \|_{r,t}$, $\mathcal{L}_{r,T}$ und $\mathcal{L}_{r,t}$ für $1 \leq r \leq \infty$ und $t \in T$ definiert. Dann gilt

<u>Satz 4.2</u> *a)* $\{f \in \bigcap_{t\in T} \mathcal{L}_{p,t} : \sum_{t\in T} (\|f\|_{p,t}^*)^p < \infty\} = \mathcal{L}_{p,T} \subset \mathcal{L}_p$

b) $\{f \in \bigcap_{t\in T} \mathcal{L}_{\infty,t} : \sup_{t\in T} \|f\|_{\infty,t}^* < \infty\} = \mathcal{L}_{\infty,T} \subset \mathcal{L}_\infty$

Beweis: Für $1 \leq p < \infty$ ist $\|\ \|_{p,T}$ starke und (unter Berücksichtigung von 2.3) auf $\mathfrak{E}'$ halbadditive Integralnorm (vgl.[6],(2),(3)); folglich gilt nach 1.3 $\mathcal{L}_{p,T} = \{f \in \mathfrak{M}_T: \|f\|_{p,T}^* < \infty\}$ und entsprechend $\mathcal{L}_{p,t} = \{f \in \mathfrak{M}_t: \|f\|_{p,t}^* < \infty\}$. a) ist dann mit
$\|\psi\|_p \leq \|\psi\|_{p,T} = (\|\psi^p\|_T)^{\frac{1}{p}} = (\sum_{t\in T} \|\psi^p\|_t)^{\frac{1}{p}} = (\sum_{t\in T} \|\psi\|_{p,t}^p)^{\frac{1}{p}}$ $(\psi \in \mathfrak{P}(\mathfrak{R}))$
und 4.1.f gegeben. Für b) genügt wegen 4.1.f der Nachweis von $\|\psi\|_\infty \leq \|\psi\|_{\infty,T} = \sup_{t\in T} \|\psi\|_{\infty,t}$ $(\psi \in \mathfrak{P}(\mathfrak{R}))$. $\|\psi\|_\infty \leq \|\psi\|_{\infty,T}$ ist klar, da jede $\|\ \|_T$-Nullmenge auch $\|\ \|$-Nullmenge ist. Nach 2.7 ist eine Teilmenge von $\mathfrak{R}$ genau dann $\|\ \|_T$-Nullmenge, wenn sie für jedes $t \in T$ $\|\ \|_t$-Nullmenge ist; dies zeigt $\|\psi\|_{\infty,T} = \sup_{t\in T} \|\psi\|_{\infty,t}$.

W.M.BOGDANOWICZ und J.N.WELCH betrachten in [3] und [4] nur den Fall, daß $\mathfrak{Y}_1 = \mathfrak{Y}_2$ und für jedes $t \in T$ $\mu_t: \mathfrak{I} \longrightarrow [0,\infty)$ (klassisches) Maß ist. Dann ist $\|\phi\|_{o,t} = i_t(\phi)$ $(\phi \in \mathfrak{E}')$, also $\|\ \|_{o,t}: \mathfrak{E}' \longrightarrow [0,\infty)$ sogar additiv; und die Annahme $\forall \phi \in \mathfrak{E}'$ $\sum_{t\in T} \|\phi\|_{o,t} < \infty$ bedeutet gerade die Konvergenz von $\sum_{t\in T} \mu_t(A)$ für A aus $\mathfrak{I}$. (in diesem Fall gilt natürlich $\|\ \| = \|\ \|_T$ und daher in 4.1 und 4.2 überall "=" statt "$\subset$".)

Literatur

[1] BOGDANOWICZ W.M., An Approach to the Theory of Lebesgue-Bochner Measurable Functions and to the Theory of Measure. Math.Ann. 164 (1966), 251-269

[2] BOGDANOWICZ W.M. und WELCH J.N., Integration Generated by Projective Volumes Over Hilbert Spaces. Bull.Acad.Polon.Sci.Sér.Sci. Math.Astronom.Phys. XIX (1971), 49-57

[3] BOGDANOWICZ W.M. und WELCH J.N., Integration Generated by a Volume Being the Sum of Volumes with an Application to Spectral Integrals. Bull.Acad.Polon.Sci.Sér.Sci.Math.Astronom.Phys. XIX (1971), 719-726

[4] BOGDANOWICZ W.M. und WELCH J.N., Integration generated by a volume which is an infinite sum of volumes. Comment.Math.Prace Mat. XVII (1974), 321-334

[5] DINCULEANU N., Integration on locally compact spaces. Noordhoff International publishing, LEYDEN (1974)

[6] HOFFMANN D., $\mathcal{L}_p$-Räume bezüglich operatorwertiger schwacher Maße. preprint (1975)

[7] SCHÄFKE F.W., Integrationstheorie II. J.Reine Angew.Math. 248 (1971), 147-171

[8] SCHÄFKE F.W., Integrationstheorie und quasinormierte Gruppen. J.Reine Angew.Math. 253 (1972), 117-137

[9] SCHÄFKE F.W., Lokale Integralnormen und verallgemeinerte uneigentliche Riemann-Stieltjes-Integrale. (1975), erscheint bei J.Reine Angew.Math.

THE SEMI-M PROPERTY FOR NORMED RIESZ SPACES

E. de Jonge and A.C. Zaanen

Katholieke Universiteit, Nijmegen and Leiden State University

It is well-known that if (Δ,Γ,μ) is a σ-finite measure space and if $1 \le p < \infty$, then the Banach dual L_p^* of the Banach space $L_p = L_p(\Delta,\mu)$ can be identified with $L_q = L_q(\Delta,\mu)$, where $p^{-1} + q^{-1} = 1$. For $p=\infty$ the situation is different; the space L_1 is a linear subspace of L_∞^*, and only in a very trivial situation (the finite-dimensional case) the subspace L_1 is equal to the whole space L_∞^*. Restricting ourselves to the real case, the space L_∞ and its Banach dual L_∞^* are <u>(real) Riesz spaces</u>, i.e., <u>vector lattices</u>, and L_1 is now a <u>band</u> in L_∞^*. The <u>disjoint complement</u> of L_1 (i.e., the set of all elements in L_∞^* disjoint to all elements in L_1) is also a band in L_∞^*, called the band of <u>singular linear functionals</u> on L_∞. It is evident that for any pair F_1, F_2 of positive elements in L_∞^* we have $\|F_1 + F_2\| = \|F_1\| + \|F_2\|$. This is due to the fact that L_∞^* is an abstract L-space.

More generally, let Φ and Ψ be a pair of conjugate and continuous Orlicz functions (also called Young functions). It is well-known that if Φ does not increase too fast, then the Banach dual L_Φ^* of the Orlicz space L_Φ can be identified with the Orlicz space L_Ψ, provided the norms in L_Φ and L_Ψ are appropriately chosen. Indeed, there are two methods according to which an Orlicz space is usually normed, giving rise to equivalent (but not necessarily equal) norms. If one wants L_Φ^* and L_Ψ isometric, and L_Φ is normed according to the first method, then L_Ψ has to be normed according to the second method, or vice versa. Norm isometry between L_Φ^* and L_Ψ occurs in particular if Φ satisfies a so-called <u>Δ_2-condition</u> (i.e., there exists a constant $M > 0$ such that $\Phi(2u) \le M\,\Phi(u)$ for all $u \ge 0$). An example is the case that $\Phi(u) = u^p$ for some p satisfying $1 \le p < \infty$. However, if Φ increases too fast (as in $\Phi(u) = e^u - 1$), then L_Ψ is a proper linear subspace of L_Φ^*. More precisely, in the real case again, L_Ψ is a band in the (real) Riesz space L_Φ^*. The disjoint complement of L_Ψ in L_Φ^* is also a band in L_Φ^*, called the band of the <u>(bounded) singular linear functionals</u> on L_Φ. It was proved by T. Ando ([1], 1960) that this subspace of all singular linear functionals on L_Φ is an

L-space, i.e., if S_1, S_2 are positive singular linear functionals on L_Φ , then $\| S_1 + S_2 \| = \| S_1 \| + \| S_2 \|$. The proof was extended to the case of discontinuous Orlicz functions by M.M. Rao ([2], 1968). His definition of singular functionals, however, did not cover all possible cases. The general situation for Orlicz spaces was discussed by the present author ([3], 1975).

Orlicz spaces are special examples of normed Köthe spaces. It may be asked, therefore, for which normed Köthe spaces we have the triangle equality $\| S_1 + S_2 \| = \| S_1 \| + \| S_2 \|$ for positive singular bounded linear functionals. More generally, we may ask for which normed Riesz spaces (i.e., normed vector lattices) this triangle equality holds.

Let L_ρ be a (real) normed Riesz space with Riesz norm ρ (i.e., ρ is a norm such that $\rho(f) = \rho(|f|)$ for all $f \in L_\rho$ and $0 \le u \le v$ implies $\rho(u) \le \rho(v)$). The notation $f_n \downarrow f_0$ means that the sequence $(f_n : n = 1,2,\ldots)$ in L_ρ is decreasing and $\inf f_n = f_0$; the notation $f_n \uparrow f_0$ is defined similarly. The Banach dual L_ρ^* of L_ρ is also a Riesz space; the positive bounded linear functionals on L_ρ form the positive cone of L_ρ^*. The element $F \in L_\rho^*$ is called an <u>integral</u> if $|F|(f_n) \downarrow 0$ holds for any sequence $f_n \downarrow 0$ in L_ρ . It is well-known that the set of all integrals is a band in L_ρ^*, which we shall denote by $L_{\rho,c}^*$. As an example, if L_ρ is the space $L_\infty = L_\infty(\Delta,\mu)$, then $L_{\rho,c}^*$ is exactly the space $L_1(\Delta,\mu)$. More generally, if L_ρ is the Orlicz space L_Φ, then $L_{\rho,c}^*$ is exactly the space L_Ψ, where Ψ is the conjugate Orlicz function of Φ. We return to the general case. The set $L_{\rho,s}^*$ of all elements disjoint to $L_{\rho,c}^*$ is also a band in L_ρ^*, and we have $L_\rho^* = L_{\rho,c}^* \oplus L_{\rho,s}^*$. The elements in $L_{\rho,s}^*$ are called the <u>singular bounded linear functionals</u> on L_ρ. The problem is now to find a necessary and sufficient condition for $L_{\rho,s}^*$ to be an abstract L-space. For the formulation of a satisfactory answer, we shall say that the normed Riesz space L_ρ is a <u>semi-M-space</u> if L_ρ satisfies the following condition: <u>If</u> u_1 <u>and</u> u_2 <u>are positive elements in</u> L_ρ <u>such that</u> $\rho(u_1) = \rho(u_2) = 1$ <u>and if</u> $\sup(u_1,u_2) \ge v_n \downarrow 0$, <u>then</u> $\lim \rho(v_n) \le 1$. It is evident that any M-space in the sense of Kakutani is semi-M. Another special case of a semi-M-space arises whenever the norm ρ is <u>absolutely continuous</u>, i.e., whenever $u_n \downarrow 0$ in L_ρ implies $\rho(u_n) \downarrow 0$ (as in L_p -spaces for $1 \le p < \infty$ and in Orlicz spaces

L_Φ if Φ satisfies a Δ_2-condition).

Our main result in this brief report is that the space L_ρ is a semi-M-space if and only if $L^*_{\rho,s}$ is an abstract L-space. Since all Orlicz spaces are semi-M-spaces, this is a meaningful generalization of the abovementioned result for Orlicz spaces. In one direction (L_ρ semi-M implies that $L^*_{\rho,s}$ is an L-space) the general result was proved by the first one of us; the proof in the converse direction was originally given for the case that L_ρ has an extra property (the principal projection property). The extra condition was removed by D.H. Fremlin (oral communication at the Oberwolfach Riesz Spaces Conference, June 1975).

For an indication of part of the proof we first note that if $F \in L^*_\rho$ is positive (notation $F \geq \Theta$, where Θ is the null functional), then according to what we observed above F has a unique decomposition $F = F_c + F_s$ with $\Theta \leq F_c \in L^*_{\rho,c}$ and $\Theta \leq F_s \in L^*_{\rho,s}$. The functionals F_c and F_s are the integral component and the singular component of F respectively. It is not difficult to prove that for any $0 \leq u \in L_\rho$ the value of $F_c(u)$ is given by the formula

$$F_c(u) = \inf(\lim F(u_n) : 0 \leq u_n \uparrow u).$$

Hence, if F is positive and singular, ~~then~~ $F_c(u) = 0$ for every $0 \leq u \in L_\rho$; i.e., for every $0 \leq u \in L_\rho$ and every $\varepsilon > 0$ there exists a sequence $(u_{n,\varepsilon}: n=1,2,\ldots)$ in L_ρ such that $0 \leq u_{n,\varepsilon} \uparrow u$ and $F(u_{n,\varepsilon}) < \varepsilon$ for all n.

Assume now that L_ρ is semi-M. Given positive elements S_1, S_2 in $L^*_{\rho,s}$, we indicate the proof that $\| S_1 + S_2 \| = \| S_1 \| + \| S_2 \|$. Let $\varepsilon > 0$. There exist positive elements u_1, u_2 in L_ρ of norm one and such that $S_i(u_i) > \| S_i \| - \frac{1}{2}\varepsilon$ ($i = 1,2$). Set $u = \sup(u_1, u_2)$ and $S = S_1 + S_2$, so $u \geq 0$ and $\Theta \leq S \in L^*_{\rho,s}$. Hence, in view of the above remark, there is a sequence $(u_{n,\varepsilon}: n=1,2,\ldots)$ in L_ρ such that $0 \leq u_{n,\varepsilon} \uparrow u$ and $S(u_{n,\varepsilon}) < \varepsilon$ for all n. Then $v_n = u - u_{n,\varepsilon}$ satisfies $u \geq v_n \downarrow 0$, so $\lim \rho(v_n) \leq 1$ since L_ρ is semi-M by hypothesis. Thus $\rho(v_n) < 1 + \varepsilon$ for every $n \geq n_0$, which implies that for $n \geq n_0$ we have

$$(1 + \varepsilon) \| S_1 + S_2 \| = (1 + \varepsilon) \| S \| \geq S(v_n) =$$

$$S(u-u_{n,\varepsilon}) > S(u) - \varepsilon \geq S_1(u_1) + S_2(u_2) - \varepsilon >$$

$$\|S_1\| + \|S_2\| - 2\varepsilon.$$

The rest is evident. For the proof that L_ρ is semi-M if $L_{\rho,s}^*$ is an L-space and for applications to rearrangement invariant Köthe spaces we refer to a forthcoming paper by the first one of us ([4]).

REFERENCES

[1] T. ANDO, Linear functionals on Orlicz spaces, Nieuw Archief voor Wiskunde (3) 8(1960), 1-16.

[2] M.M. RAO, Linear functionals on Orlicz spaces: General theory, Pacific J. of Math. 25 (1968), 553-585.

[3] E. de JONGE, The triangle equality for positive singular linear functionals on some classes of normed Köthe spaces, Proc.Kon.Ned.Akad.Wet. 78(1975), 48-69 (Indágationes Math.).

[4] E. de JONGE, The semi-M property for normed Riesz spaces, to appear in Proc. Kon.Ned.Akad.Wet.

ON R. PALLU DE LA BARRIÈRE'S CHARACTERIZATION OF NORMAL STATES

W. A. J. Luxemburg 1)

California Institute of Technology, Pasadena, California

1. Introduction. Let M be a commutative von Neumann algebra of operators acting in a complex Hilbert space H. By Re M we shall denote the real linear space of the self-adjoint elements of M. Under the natural order induced on Re M by the cone of the positive operators, Re M is a Dedekind complete Riesz space. The operator norm is a an absolute and monotone increasing norm on Re M under which Re M is an abstract L^{∞}-space with the identity operator as a strong order unit. A linear functional φ on M is called positive if it is non-negative on the cone of the positive operators, or in other words, if its restriction to the Riesz space Re M is a positive linear functional. A positive linear functional φ on M is called normal (see [1]) whenever for every directed system $\{A_\tau, \tau \in \{\tau\}\} \subset \text{Re } M$ which converges to an operator A in the weak operator topology, $\lim_\tau \varphi(A_\tau) = \varphi(A)$. It is evident that the positive linear functional $\varphi(A) = (Ax, x)$, where $x \in H$, is normal. It was shown conversely by R. Pallu de la Barrière [7] that every positive linear functional φ which is normal is of the form $\varphi(A) = (Ax, x)$ for some $x \in H$. The original proof of this surprising result was obtained, via the Gelfand theory, by representing M as the space of real continuous function on a compact extremely disconnected space and then by applying standard results of measure theory such as the Riesz representation theorem and the Radon-Nikodym theorem. Since Re M is a Riesz space of a very special type an alternative approach to the theorem of Pallu de la Barrière via the theory of Riesz spaces is possible. A proof of the theorem along such lines appeared recently in a paper by Peter G. Dodds [4]. The purpose of the present note is to show that there is still another way of proving the theorem of Pallu de la Barrière namely via a Riesz space version of the classical Radon-Nikodym theorem.

---

1) This work was supported in part by NSF Grant MPS-74-17845.

2. A Radon-Nikodym theorem. In this section as well in the rest of the paper we shall adhere to the notation and terminology from the theory of Riesz spaces as developed in [5]. If L is a Riesz space, then we write $f^+ = \sup(f,0), f^- = \sup(-f,0)$, $|f| = \sup(f,-f)$, from which $f = f^+ - f^-$ and $|f| = f^+ + f^-$ follows. A pair of elements f,g are called disjoint whenever $\inf(|f|,|g|) = 0$ and this is denoted by $f \perp g$. If D is an arbitrary subset of L, then we put $D^d = \{f: f \perp D\}$. A Riesz space L is called Dedekind complete whenever every non-empty subset of L which is bounded above has a least upper bound. A linear subspace K of L is called an ideal of L if $f \in K$ and $|g| \leq |f|$ implies $g \in K$. An ideal is called a band whenever $0 \leq f_\tau \uparrow f$, $f_\tau \in K$ for all τ implies $f \in K$. If L is Dedekind complete, then every band is a projection band, that is, if B is a band, then $L = B \oplus B^d$.

A real linear functional φ on a Riesz space L is called order bounded whenever $\sup(|\varphi(g)| : 0 \leq |g| \leq f)$ is finite for all $0 \leq f \in L$. The family of all order bounded linear functionals on L is denoted by $L^\sim$. Under the ordering defined by the cone of the positive linear functionals (a real linear functional φ is called positive whenever $0 \leq f \in L$ implies $\varphi(f) \geq 0$), $L^\sim$ is always a Dedekind complete Riesz space. An orderbounded linear functional $\varphi \in L^\sim$ is called normal whenever $0 \leq f_\tau \downarrow 0$, $f_\tau \in L$ for all τ, implies $\inf_\tau |\varphi(f_\tau)| = 0$. The set of normal integrals is a band in $L^\sim$ and is denoted by $L_n^\sim$. We have $L^\sim = L_n^\sim \oplus L_s^\sim$, where $L_s^\sim$ denotes the band of the singular functionals. For any element $\varphi \in L^\sim$ we set $N_\varphi = \{f: |\varphi|(|f|) = 0\}$. N_φ is an ideal of L, and is called the null ideal of φ. The disjoint complement N_φ^d of N_φ is called the carrier of φ, and is denoted by C_φ. The carrier of φ is a band and in fact the largest band on which φ is strictly positive. If σ is normal, then N_φ is a band and $C_\varphi \neq \{0\}$. If L is Dedekind complete, then for each normal linear functional φ we have $L = N_\varphi \oplus C_\varphi$. An element $\psi \in L^\sim$ is said to be absolutely continuous with respect to an element $\varphi \in L^\sim$ whenever $N_\psi \supset N_\varphi$ and is denoted by $\psi \ll \varphi$. If $\psi, \psi \in L_n^\sim$, then $\psi \ll \varphi$ if and only if ψ is in the band generated by ψ, in symbols, $\psi = \sup_n \inf(\psi, n|\varphi|)$. This result can be looked upon as a weak Riesz space version of the Radon-Nikodym theorem. It is possible, however, to strengthen this result. To this end, we shall need to introduce a few more notions of the theory of Riesz spaces. A linear transformation h of a Riesz space L into itself is called a Riesz homomorphism whenever h preserves also the lattice operations. We are particularly interested in a special kind of Riesz homomorphisms namely the so-called <u>orthomorphisms</u> which are those homomorphisms h which leave the bands invariant, in symbols, $h(B) \subset B$ for all bands $B \subset L$. It turns out that every orthomorphism h is always normal, that is, $0 \leq f_\tau \downarrow 0$ implies $h(f_\tau) \downarrow 0$ in L. Furthermore, two orthomorphisms h_1 and h_2 are equal if they coincide on a subset D of L which is dense in L in the sense that $D^d = \{0\}$. From this it follows, in particular, that if the Riesz space L has an order unit or a weak unit, then it is completely determined by its value on the order unit. So for instance, if L is the Riesz space $C(X)$ of all real continuous functions on a compact Hausdorff space, then every orthomorphism h of C into itself is of the form $h(f)(x) = h(1)(x)\cdot f(x)$ for all $x \in X$. For further details about the form of orthomorphisms on special Riesz spaces we refer the reader to [8].

If φ is a normal integral on a Riesz space L, then by Φ we denote the band in $L^\sim$ generated by φ. The Riesz space of all the normal order bounded linear functional on Φ we denote by $\hat{C}_\varphi$. Then obviously $C_\varphi \subset \hat{C}_\varphi$. Furthermore, $\hat{C}_\varphi$ is perfect in the sense that if $0 \leq \hat{f}_\tau \in \hat{C}_\varphi$, $\hat{f}_\tau \uparrow$ such that $\sup_\tau \psi(\hat{f}_\tau) < \infty$ for all $\psi \in \Phi$, there is an element $\hat{f} \in \hat{C}_\varphi$ such that $\sup_\tau \hat{f}_\tau = \hat{f}$ in $\hat{C}_\varphi$. We are now in a position to announce the following Riesz space version of the Radon-Nikodym theorem.

(2.1) THEOREM. <u>If L is a Dedekind complete Riesz space and if $\varphi, \psi \in L_n^\sim$ is a pair of normal order bounded linear functionals on L, then $\psi \ll \varphi$ if and only if there exists an orthomorphism h_φ of $\hat{C}_\varphi$ into $\hat{C}_\varphi$ such that $\psi(f) = \varphi(h_\varphi(f))$ for all $f \in C_\varphi$.</u>

For a complete discussion of the theory of orthomorphisms and Theorem 2.1 the reader is referred to [7]. It is also of interest to compare the above result with the abstract Radon-Nikodym type results contained in [3].

We are now ready to fulfill the promise of deducing the theorem of R. Pallu de la Barrière from Theorem 2.1.

3. <u>R. Pallu de la Barrière's representation theorem for normal states.</u>

Let M be an Abelian von Neumann algebra of operators acting on a Hilbert space H. The Dedekind complete normed Riesz space Re M of all the self-adjoint elements of

M is precisely the Banach dual of the band of all normal linear functionals on Re M. This follows from the fact that an Abelian von Neumann algebra is a closed subalgebra of the Banach algebra of all bounded operators on H in the weak operator topology and the fact that for each $x \in H$ the functional $\varphi_x(A) = (Ax,x)$ is normal. Furthermore, from this result it follows that Re M is also perfect.

Since for any operator $A \geq 0$ the transformation $h_A(T) = AT$, $T \in \text{Re } M$, is obviously an orthomorphism and the identity operator E on H is a weak order unit of Re M that any orthomorphism is of this form.

If $0 \leq \varphi \in (\text{Re } M)_n^{\sim}$, then its null ideal N_φ is a band in Re M and $\text{Re } M = N_\varphi \oplus C_\varphi$. By E_φ we denote the components of the identity E in the band C_φ. The projection operator E_φ on H will be called the carrier of φ. It is easily shown from the perfectness of Re M that, in this case, $\hat{C}_\varphi = C_\varphi$. Furthermore, $\varphi(A) = (Ax,x)$ for some $x \in H$, then $E_\varphi = E_x^{M'}$, where M' denotes the commutant of M, and $E_x^{M'}$ denotes the orthogonal projection in M onto the closure in H of the linear subspace generated by the orbits $\{Ax;\ A \in M'\}$.

We are now, finally, in a position to prove the theorem of Pallu de la Barrière [7] referred to in the introduction.

(3.1) THEOREM (R. Pallu de la Barriere). Let $0 \leq \varphi$ be a positive linear functional on M. Then φ is normal if and only if there is an element $y \in H$ such that $\varphi(A) = (Ay,y)$ for all $A \in M$.

Proof. Let E_φ be the carrier of φ. By [2], p. 19, there exists an element $x \in H$ such that $E_x^{M'} = E_\varphi$. Then φ is absolutely continuous with respect to the functional $\varphi_x(A) = (Ax,x)$, $A \in \text{Re } M$. Hence, by the Radon-Nikodym Theorem (Theorem 2.1), there exists an orthomorphism h such that $\varphi(A) = (h(A)x,x)$ for all $A \in \text{Re } M$. Since $h(A) = h(E)A$, $A \in \text{Re } M$, with $0 \leq h(E) = B \in \text{Re } M$, we obtain that $\varphi(A) = (BAx,x) = (AB^{\frac{1}{2}}x, B^{\frac{1}{2}}x) = (Ay,y)$, where $y = B^{\frac{1}{2}}x$; and the proof is finished.

REMARK. A normal state is a normal positive linear functional φ such that $\varphi(E) = 1$, where E is the identity operator.

REFERENCES

[1] J. Dixmier, Sur certaines espaces considérés par M. H. Stone, Summa Brasil Math., 2(1951), 151-182.

[2] J. Dixmier, Les algebres d'operateurs dans l'espace hilbertien, Gauthier-Villars, Paris 1957.

[3] Jean Dieudonné, Sur les théorème de Lebesgue-Nikodym III, Ann. Univ. Grenoble Sect. Sci. Math. et Phys.,23 (1948), 25-53.

[4] Peter G. Dodds, The order dual of an Abelian von Neumann algebra, Journ. Austr. Math. Soc. 18(1974), 153-160.

[5] W. A. J. Luxemburg and A. C. Zaanen, Riesz Spaces I, North-Holland Mathematical Library, Amsterdam, 1972.

[6] W. A. J. Luxemburg, Riesz homomorphisms, To appear in Ind. Math.

[7] R. Pallu de la Barrière, Sur les algèbres d'operateurs dans les espaces hilbertiens, Bull. Soc. Math. de France, 82(1954), 1-51.

[8] A. C. Zaanen, Examples of orthomorphisms, Journ. of Approx. Theory, 13(1975), 192-204.

NON COMMUTATIVE INTEGRATION IN SPECTRAL THEORY

By W. Hackenbroch
University of Regensburg (Germany)

Commutative spectral decomposition is, by Gelfand's isomorphism, essentially reduced to the approximation of bounded measurable functions by step functions or, from a slightly different point of view, to an integration procedure with respect to the spectral measure "multiplication by characteristic functions of measurable sets". In non-commutative W*-algebras we still have an abundance of projections, but now they form a non-distributive lattice rather than a Boolean algebra. On the other hand, such a "logic" is more than the collection of all its Boolean subalgebras (for the latter point of view compare [5,8]), as a W*-algebra is more than the collection of its commutative subalgebras. As is well-known, it is just this relation of incompatibility between various Boolean subalgebras which gives rise to the non-classical effects in quantum theory.

In this paper we study these phenomena in a rather simple function space model, namely an order unit space A of real functions together with its lattice $\mathfrak{F}$ of faces $a^{-1}(0)$, $0 \leq a \in A$. We shall assume that each face F gives rise to a linear projection in A generalizing the "multiplication by χ_F" - mapping of the commutative case.
In section 1. we show that these assumptions make the lattice $\mathfrak{F}$ a logic and characterize compatibility as well as the intimate relations between the lattice ordering in $\mathfrak{F}$ and the usual (pointwise) ordering in A. In section 2. we discuss the spectral theorem and its connection with measurability with respect to $\mathfrak{F}$. In 3. it is shown how W*-algebras easily fit into our model.
Questions of non-commutative probability theory on $\mathfrak{F}$ will be dealt with in a separate paper.

The predominance of the aspect of order in this note is strongly influenced by a recent series of papers by Alfsen and Shultz [1,2,3]. In particular, the measurability concept in section 2. is essentially a generalization of the well-known Stone condition (see [11]) following the discussion in [1].

Also G. Ludwig's approach to axiomatic quantum mechanics [9,10] very soon leads to a logic of faces, although the projectivity of these faces, which is the starting point of our discussion, in his treatment only indirectly comes out after posing several further axioms to make the logic a standard one. So it might be interesting to have a direct physical interpretation of projectivity in quantum theory.

1. The logic of faces.

Let Ω be a non-void set and A a linear space of bounded real-valued functions containing the constants and closed with respect to the sup-norm $\|\ \|$. In A we consider the pointwise ordering $\leq$ with positive cone A_+. Assume that the set $\mathfrak{F} = \{a^{-1}(0) : a \in A_+\}$ of all A-<u>faces</u> is provided with an orthocomplementation $^\perp : \mathfrak{F} \to \mathfrak{F}$ (i.e. $F \cap F^\perp = \emptyset$, $F^{\perp\perp} = F$ and $F \subset G \Rightarrow G^\perp \subset F^\perp$ for all $F,G \in \mathfrak{F}$), and furthermore, that every face $F \in \mathfrak{F}$ is <u>projective</u> in the following sense:
To every $a \in A_+$ there exists exactly one element $a_F \in A_+$ such that $a_F(t) = a(t)$ for $t \in F$ and $a_F(t) = 0$ for $t \in F^\perp$; in addition for the constant function one we require $1_F \leq 1$.

<u>Remark:</u> i) $\mathfrak{F}$ with respect to indusion ordering "$\subset$" is a σ-complete lattice, in fact, for the lattice operations in $\mathfrak{F}$ we have

$$\bigwedge_1^\infty F_n = \bigcap_1^\infty F_n \ ; \ \bigvee_1^\infty F_n = (\bigwedge_1^\infty F^\perp)^\perp \text{(in particular } F \vee F^\perp = \Omega).$$

(For since $^\perp$ is an order-reversing bijection we have only to show that $\bigcap_1^\infty F_n \in \mathfrak{F}$ for any sequence (F_n) in $\mathfrak{F}$. But $F_n = a_n^{-1}(0)$ with $a_n \in A_+$, so $\bigcap_1^\infty F_n = a^{-1}(0)$, where $a := \sum_1^\infty \frac{1}{2^n} \frac{a_n}{1+\|a_n\|}$ belongs to A_+ by completeness of A).

ii) For every $F \in \mathfrak{F}$ the mapping $A_+ \ni a \mapsto a_F \in A$ extends uniquely to a (monotone, contractive) linear projection $P_F : A \to A$, which in turn determines F according to $F^\perp = (1_F)^{-1}(0)$.
(For obviously $a \mapsto a_F$ is additive and positively homogeneous and therefore extends linearly to $A = A_+ - A_+$. By definition we have $F^\perp \subset (1_F)^{-1}(0)$. Conversely there is $a \in A_+$ with $0 \leq a \leq 1$ and $F^\perp = a^{-1}(0)$, so $a = a_F \leq 1_F$ and thus: $(1_F)^{-1}(0) \subset F^\perp$).

iii) For any $F \in \mathfrak{F}$ we have (with $\ker^+ P_F = (\ker P_F) \cap A_+$, $\operatorname{im}^+ P_F = (\operatorname{im} P_F) \cap A_+$

$$\ker^+ P_F - \ker^+ P_F \subset \ker P_F = \{a \in A : a(t) = 0 \text{ for } t \in F\}$$

$$\operatorname{im}^+ P_F - \operatorname{im}^+ P_F = \operatorname{im} P_F = A_{1_F} \quad \text{(order ideal in A generated by } 1_F)$$

$$\ker^+ P_F = \operatorname{im}^+ P_{F^\perp}$$

(Because for $a \in A$ with $\|a\| \leq 1$ we have $a - \frac{1}{2}(1+a) - \frac{1}{2}(1-a) =: a' - a''$ with $a', a'' \geq 0$, so $P_F a = 0 \iff a'_F = a''_F \iff a'|F = a''|F \iff a|F = -a|F \iff a|F = 0$, which proves the first statement. $A = A_+ - A_+$ implies that im P_F is positively generated; also for $a \in im^+ P_F$ we have $0 \le a = a_F \le \|a\| 1_F, \Rightarrow$ im $P_F \subset A_{1_F}$, and conversely for $0 \le a \le 1_F$ we have $a|F^\perp = 0$ and therefore $a = a_F \in$ im P_F. $im^+ P_F^\perp \subset ker^+ P_F$ is obvious; if on the other hand $a \in ker^+ P_F$, i.e. $a \ge 0$ and $a|F = 0$, we have $a = a_{F^\perp} \in im^+ P_F^\perp$).

By remark ii) we have a bijective mapping $F \mapsto P_F$ from $\mathfrak{F}$ onto a certain set $\mathcal{P}$ of positive projections in A. $\mathcal{P}$ becomes an orthocomplemented σ-complete lattice (with smallest element O the null-projection and largest element I the identity) under the lattice ordering and operations induced from $\mathfrak{F}$ via this bijection, which we denote again by $\le, \vee, \wedge, ^\perp$ (with an additional Index if there is any danger of confusion with corresponding notations in A). Also we write $F = F_P$ if $P = P_F$ (by abuse of notation). Finally we consider the orthogonality relation $P \perp Q$ in $\mathcal{P}$ when $P \le Q^\perp$ (or equivalently $Q \le P^\perp$).

<u>Proposition:</u> For $P \in \mathcal{P}$ and $a \in A_+$ the following are equivalent:

i) $Pa \le a$ (or equivalently: $a = Pa + P^\perp a$ or else $P^\perp a \le a$).

ii) $Pa = \max_A \{b \in A : 0 \le b \le a \text{ and } b|F_P^\perp = 0\}$

<u>Proof:</u> ii) $\Rightarrow$ i) is trivial

i) $\Rightarrow$ ii) Obviously, assuming i), $Pa \in \{b \in A : 0 \le b \le a \text{ and } b|F_P^\perp = 0\}$
Conversely for all such b we have $b = Pb \le Pa$. ■

In the sequel for any $a \in A$ we write

$$\mathcal{P}(a) = \{P \in \mathcal{P} : a = Pa + P^\perp a\}.$$

<u>Corollary:</u> Let $a \in A_+$ and $P, Q \in \mathcal{P}(a)$. Then

i) $P \le_{\mathcal{P}} Q \Rightarrow Pa \le_A Qa$

ii) $P \perp Q \Rightarrow (P \vee Q)a = Pa + Qa$

<u>Proof:</u> i) follows immediately from the Prop. ii).

ii) By i) we have $(P+Q)a \le (P+P^\perp)a = a$. On the other hand $(P+Q)a = 0$ on $F_P^\perp \cap F_Q^\perp$, and $= a$ on $F_P \cup F_Q$; but then also $(P+Q)a = a$ on

$F_P \vee F_Q$, since any $b \in A_+$ vanishing on the union $F \cup G$ of two faces already vanishis on their supremum $F \vee G$. ∎

Since $P = P(1)$, in particular we have $P1 \leq Q1$ for all $P,Q \in P$ such that $P \leq Q$. In fact the following stronger formula can easily be deduced from Prop. ii) for any $P,Q \in P$:

$$(*) \quad (P \wedge Q)(1) = \max_A \{a \in A : 0 \leq a \leq 1 \text{ and } a | F_P^\perp \cup F_Q^\perp = 0\} = P1 \wedge_{A_+} Q1.$$

An orthocomplemented σ-lattice L is called a <u>logic</u> whenever to any $u,v \in L, u \leq v$, there is $w \in L$ such that $u \perp w$ and $u \vee w = v$. Then neccesarily $w = v \wedge u^\perp$ (see [12] p. 105).

<u>Theorem 1:</u> P is a logic. For any $P,Q \in P$ the following are equivalent:

i) $P \leq_P Q$ ($\Longleftrightarrow$ $P1 \leq_A Q1$ by (*))

ii) $\text{im } P \subset \text{im } Q$ ($\Longleftrightarrow$ $QP = P$)

iii) $PQ = P$

<u>Proof:</u> To show that P is a logic, for $P \leq Q$ in P and $R = Q \wedge P^\perp$ we must prove $P \vee R = Q$, i.e. $Q1 = (P \vee R)1$, $= P1+R1$ by Cor. ii). But $Q1-P1 = 1$ on $F_Q \cap F_P^\perp$, and $= 0$ on $F_P \cup F_Q^\perp$ and thus on $F_P \vee F_Q^\perp$. To prove the equivalences, we note first of all that i) and ii) are equivalent by remark iii) above. i) implies iii) because for any $a \in A_+$ we have $PQa = Qa = a$ on F_P, and $= 0$ on $F_P^\perp$. Finally iii) implies i), for $\text{im}^+ Q^\perp = \ker^+ Q \subset \ker^+ P = \text{im}^+ P^\perp$ hence $\text{im } Q^\perp \subset \text{im } P^\perp$ and so $Q^\perp \leq P^\perp$ by the equivalence of i) and ii). Thus $P \leq Q$. ∎

The next theorem describes <u>compatibility</u> in the logic P in different forms, in particular through the order structure of A and through the multiplicative structure in the space of linear operators on A. We use the (symmetric) definition given in [12], p. 118:
$P \leftrightarrow Q$ iff there are pairwise orthogonal $R,S,T \in P$ such that $P = R \vee S$ and $Q = S \vee T$.

<u>Theorem 2:</u> For $P,Q \in P$ the following properties are equivalent:

i) $P \leftrightarrow Q$

ii) $P \in P(Q1)$

iii) $PQ = P\wedge_{\mathcal{P}}Q$

iv) $PQ \in \mathcal{P}$

v) $PQ = QP$

Proof: i) ⇒ ii): Writing $P = R \vee S$, $Q = S \vee T$ with $R,S,T\in\mathcal{P}$ pairwise orthogonal we have $PT = 0$ and $PS = S$, so

$$(PQ)1 = P(S1+T1) = (PS)1 = S1 \leq Q1.$$

ii) ⇒ iii): For any $a\in A_+$ $PQa = a$ on $F_P \cap F_Q$ and $0 \leq (PQ)a \leq \|a\|(PQ1) =$ on $F_P^\perp \cup F_Q^\perp$ and thus on $F_P^\perp \vee F_Q^\perp$. iii) ⇒ iv) is trivial. iv) ⇒ i): Put $R = PQ$; $S = P \wedge R^\perp$, $T = Q \wedge R^\perp$. Then by definition $P = R \vee S$ and $Q = R \vee T$; also $S \perp R \perp T$ is obvious. We must show $S \perp T$, i.e. $S1 \leq T^\perp 1$. Now

$$S1 = P1-PQ1 = 1-P^\perp 1 - (PQ)1 \leq 1-(PQ)1 - (P^\perp Q1) = 1-Q1,$$

the last equality by $(PQ)1 = R1$ and $R = RQ \leq Q$, i.e. $P\in\mathcal{P}(Q1)$; so $S1 \leq Q^\perp 1$, and obviously $Q^\perp \leq T^\perp$. v) ⇒ ii): $(PQ)1 = (QP)1 \leq Q1$. Finally the equivalent properties i) and iii) together with the symmetry of the compatibility relation imply v). ∎

2. Spectral theory

In this paragraph we strengthen the assumption of norm completeness of A assuming A to be monotone-σ-complete, i.e. every increasing sequence (a_n) in A, bounded above by some element of A, has a least upper bound $\overset{\infty}{\underset{1}{\vee}} a_n$ in A.

Theorem 1: Let $a\in A_+$ and consider $\mu : \mathcal{P}(a) \to A$ with $\mu(P) = Pa$. Then for any sequence (P_n) in $\mathcal{P}(a)$ of pairwise orthogonal elements the $\mathcal{P}$-supremum $\overset{\infty}{\underset{1}{\vee}} P_n$ is contained in $\mathcal{P}(a)$, and μ is σ-additive in the sense that

$$\mu\Big(\overset{\infty}{\underset{1}{\vee}} P_n\Big) = \overset{\infty}{\underset{k=1}{\vee}} \sum_{n=1}^{k} \mu(P_n).$$

<u>Proof:</u> If $P \perp Q$ in $\mathcal{P}(a)$, by the corollary in section 1 we have

$$(P \vee Q)a = Pa + Qa \leq Pa + P^{\perp}a = a$$

and therefore $P \vee Q \in \mathcal{P}(a)$. Furthermore, if Q_n is an increasing sequence in $\mathcal{P}(a)$ with $\overset{\infty}{\underset{1}{\vee}} Q_n = Q$ (in $\mathcal{P}$), $(Q_n a)$ is increasing in A, dominated by a. So $\overset{\infty}{\underset{1}{\vee}} Q_n a = b$ exists in A, and we have to show $Qa = b$ ($\leq a$, so in particular $Q \in \mathcal{P}(a)$). Now $b \leq Qa$ because $Q_n a = (QQ_n)a \leq Qa$; in particular $b | F_Q^{\perp} = 0$. On the other hand

$$Qa \geq b \geq Q_n a = Q_n Qa = Qa \text{ on } F_{Q_n}$$

and therefore $b = Qa$ on $\overset{\infty}{\underset{1}{\cup}} F_{Q_n}$ and thus on $\overset{\infty}{\underset{1}{\vee}} F_{Q_n} = F_Q$. ∎

For any $a \in A$ let us denote

$$\mathcal{B}(a) = \{P \in \mathcal{P}(a) \;:\; P \leftrightarrow Q \text{ for all } Q \in \mathcal{P}(a)\} .$$

<u>Corollary:</u> $\mathcal{B}(a)$ is a σ-distributive sublogic of $\mathcal{P}$.

<u>Proof:</u> As a compatible set, $\mathcal{B}(a)$ is contained in some σ-distributive sublogic ([12], Cor. 6.15. p. 123). On the other hand, it is closed under orthocomplementation (obviously) and under countable suprema in $\mathcal{P}$. For $P \leftrightarrow P_n$ (for $n = 1,2,\dots$) implies $P \leftrightarrow \overset{\infty}{\underset{1}{\vee}} P_n$ ([12], Lemma 6.10, p. 120); so in view of the theorem above we only have to show that $\mathcal{B}(a)$ is closed under finite suprema (or infima by orthocomplementation). Since $\mathcal{B}(a) = \mathcal{B}(a + \|a\| 1)$ we may assume $a \geq 0$. But then for $P, Q \in \mathcal{B}(a)$, we have

$$(P \wedge Q)a = (PQ)a \leq Pa \leq a,$$

so $P \wedge Q \in \mathcal{P}(a)$ and thus in $\mathcal{B}(a)$ by the lemma mentioned above. ∎

Let us denote, for S any locally compact Hausdorff space, a Baire set function $\pi : \text{Baire}(S) \to A_+$ a <u>spectral measure,</u> whenever there exists a σ-homomorphism $\lambda : \text{Baire}(S) \to \mathcal{P}$ such that $\pi = \lambda(\cdot)1$.
An example of a spectral measure is furnished (for any $a \in A$) by the Loomis-homomorphism $\lambda : \text{Baire}(S(a)) \to \mathcal{B}(a)$ with $S(a)$ denoting the Stone re-

presentation space for the σ-algebra $\mathcal{B}(a)$ (cf. [7], p. 102).

<u>Remark:</u> Let $\pi = \lambda(\cdot)1$: Baire(S) $\to A_+$ be any spectral measure.

i) By theorem 1. π is an A-valued measure in the sense of J.M. Wright [13], so in particular we can integrate bounded realvalued Baire functions with respect to π. The integration mapping $\pi : L^\infty(\pi) \to A$ evidently is a positive contraction.

ii) By theorem 2. of section 1 we have the multiplicativity property

$$\lambda(M\cap N) = \lambda(M)\lambda(N) \qquad (\Rightarrow \pi(M\cap N) = \pi(M)\wedge_{A_+} \pi(N) \text{ by } (*) \text{ in sect.1}).$$

From this (and the norm continuity of the integral) we obtain by integration with respect to N

$$\pi(\chi_M \cdot f) = \lambda(M)\pi(f) \qquad (M \text{ Baire set, } f\in L^\infty(\pi)).$$

iii) $\pi : L^\infty(\pi) \to A$ is bi-positive.
For if $\pi(f) \geq 0$ also

$$\lambda([f<0])\ \pi(f) = \pi(\chi_{[f<0]}\cdot f) \geq 0 \text{ and thus } = 0.$$

iv) The Baire measurability of f is reflected in the following "non-commutative" measurability property of $\pi(f)$: If for $t \in \mathbb{R}$ F_t denotes the face corresponding to $\lambda([f<t])$ we have

$$\pi(f)|F_t \leq t \quad \text{and} \quad \pi(f)|F_t^\perp \geq t.$$

For, from ii) we obtain

$$\pi(f)|F_t = \lambda([f<t])\pi(f)|F_t = \pi(\chi_{[f<t]}\cdot f)|F_t \leq t1|F_t\ ,$$
$$\pi(f)|F_t^\perp = \lambda([f\geq t])\pi(f)|F_t^\perp = \pi(\chi_{[f\geq t]}\cdot f)|F_t^\perp \geq t1|F_t^\perp\ .$$

This shows immediately (taking for $\mathcal{B}(a)$ e.g. the range of λ), that every $\pi(f)$ is measurable in the following sense (compare the notion of "spectral duality" in [1]):

<u>Definition:</u> $a\in A$ is called <u>measurable</u> if there exists a σ-distributive sublogic $\mathcal{B}(a) \subset \mathcal{F}(a)$ such that for any $P \in \mathcal{B}(a)$ and $s \in \mathbb{R}$ the following

property $*(P,s)$ is fulfilled:
To every $t \in \mathbb{R}$ there is a $P_t \in \mathcal{B}(a)$ with

$$P(a+s1)|F_{P_t} \le t \text{ and } P(a+s1)|F^{\perp}_{P_t} \ge t.$$

<u>Theorem 2:</u> In order that $a \in A$ admits a spectral decomposition

$$a = \int_{-\infty}^{\infty} t\, \pi_a(dt) \qquad (\pi_a \text{ a spectral measure on } \mathbb{R})$$

it is necessary and sufficient that a be measurable.

<u>Proof:</u> Necessity has just been shown in remark iv) above, so we prove sufficiency in several steps. Let $\mathcal{B}(a)$ be the Boolean σ-algebra from the definition of the measurability of a, S its Stonean space and λ : Baire(S) $\to \mathcal{B}(a) \subset P$ its Loomis representation, $\pi = \lambda(\cdot)1$ the corresponding spectral measure. It suffices to show $a = \pi(f)$ for some $f \in L^{\infty}(\pi)$, since then also

$$a = \int_S f d\pi = \int_{-\infty}^{\infty} t(\pi \circ f^{-1})(dt).$$

i) $s < t \Rightarrow P_s \le P_t$ for all $P_s, P_t \in \mathcal{B}(a)$ satisfying $*(I,O)$.
For otherwise $P_s \wedge P_t < P_s$ and therefore

$$0 \neq Q := P_s \wedge (P_s \wedge P_t)^{\perp} = P_s \wedge (P_s^{\perp} \vee P_t^{\perp}) = P_s \wedge P_t^{\perp}.$$

But $Q \le P_s \Rightarrow a|F_Q \le s$ and $Q \le P_t^{\perp} \Rightarrow a|F_Q \ge t > s$.

ii) For any $P \in \mathcal{B}(a)$ we have $a|F_P \ge s$ iff $Pa \ge sP1$.
For clearly $Pa \ge sP1$ implies $a \ge s$ on F_P. For the converse direction (by considering a-s1 which is also measurable with respect to $\mathcal{B}(a)$) we may assume $s = 0$. Now $a|F_P \ge 0 \Rightarrow Pa|F_P \cup F_P^{\perp} \ge 0$. Take $Q_n = P_{-\frac{1}{n}}$ according to $*(P,0)$. Then $F_{Q_n} \cap (F_P \cup F_P^{\perp}) = \emptyset$ and so

$$\emptyset = (F_{Q_n} \cap F_P) \vee (F_{Q_n} \cap F_P^{\perp}) = F_{Q_n} \wedge (F_P \vee F_P^{\perp}) = F_{Q_n}.$$

Therefore $Pa \ge -\frac{1}{n}$ on Ω for all $n \in \mathbb{N}$.

iii) $s < t \Rightarrow s(P_t - P_s)1 \le (P_t - P_s)a \le t(P_t - P_s)1$ for all

P_s , $P_t \in \mathcal{P}(a)$ according to *(I,O).

For, let F be the face corresponding to $Q = P_t \wedge P_s^{\perp}$. Since $a-s1|F \geq 0$ and $t1-a|F \geq 0$, from ii) we infer $sQ1 \leq Qa \leq tQ1$. But $P_t = P_s \vee Q$ and $P_s \perp Q$; therefore by the corollary of section 1

$$Q1 = (P_t-P_s)1 \ , \ Qa = (P_t-P_s)a \ .$$

iv) Taking a finite sequence $t_1 < t_2 <\ldots< t_r$ of real numbers with $t_1 < \inf a(\Omega)$ and $t_r > \sup a(\Omega)$, and corresponding $P_{t_i} \in \mathcal{P}(a)$ according to *(I,O), we obtain from iii)

$$\sum_{i=1}^{r-1} t_i(P_{t_{i+1}}-P_{t_i})1 \leq \sum_{i=1}^{r-1} (P_{t_{i+1}}-P_{t_i})a = a \leq \sum_{i=1}^{r-1} t_{i+1}(P_{t_{i+1}}-P_{t_i})1 \ .$$

From this it becomes obvious that a is in the image of the integration mapping $\pi : L^{\infty}(\pi) \to A$ for the spectral measure π described above. ∎

The measure π_a constructed in the proof obviously has compact support. By remark iii) above the image of its integration mapping is a vector lattice A(a) with respect to pointwise ordering, isomorphic to $L^{\infty}(\pi_a)$. As shown in [6] p. 128 , this implies a Freudenthal's formula

$$\pi_a(-\infty,t) = \bigvee_{n=1}^{\infty} (1 \wedge n(t1-a)_+)$$

with lattice operations taken in A(a).

As a corollary we determine the <u>center</u> of the logic $\mathcal{P}$ ([12] p.127),

$$\mathbb{C}(\mathcal{P}) = \{P \in \mathcal{P} : P \leftrightarrow Q \text{ for all } Q \in \mathcal{P}\} = \mathcal{B}(1):$$

<u>Corollary:</u> Assume that every $a \in A_+$ be measurable. Then $P \in \mathcal{P}$ is in the center of $\mathcal{P}$ iff P is an order-direct projection in A, i.e. $0 \leq Pa \leq a$ for all $a \in A_+$.

<u>Proof:</u> If $P \in \mathcal{P}$ is an order-direct projection we have in particular $P \in \cap\{\mathcal{P}(Q1) : Q \in \mathcal{P}\}$, which equals $\mathbb{C}(\mathcal{P})$ by theorem 2 of section 1. Converse suppose $P \in \mathbb{C}(\mathcal{P})$. Then the set $\{a \in A_+ : Pa \leq a\}$ is closed with respect to addition, multiplication by non-negative scalars and $\| \ \|$-limits. It contains the ranges of every spectral measure because $P \in \mathbb{C}(\mathcal{P})$. Thus it is all of A_+ by the spectral theorem, since the corresponding integrals are $\| \ \|$-convergent. ∎

3. Example: W*-algebras

Let $\mathfrak{A}$ be a von Neumann algebra of operators acting on some Hilbert space $(H;\langle,\rangle)$, containing the identity operator I for simplicity. Take for Ω the unit sphere of H and consider the mapping

$$\underline{a} \mapsto a : a(\xi) = \langle \underline{a}\xi,\xi\rangle \qquad (\xi\in\Omega)$$

of the self-adjoint part $\mathfrak{A}^s$ of $\mathfrak{A}$ onto some real vector space A of bounded real-valued functions. Obviously this mapping is bi-positive and carries I onto 1, so A is isometrically order-isomorphic to $\mathfrak{A}^s$; in particular A is (even pointwise) monotone complete.
As for the set $\mathfrak{F}$ of faces we have:

Proposition 1: The mapping $\underline{p} \mapsto F(p) := \mathrm{im}(\underline{p}) \cap \Omega$ is an order-isomorphism from the logic $\mathfrak{A}^{pr}$ of all projections in $\mathfrak{A}$ onto $\mathfrak{F}$.

Proof: Evidently we have a strictly bi-monotone injection of $\mathfrak{A}^{pr}$ into $\mathfrak{F}$ (note that $F(p) = \{\xi\in\Omega : \langle(I-\underline{p})\xi,\xi\rangle = 0\} \in \mathfrak{F}$). It is surjective since for $a\in A_+$ the selfadjoint projection $\underline{p}$ onto the kernel of $\underline{a}$ belongs to $\mathfrak{A}^{pr}$ and obviously $a^{-1}(0) = F(p)$. ■

Next, in $\mathfrak{F}$ we take the orthocomplementation

$$F(p) \mapsto F(p)^{\perp} := F(1-p)$$

induced by the ordinary orthocomplementation in the logic $\mathfrak{A}^{pr}$, so that $\mathfrak{F}$ becomes a logic isomorphic to $\mathfrak{A}^{pr}$.

Proposition 2: Every $F = F(p)$ is projective; the corresponding projection is given by

$$(P_F a)(\xi) = \langle \underline{p}\,\underline{a}\,\underline{p}\,\xi,\xi\rangle \qquad (a\in A;\ \xi\in\Omega).$$

Proof: To $a\in A_+$ consider $a_F : a_F(\xi) = \langle \underline{p}\,\underline{a}\,\underline{p}\,\xi,\xi\rangle$. Then $a_F\in A_+$ and $a_F = a$ on F, and $= 0$ on $F^{\perp}$. Let also $b\in A_+$ be $= a$ on F, and $= 0$ on $F^{\perp}$; then $b = a_F$, that is $\underline{b} = \underline{p}\,\underline{a}\,\underline{p}$, because $\langle\underline{b}\xi,\xi\rangle = \langle\underline{a}\xi,\xi\rangle$ for $\xi\in\mathrm{im}\,\underline{p}$, and $= 0$ on $\mathrm{im}(I-\underline{p})$. ■

Proposition 3: For $\underline{p}, \underline{q} \in \mathfrak{A}^{pr}$ consider $P = P_{F(p)}$, $Q = P_{F(q)}$.

i) $P \in \mathcal{P}(a)$ iff $\underline{p}$ commutes with $\underline{a}$

ii) $P \leftrightarrow Q$ iff $\underline{p}$ commutes with $\underline{q}$.

Proof: ii) is a special case of i), and i) is seen as follows:

$$P \in \mathcal{P}(a) \longleftrightarrow a = Pa + P^{\perp}a \overset{\text{(Prop.2)}}{\longleftrightarrow} \underline{a} = \underline{p}\,\underline{a}\,\underline{p} + (I-\underline{p})\underline{a}(I-\underline{p})$$

$$\longleftrightarrow 2\underline{p}\,\underline{a}\,\underline{p} = \underline{p}\,\underline{a} + \underline{a}\,\underline{p} \longleftrightarrow \underline{p}\,\underline{a} = \underline{p}\,\underline{a}\,\underline{p} = \underline{a}\,\underline{p}\,. \blacksquare$$

Finally, let $\underline{a} = \int_{\mathbb{R}} t\, \underline{\alpha}(dt)$ be the spectral decomposition of some $\underline{a} \in \mathfrak{a}^s$.

Since the values $\underline{\alpha}(M)$ ($M \in \text{Baire}(\mathbb{R})$) double-commute with $\underline{a}$, proposition 3. shows that $P_{F(\alpha(M))} \in \mathcal{B}(a)$. In fact,

$$\lambda_a : \text{Baire}(R) \to \mathcal{B}(a) \;:\; \lambda_a(M) = P_{F(\alpha(M))}$$

is a σ-homomorphism of Boolean algebras (since $\underline{\alpha}$ is). For the corresponding spectral measure $\pi_a = \lambda_a(\cdot)1$ we have

$$\pi_a(M)(\xi) = (\lambda_a(M)1)(\xi) = \langle \underline{\alpha}(M)\xi, \xi \rangle \,.$$

Therefore, $\underline{a} = \int_{\mathbb{R}} t\underline{\alpha}(dt)$ immediately translates into $a = \int_{\mathbb{R}} t\pi_a(dt)$. In particular every $a \in A$ is measurable.

Literature

[1] Alfsen, E. and F.W. Shultz: Non-commutative spectral theory for affine function spaces on convex sets I and II (Preprint)

[2] — : Geometric notions related to self-adjoint projections and one-sided ideals in operator algebras (Preprint)

[3] — : On the geometry of non-commutative spectral theory (Preprint)

[4] Fremlin, D.H.: Topological Riesz space and measure theory, University Press, Cambridge 1974

[5] Gudder, S.P.: Spectral methods for a generalized probability theory, Transactions Am. Math. Soc. 119 (1965) 428-442

[6] Hackenbroch, W.: Zur Darstellungstheorie σ-vollständiger Vektorverbände, Math. Z. 128 (1972) 115-128

[7] Halmos, P.R.: Lectures on Boolean algebras, Van Nostrand, New York 1963

[8] Kappos, D.A.: Generalized probability with applications to the quantum mechanics, The Greek Mathematical Society C. Caratheodory Symposium (1973) 253-270

[9] Ludwig, G.: Deutung des Begriffs "physikalische Theorie" und axiomatische Grundlegung der Hilbertraumstruktur der Quantenmechanik durch Hauptsätze des Messens, L.N. in Physics 4 (1970), Berlin-Heidelberg-New York

[10] Neumann, H.: The structure of ordered Banach spaces in axiomatic quantum mechanics L.N. in Physics 29 (1974) 116-121, Berlin-Heidelberg-New York

[11] Stone, M.H.: Notes on integration I and II, Proc. Nat. Acad. Sci. USA 34 (1948) 336-342 and 447-455

[12] Varadarajan, V.S.: Geometry of quantum theory I, Van Nostrand, Princeton 1968

[13] Wright, M.J.D.: Measures with values in partially ordered vector space, Proc. London Math. Soc. 25 (1972) 675-688

MEASURE THEORY ON ORTHOMODULAR POSETS AND LATTICES

by

Demetrios A. Kappos

0. Introduction

0.1. The mathematical description of statistical models for a classical mechanical system Σ can be made in terms of probability spaces (Ω,L(Σ),P), where Ω is the so called phase space associated with Σ, L(Σ)a Boolean σ-algebra of subsets of Ω, the so called statements (events), which can be made about Σ and P a probability measure an L(Σ). The points of Ω are considered as the states of the system Σ. The observables or physical quantities, in which the observer is interested, are then represented by L(Σ)-measurable functions (random variables) defined on Ω. Such a mathematical description is inadequate if Σ is a quantum mechanical system. The reason is that a phase space Ω can not be associated with a quantum mechanical system. Therefore the quantum mechanical statements about Σ fail to form a Boolean σ-algebra of subsets. They even fail to form in general an abstract Boolean algebra, if we consider these as propositions of a logic of Σ, because only experimentally verifiable statements are to be regarded as members of the logic L(Σ) of Σ. Such a logic is not a distributive lattice not even a modular lattice. Modern quantum theory works with the assumption that the logic L(Σ) is an orthomodular lattice [11β] or more generally on orthomodular poset [5β,11a].

0.2. A typical example of such an algebraic structure is the lattice L(H) of all closed subspaces of a separable infinite dimensional

Hilbert space H over the real or complex numbers. The logic $L(\Sigma)$ of a quantum mechanical system Σ is said to be standard, if it is isomorphic to this lattice $L(H)$. In quantum theory also not standard logics are considered, which are isomorphic to sublogics of a standard logic.

0.3. Let now Σ be a quantum mechanical system and ξ be a physical quantity (observable) associated with Σ. Then ξ can not be represented by a point function. In this case we have not a phase space Ω. But the statements that can be made concerning ξ in any experiment are of the type, which asserts that the value of ξ lies in some Borel subset A of the real line. Let $x_\xi(A)$ denote the statement that the value of ξ lies in the Borel set $A \subseteq R$, then we have a mapping:

$$(1) \qquad x_\xi \colon B(R) \ni A \to x_\xi(A) \in L(\Sigma).$$

of the Boolean σ-Algebra $B(R)$ of all Borel subsets of the real line R into the orthomodular σ-lattice or σ-poset $L(H)$ (= the logic of all statements (events) which can be made about Σ). In the classical case of a mechanical system Σ a mapping:

$$(2) \qquad h_\xi := \xi^{-1} \colon B(R) \ni A \to h_\xi(A) := \xi^{-1}(A) \in L(\Sigma)$$

of $B(R)$ into the Boolean σ-algebra of subsets of the phase space Ω, corresponds also to any observable (= $L(\Sigma)$-measurable function)ξ. This mapping is a σ-homomorphism of $B(R)$ into $L(\Sigma)$. the mapping $h_\xi := \xi^{-1}$ can replace the observable ξ in probabilistic problems about Σ. For example $P(h_\xi(A))$, $A \in B(R)$ defines the probability distribution of ξ on the real line R. Therefore if in the case of a quantum mechanical system Σ, the mapping (1) is defined as a σ-homomorphism of $B(R)$ in to the logic $L(\Sigma)$, then this σ-homomorphism x can be considered as a representation of an observable ξ. If moreover a kind of probability p can be defined on $L(\Sigma)$ [116,6], then p (x(A)), $A \in B(R)$ will be the probability distribution of the observable ξ on the real line R. The class $O(\Sigma)$ of all σ-homomorphisms of B(R) into the σ-lattice or σ-poset $L(\Sigma)$ can then be considered as the class of all observables associated with Σ.This is a pointfree way to introduce the notion of measurable functions over an abstract orthomodular σ-lattice or σ-poset.

0.4. C. Caratheodory had introduced 1938 in a paper [1a] also a pointfree way to define the notion of measurable functions, socalled place functions (= Ortsfuntionen) over an abstract Boolean σ-algebra and in this way had explained an algebraic measure and integration theory [1β]. This theory was later applied by me [7α,7β] in the probability theory, in order to study the algebraic structure of probability algebras and spaces of random variables. This algebraic theory was not taken in to consideration by quantum theory as far as I know. In quantum theory the knowledge of such an algebraic theory is more important, because there we have not a representation theorem for orthomodular lattices corresponding to the Stone representation theorem for Boolean algebras. In order to define the notion of states, a measure theory, especially for probability measures on orthomodular lattices or posets, is introduced and studied in quantum theory [11a,11β,5a,5β]. In this paper our aim is to give a brief sketch of such a generalized algebraic measure and integration theory using as a model our pointfree theory explained in the book of Carathèodory [1β] and in our book [7β]. In such a generalized theory lots of problems corresponding to the classical theory remain yet open. Some of them will be pointed out.

1. Orthomodular posets and lattices

1.1. A <u>partially ordered set</u> (briefly <u>poset</u>) $(P,\leq)$ is said to be an <u>orthocomplemented poset</u> (briefly <u>orthoposet</u>) if

1.1.1 There are two elements 0 (= zero) and e (= unit) in P such that

$$0 \leq a \leq e, \qquad \forall\, a \in P.$$

1.1.2 There exists a mapping $\perp: a \to a^{\perp}$ of P into itself such that

a) $(a^{\perp})^{\perp} = a, \quad \forall\, a \in P$

b) if $a \leq b$, then $b^{\perp} \leq a^{\perp}$

c) there is $a \vee a^{\perp} = e, \quad \forall\, a \in P$.

If moreover P is a lattice, then P is said to be an <u>ortholattice</u>

It is easy to show that this mapping is one-one and onto and that

d) $0^{\perp} = e$ and $e^{\perp} = 0$.

1.1.3 <u>Definition</u>. Two elements a and b in the orthoposet P are said to be <u>orthogonal</u> (<u>disjoint</u>), written

$$a \perp b \quad \text{if} \quad a \leq b^{\perp}.$$

We notice that $a \leq b^{\perp}$ is equivalent to $b \leq a^{\perp}$ i.e

$$a \perp b \quad \text{if} \quad b \perp a, \text{ i.e.}$$

Te relation $\perp$ is symmetric. It is easy to prove.

1.1.4 $0 \perp a$ and $a \perp a^{\perp}, \quad \forall\, a \in P$.

1.1.5 If $a \vee b$ resp. $a \wedge b$ exists, then $a^{\perp} \wedge b^{\perp}$ resp. $a^{\perp} \vee b^{\perp}$ exists and $a^{\perp} \wedge b^{\perp} = (a \vee b)^{\perp}$ resp. $a^{\perp} \vee b^{\perp} = (a \wedge b)^{\perp}$ and in general

$$(\vee a_i)^c = \wedge a_i^c \quad \text{resp.} \quad (\wedge a_i)^c = \vee a_i^c\,,$$

in the sense: if either side exists, so does the other and the two are equal

1.1.6 It is $a \wedge a^{\perp} = 0$.

1.1.7 $a \perp b$ implies $a \wedge b = 0$, although the inverse is not true.

1.1.8 If $\bigvee_{i\in I} a_i$ and $\bigvee_{j\in J} b_j$ exist and $a_i \perp b_j$, $\forall\,(i,j) \in I \times J$,

then $\bigvee_{i\in I} a_i \perp \bigvee_{j\in J} b_j$.

1.2. An orthoposet P is said to be an orthomodular poset if:

e) If $a \perp b$, then $a \vee b$ exists in P, We shall write then $a + b$ for $a \vee b$.

f) If $a \leq b$, then there is $d \in P$ such that $a \perp d$ and $a + d = b$.

We have then: [11a]:

1.2.1 The element d in f is unique and $d = b \wedge a^{\perp}$. We write $b - a := d = b \wedge a^{\perp}$ if $a \leq b$, the socalled (relative) orthocomplement of a relative to b.

1.2.2 In the presence of condition e) the condition f) is equivalent to:

f*) if $a \leq b$ then $a \vee (b \wedge a^{\perp}) = b$.

1.2.3 Then subset $P_b := \{a \in P : a \leq b\}$ forms an orthomodular poset with the same zero o and as unit the element b. The orthocomplement of a is then given by $b - a$ for any $a \in P_b$.

1.3. An orthomodular lattice is an orthomodular poset P in which for every pair $a,b \in P$ we have $a \vee b \in P$. Then there exist also $a \wedge b \in P$ and P is a lattice.

We have [s. 11b, Ch VI]:

1.3.1 Let P be an orthomodular lattice and a,b,c three elements in P such that $a \perp b$ and $b \leq c$, then

$$(a \vee b) \wedge c = (a \wedge c) \vee b,$$

This property is equivalent to 1.2.f).

1.3.2 We remark that (1.3.1) is a special case of the usual modular law, the socalled weakly modular law. A modular ortholattice is hence a special case of an orthomodular lattice. A distributive ortholattice is a Boolean algebra. We have for an ortholattice: distributive $\Rightarrow$ modular $\Rightarrow$ orthomodular the converse is not always true.

1.4. An orthomodular poset P resp. orthomodular lattice L is said to be an orthomodular σ-poset resp. orthomodular σ-lattice iff

e_σ) If $a_i \in P$, $i = 1,2,\dots$ are pairwise orthogonal (i.e. $a_i \perp a_j$, $i \neq j$) then $\bigvee_{i=1}^{\infty} a_i$ exists in P, we shall write then:

$$\sum_{i=1}^{\infty} a_i \quad \text{for} \quad \bigvee_{i=1}^{\infty} a_i$$

resp.

1) If $a_i \in L$, $i = 1,2,\dots$, then $\bigvee_{i=1}^{\infty} a_i$ exists in P, then $\bigwedge_{i=1}^{\infty} a_i$ exists also in L.

We have:

1.4.1 If $\sum_{i\in I} a_i$ and $\sum_{j\in J} b_j$ exist in a orthoposet and $a_i \perp b_j$ for every $(i,j) \in I \times J$, then $\sum_{i\in I} a_i \perp \sum_{j\in J} b_j$.

2. Compatibility in orthomodular posets.

2.1. Let P be an orthomodular poset then we say that an element $a \in P$ is compatible with an element $b \in P$ *) in symbols $a \leftrightarrow b$ if there are elements a_1, b_1, c in P which are pairwise orthogonal such that

$$a = a_1 + c \quad \text{and} \quad b = b_1 + c.$$

*) or a commutes with b or a and b are simultaneously verifiable. Least designation is used in the quantum mechanics for characterisation of observables, whose values can be simultaneously measured.

we have [11α,β, and 5β]:

2.1.1 If $a \leftrightarrow b$, then $b \leftrightarrow a$

2.1.2 If $a \perp b$, then $a \leftrightarrow b$

2.1.3 If $a \leq b$, then $a \leftrightarrow b$

2.1.4 If $a \leftrightarrow b$, any two of $a, b, a^{\perp}, b^{\perp}$ are compatible and $a \vee b$, $a \wedge b$ exist in P.

2.1.5 By the definition of $a \leftrightarrow b$ the elements a_1, b_1, c are unique and $c = a \wedge b$, $a_1 = a - (a \wedge b)$, $b_1 = b - (a \wedge b)$
$a^{\perp} = [b - (a \wedge b)] + (a \vee b)^{\perp}$, $b^{\perp} = [a - (a \wedge b)] + (a \vee b)^{\perp}$.

2.1.6 $a \leftrightarrow b$ iff $a \wedge b$ exists in P and $(a - (a \wedge b) \perp b$.

2.2. The subset $C(P) := \{a \in P : a \leftrightarrow x \text{ for every } x \in P\} \subseteq P$ is called the center of P. Obviously $C(P) \neq \emptyset$, namely $C(P) \supseteq \{o, e\}$.

2.3. Let P_1 and P_2 be to orthomodular posets then a mapping: $h: P_1 \to P_2$ is said to be a homomorphism of P_1 into P_2 if:

2.3.1 $h(e_1) = e_2$, Here o_i resp. e_i is the zero resp. the unit of P_i, $i = 1,2$.

2.3.2 If $a \perp b$, then $h(a) \perp h(b)$ and $h(a+b) = h(a) + h(b)$.

In the case of orthomodular σ-posets P_1 and P_2 an homomorphism is said to be σ-homomorphism if

2.3.2$_\sigma$ If $a_i \in P_1$, $i = 1,2,\ldots$ are pairwise orthogonal, then

$$h\left(\sum_{i=1}^{\infty} a_i\right) = \sum_{i=1}^{\infty} h(a_i).$$

It follows:

2.3.3 $h(O_1) = O_2$ and $h(a^{\perp}) = [h(a)]^{\perp}$

In fact: $e_2 = h(e_1 + O_1) = h(e_1) + h(O_1 = e_2 + h(O_1)$, hence $h(O_1) = O_2$
$e_2 = h(e_1) = h(a + a^{\perp}) = h(a) + h(a^{\perp})$, $e = h(a) + h(a^{\perp})$ with $h(a) \perp h(a^{\perp})$, hence $h(a^{\perp}) = [h(a)]^{\perp}$.

2.3.4 Let P_1 be an orthomodular lattice and h a homomorphism of P_1 into an orthomodular poset P_2, then $h(a) \vee h(b)$ and $h(a) \wedge h(b)$ exists in P_2 and we have $h(a \vee b) = h(a) \vee h(b)$ and $h(a \wedge b) = h(a) \wedge h(b)$.

Proof: In view of the de Morgans's law, the second equation will follow if we prove the first:
we put $a \vee b = d$, then $h(a) \leq h(d)$ and $h(b) \leq h(d)$. We put $a_1 = a$ and $a_2 = d - a_1$, then $d = a_1 + a_2$ and $a_1 \leq a$, $b_1 \leq b$ and $h(d) = h(a_1) \subseteq h(a_2)$. Let $x \in P_2$ with $h(a) \leq x$ and $h(b) \leq x$, we have $h(a_1) \leq h(a)$ and $h(a_2) \leq h(b)$ hence $h(a_1) \leq x$ and $h(a_2) \leq x$, $h(d) = h(a_1) + h(a_2 \leq x$, i.e $h(d) = h(a) \vee h(b)$ und $h(a \vee b) = h(a) \vee h(b)$.

2.3.5 If a homomorphism resp. σ-homomorphism of P_1 into P_2 is one - one then h is called an isomophism resp. σ-isomorphism. If the isomorphism is from P_1 onto P_2, then the two orthomodular posets are said to be isomorph resp. σ-isomorph. Let $P_1 \subseteq P_2$, then P_1 is called an orthomodular subposet resp. σ-poset of P_2 if the identity i.e the mapping $h(a) = a$ for every $a \in P_1$ is an isomorphism resp. σ-isomorphism from P_1 into P_2.

2.4. Let P an orthomodular σ-poset. If an orthomodular subposet resp. σ-subposet $B \subseteq P$ is a distributive lattice resp. σ-Lattice i.e a Boolean algebra resp. σ-algebra, then we shall say: B is a Boolean algebra resp. σ-algebra in P. There are Boolean algebras resp. σ-algebras in P. For example the subset $\{0,a,a^{\perp},e\}$ of P for every $a \in P$ forms a Boolean algebra in P.

2.4.1 Equivalent to the definition of the relation $a \leftrightarrow b$ in P is: $a \leftrightarrow b$, if there exists a Boolean algebra B in P containing a and b.

2.4.2 We say the elements of a subset $A \subseteq P$ are compatible briefly A is a compatible subset of P iff for every pair $a_i \in A$, $i = 1,2$ we have $a_1 \leftrightarrow a_2$.

2.4.3 Let $\{a_1,a_2\} \subseteq P$ i.e a subset of P with two elements, then if $\{a_1,a_2\}$ is compatible according to 2.4.1 there exists a Boolean algebra B in P containing $\{a_1,a_2\}$. This theorem does not hold in general for subsets of more than two elements *). But if an orthomodular poset satisfies the condition:

(c) If any three elements a, b, c in P are pairwise compatible, then we have also $a \leftrightarrow b \vee c$.

Then one can prove [5β]:

2.4.4 Any compatible subset A of an orthomodular poset resp. σ-poset P is contained in a Boolean algebra resp. σ-algebra in P iff P satisfies condition (c).

2.4.5 We remark that Condition (c) is always satisfied if P is an orthomodular lattice

2.5 In the following P is an orthomodular poset, which satisfies condition (c). A Boolean algebra B in P is called <u>maximal</u> if it is not properly contained in any other Boolean algebra in P. If P is moreover a σ-poset, then according to 2.4.4 every maximal Boolean algebra B in P is a σ-algebra. It is easy to prove by transfinite induction:

2.5.1 There are maximal Boolean algebras in any orthomodular poset P and if B_i, $i \in I$, is the not empty class of all maximal algebras in P, then we have

(I) $$P = \bigcup_{i \in I} B_i$$

if P is a σ-poset, then every B_i is a maximal Boolean σ-algebra in P. It is easy to prove:

(II) $$C(P) = \bigcap_{i \in I} B_i \geq \{o,e\}$$

*) Ramsey [9] gave an example of a subset of three elements

Therefore the center C(P) of P is a Boolean algebra and if P is a σ-poset, a Boolean σ-algebra in P.

2.5.2 We say that an orthomodular poset P is <u>irreducible</u> if $C(P) = \{o,e\}$.

Theorem 2.5.1 is very important for the introduction of measure theoretical concept in an orthomodular σ-poset.
Finch [3] has shown that an orthomodular σ-poset can be defined as an indexed set B_i, $i \in I$, of Boolean σ-algebras which are so related, that one can define orthocomplement and partially order relation in

$\bigcup_{i \in I} B_i$, such that $P := \bigcup_{i \in I} B_i$ is an orthomodular σ-poset.

2.5.3 An open problem is now: can one consider an indexed set (B_i, m_i), $i \in I$, of measure algebras, which are so related that one can define a measure on the Finch σ-poset $P := \bigcup_{i \in I} B_i$ such that the restriction of this measure from P to B_i is equal to m_i, $i \in I$ *).

3. <u>Carathèodory place functions</u> over an orthomodular <u>σ-poset</u>.

3.1 In this section 3 and in the following 4 we assume that the orthomodular poset satisfy condition (c). We shall then show that a reasonable algebraic measure and integration theory can be esplained on an orthomodular σ-poset or on an orthomodular σ-lattice. However it is interesting to study when an orthomodular σ-poset can be embedded isomorphically into an orthomodular σ-lattice. The reason is that a measure theory can be explained better on an orthomodular σ-lattice, since the

---

*) For the definition of a measure on an orthomodular poset s. section 4 of the present paper

lattice is closed for infima and suprema. This problem is open.

3.2 Let now P be an orthomodular σ-poset, then there exists a family of maximal Boolean σ-algebras B_i, $i \in I$, in P such that

$$(M) \qquad P = \bigcup_{i\in I} B_i .$$

Let now B be any Boolean σ-algebra and h be a σ-homomorphism of B into P. The image

$$h(B) = \{a \in B: \text{there exists } b \in B \text{ such that } b = h(a)\}$$

of B in P by this σ-homomorphism is a Boolean σ-algebra in P. In fact, the elements of h(B) are compatible. Let h(a) und h(b) $\in$ h(B), then obviously $a \leftrightarrow b$, hence there exists a_1, b_1, c pairwise disjoint (= orthogonal) such that $a = a_1 + c$, $b = b_1 + c$, then $h(a) = h(a_1) + h(c)$ $h(b) = h(b_1) + h(c)$ and $h(a_1)$, $h(b_1$, h(c) are pairwise orthogonal, therefore $h(a) \leftrightarrow h(b)$ for every pair h(a), h(b) i.e h(B) is contained in a maximal Boolean σ-algebra B_i in P. The homomorphism h can then be considered as a Boolean σ-homomorphism of B into the Boolean σ-Algebra B_i. Hence h(B) is a Boolean σ-algebra in P contained in B_i.

3.3 Let now $\mathbb{B} = \mathbb{B}(R_1)$ the Boolean σ-algebra of all Borel subset of the real line R_1 and any orthomodular σ-poset P be defined as the logic of a quantummechanical system. Then Varadarajan [11a,11b] Gudder [5a,5b] and other define the notion of an observable associated with P as a σ-homomorphism from $\mathbb{B}$ into P, i.e as a mapping

$$x : \mathbb{B} \to P$$

such that the following conditions are satisfied:

3.3.1 $x(\emptyset) = 0$, $x(R_1) = e$

3.3.2 if $A_1, A_2 \in \mathbb{B}$ mit $A_1 \cap A_2 = \emptyset$, then $x(A_1) \perp x(A_2)$

3.3.3 if $A_i \in \mathbb{B}$, $i = 1,2,\ldots$ are pairwise disjoint, then

$$x\Big(\bigcup_{i\geq 1} A_i\Big) = \sum_{i\geq 1} x(A_i).$$

Let O(P) be the set of all observables associated with P, then for every $x \in O(P)$, there is a maximal Boolean σ-algebra B_i in P, such that $x(\mathbb{B}) \subseteq B_i$. Therefore the observable x can be considered as a σ-homomorphism of $\mathbb{B}$ in B_i, i.e as an observable associated with B_i. We put now:

$$O(B_i) := \{x \in O(P) : x(\mathbb{B}) \subseteq B_i\}.$$

Then obviously we have the equation

$$\text{(o)} \qquad O(P) = \bigcup_{i\in I} O(B_i)$$

which corresponds to the equation (M) of section 3.2.

3.4 A mapping s resp. $\tau: R_1 \to P$, such that the following conditions are satisfied:

3.4.1 if $\xi < \xi'$ then $s(\xi) \leq s(\xi')$ resp. $\tau(\xi) \geq \tau(\xi')$

3.4.2 if $\xi_n \uparrow \xi$ resp. $\xi_n \downarrow \xi$, then $\vee\, s(\xi_n) = s(\xi)$ resp. $\wedge\, \tau(\xi_n) = \tau(\xi)$*)

3.4.3 $\bigvee_{n=1}^{\infty} s(n) = e$ and $\bigwedge_{n=1}^{\infty} s(-n) = 0$ resp. $\bigwedge_{n=1}^{\infty}$ (n) = 0 and $\bigvee_{n=1}^{\infty} \tau(-n) = e$.

is called an increasing left continuous resp. decreasing right continuous spectral chain in P.

We put

$$\gamma^* := \{(-\infty,\xi) := \delta_\xi, \text{ for all } \xi \in R_1\} \subseteq \mathbb{B}$$

resp.

$$\gamma_* := \{(\xi, +\infty) := \gamma_\xi, \text{for all } \xi \in R_1\} \subseteq \mathbb{B}$$

---

*) The elements $s(\xi_n)$ resp. $\tau(\xi_n)$, $n = 1,2,\ldots$ are pairwise compatible i.e $\vee\, s(\xi_n)$ resp. $\wedge\, \tau(\xi_n)$ exist in L.

and define the mapping:

$$x^*: \gamma^* \ni \delta_\xi \to s(\xi) := x^*(\delta_\xi) \in P.$$

resp.

$$y_*: \gamma_* \ni \gamma_\xi \to \tau(\xi) := y_*(\gamma_\xi) \in P$$

It is easy to prove that the mappings x^* resul x_* can be extended in both cases uniquely to a σ-homomorphism x resp. y from $\mathbb{B}$ into P such that:

$$x(\delta_\xi) = x^*(\delta_\xi) = s(\xi) \text{ resp. } y(\gamma_\xi) = y_*(\gamma_\xi) = \tau(\xi).$$

Conversely if $x \in O(P)$, then the restriction of x on γ^* resp. on γ_* defines a mapping:

$$s(\xi) = x(\delta_\xi) \text{ resp. } \tau(\xi) = x(\gamma_\xi) \text{ for all } \delta_\xi \in \gamma^* \text{ resp. } \gamma_\xi \in \gamma_*,$$

of R_1 into P,which is an increasing left continuous res. decreasing right continuous spectral chain in P.

3.5 Let $C^*(P)$ resp. $C_*(L)$ be the set of all increasing left continuous, resp. decreasing right continuous, spectral chains in L. The image $s(R_1)$ resp. $\tau(R_1)$ of an $s \in C^*(P)$ resp. $r \in C_*(P)$ lies in a maximal Boolean σ-algebra B_i in P. In other words the chain s resp. τ can be considered as a chain s resp. τ from R_1 into B_i. Hence we have the equation

(c) $\quad C^*(P) = \bigcup_{i\in I} C^*(B_i)$ resp. $C_*(P) = \bigcup_{i\in I} C_*(B_i)$

where

$$C^*(B_i) = \{s \in C^*(P) : s(R_1) \subseteq B_i\}$$

resp

$$C_*(B_i) = \{\tau \in C_*(P) : \tau(R_1) \subseteq B_i\}$$

we remark that:

3.5.1 Bodiou [2] defines $C^*(P)$ as the set of all observables (variables alètoires) in P where $P := L(H)$ i.e the orthomodular and atomic lattice of all closed subspaces of a Hilbert space H.

3.5.2 J.M. Olmsted [8] has defined $C_*(P)$ as the set of all real valued measurable functions over P, if P is any abstract Boolean measure σ-algebra and has introduced in $C_*(P)$ the algebraic structure of vector lattice. Obviously one can define now in the case of an orthomodular σ-lattice P the set $C_*(P)$ as the set of all real valued measurable functions over P. Now using the one to one correspodence between $C_*(B_i)$ and $O(B_i)$ we can introduce an algebraic and topological structure in every $O(B_i)$, $i \in I$ and can consider O(P) or $C_*(P)$ equiped with this structure.

3.6 However our aim is to give another way to define directly the notion of the socalled place functions, i.e measurable functions over P and to equip the space of these place functions with an algebraic and topological structure. We shall follow the corresponding theory of our book [7b] to define the place functions as random variables over a generalized probability algebra. We mention that the maximal Boolean σ-algebras B_i, $i \in I$, in P are probability σ-algebras in the case in which P := L(H) = the ortomodular σ-lattice of all closed subspaces of a separable infinite dimensional Hilbert space H. In fact in this case one can define on L(H) probabilities which are strictly positive [2]. Hence the restriction of such a probability on the maximal Boolean σ-algebras in L(H) is strictly positive, i.e every maximal Boolean σ- algebra in L(H) is a probability σ-algebra and that with a character $\leq \aleph_0$ *).

3.7 Let P be any orthomodular σ-poset as in section 3.2.**) A trial a in P is an at most countable subset of P:

$$\underline{a} = \{a_1, a_2, \ldots\},$$

where the elements a_i, $i = 1,2,\ldots$ are pairwise orthogonal, different

*) compare Kappos [7b] and sect. 4 of the present paper

of O and with $e = a_1 + a_2 + \dots$. Obviously this subset $\underline{a}$ is compatible. In fact, for every pair of elements we have $a_i \leftrightarrow a_j$ according to 2.1.3. Hence there exists a maximal Boolean σ-algebra B_i in P such that $\underline{a} \subseteq B_i$, i.e the trial $\underline{a}$ can be considered as a trial in B_i. An elementary place function or elementary random variable (briefly: e.r.v.) X over L is now defined as a real valued function X on a trial $\underline{a}$, i.e

3.7.1 $X: \underline{a} \ni a_i \to X(a_i) = \xi_i \in R, \ i = 1,2,\dots$

If the trial $\underline{a}$ is finite i.e. $\underline{a} = \{a_1, a_2, \dots, a_k\}$ then X is said to be a simple random variable (briefly: s.r.v.). Let $\mathbb{E}(P)$ resp. $S(P)$ the set of all e.r.v.'s resp. s.r.v.'s over P, then we have an equation of the form:

(e) $$\mathbb{E}(P) = \bigcup_{i \in I} \mathbb{E}(B_i) \quad \text{resp. (s)} \quad S(P) = \bigcup_{i \in I} S(B_i),$$

where

$$\mathbb{E}(B_i) = \{X \in \mathbb{E}(P): \text{if the trial } \underline{a} \text{ on which X is defined is a subset of } B_i\}$$

analogously is $S(B_i)$ defined. Moreover every $\mathbb{E}(B_i)$ resp. every $S(B_i)$ can be considered as the set of all e.r.v's resp. s,r.v's over the Boolean σ-algebra B_i in the meaning of my book [7β, ch.IV]. We can also define the indicator of every $a \in P$ as follows:

3.7.2 $1_a: \begin{matrix} a \to 1 \\ a \to 0 \end{matrix}$

and if $J(P)$ is the set of all indicator over O, then we have an equation

(i) $$J(P) = \bigcup_{i \in I} J(B_i), \text{ where } J(B_i) = \{1_a \in J, \ a \in \mathbb{B}_i\}, \ i \in I,$$

we have $J(P) \subseteq S(P) \subseteq \mathbb{E}(P)$.

An algebraic structure can be defined in every $\mathbb{E}(B_i)$, $i \in I$ as in

**) In this general case are B_i not always probability σ-algebras $i \in I$. However we shall explain the theory without this assumption.

[7β,ch.IV]. $E(B_i)$ can then be considered as a vector lattice in which a multiplication is also defined.

3.8 Let now x be an observable associated with P, i. $x \in O(P)$. We define

$$\sigma(x) = \cap K$$

over all K = closed subsets of R_1 with $x(K) = e$,

then $\sigma(x)$ is a closed subset of K and in particular the smallest closed subset $K \subseteq R_1$ such that $x(K) = e$. An observable is said to be discrete if $\sigma(x)$ is at most a countable subset of R_1. Let D(L) be the set of all discrete observables. Then we have an equation

(D) $$D(P) = \bigcup_{i \in I} D(B_i) \text{ with } D(B_i) = \{x \in D(P) : x(B_1) \subseteq B_i\}.$$

For an observable $x \in D(P)$ with $\sigma(x) = \{\xi_1, \xi_2, \ldots\} \subseteq R_1$ we have $\{\xi_i\} \cap \{\xi_j\} = \emptyset$, $i \neq j$, hence $x(\{\xi_i\}) \perp x(\{\xi_j\})$ and $\sum_{\xi_i \in \sigma(x)} x(\{\xi_i\}) = e$, we put $x(\{\xi_i\}) = a_i$, then $\underline{a} = \{a_1, a_2, \ldots\}$ is a trial. In this way one can correspond to every $x \in D(P)$ an e.r.v.

$$X : \underline{a} \ni a_i \to \xi_i, \; i = 1,2,\ldots$$

conversely to every e.r.v. $X: \underline{a} \ni a_i \to X(a_i) = \xi_i \in R_1$ corresponds a discrete observable x given by

$$x(A) := \sum_{\xi_i \in A} a_i \quad \text{for every} \quad A \in \mathbb{B}_1 .$$

3.9 A sequence $X_n \in \mathbb{E}(P)$, $n = 1,2,\ldots$ is said to be o-convergent to $X \in \mathbb{E}(L)$ if there exists a maximal Boolean σ-algebra B_i in P such that X and $X_n \in \mathbb{E}(B_i)$, $n = 1,2,\ldots$ and $\text{o-lim}\, X_n = X$ in $\mathbb{E}(B_i)$, as this Limit is defined in [7β]. Analogously we can define an o-fundamental sequence in $\mathbb{E}(P)$. The space $\mathbb{E}(L)$ of all e.r.v's over P can then be extented to the socalled stochastic space V(P) of all random variables over P*). Every space $\mathbb{E}(B_1)$ will then be

*) compare Extension theory in [7β,ch,IV]

extented to the stochastic space $V(B_i)$ of all random variables over B_i, $i \in I$ and we have the equation:

$$(v) \qquad V(P) = \bigcup_{i \in I} V(B_i) .$$

The space $V(P)$ is a poset and every subspace $V(B_i)$ is a vector-lattice with a multiplication and considered as a lattice, is conditionally σ-complete. We can $V(P)$ consider also as the space of all measurable functions (so called Carathèodory place functions) over P. One can prove that the theory of place functions introduced by Carathèodory [1b in ch. III and IV) is equivalent to our theory, if one defines the place functions over any Boolean σ-algebra. If every maximal Boolean σ-algebra B_i, $i \in I$ in P is a probability σ-algebra, then every $V(B_i)$ considered as a lattice is conditionally complete (s.7β ch.IV, theorem 5.3).

3.10. To every $X \in V(L)$ there corresponds exactly a chain $[X<\xi] \in P$, $-\infty < \xi < +\infty$, which is an increasing left continuous spectral chain in P*). To this chain there corresponds exactly a σ-homomorphism x from B into B_i, if $X \in V(B_i)$, i. e an observable $x \in O(B_i) \subseteq O(P)$ such that $[X < \xi] = x(\delta_\xi)$, $-\infty < \xi < +\infty$ conversely to every $x \in O(B_i) \subseteq O(P)$ there corresponds exactly a place function $X \in V(B_i) \subseteq V(P)$ such that $x(\delta_\xi) = [x < \xi]$, $-\infty < x < +\infty$. We have hence an one-one correspondence between $O(P)$ and $V(P)$ respectively between $O(B_i)$ and $V(B_I)$ for every $i \in I$. The avantage of our definition is that we have by the definition of $V(P)$ simultaneously an algebraic structure and convergence notion in this space. Moreover we remark that it is not difficult to explain in this way an analogous theory in the case in which the place functions are valued in a separable Banach-space or Banach-lattice (s.7β,ch.VII).

---

*) compare Extension theory in [7β,Ch.IV]

4. Measure on an orthomodular σ-poset

4.1 In quantum theory is defined a probability measure on $P = L(H) =$ = orthomodular lattice of all closed subspaces of an infinite dimensional separable Hilbertspace H, in order to define the concept of a state in P. This definition can be introduced to define a measure or particularly a probability on any orthomodular σ-poset [s.11a,5b] P. Namely, a measure μ on P is a map from P into $\bar{R}_+ = [0,+\infty]$ which satisfies:

4.1.1. $\mu(0) = 0$

4.1.2. If $a_n \in L$, $n = 1,2,\ldots$ pairwise orthogonal, then

$$\mu(\sum_{n=1}^{\infty} a_n) = \sum_{n=1}^{\infty} \mu(a_n).$$

A probability on P is a measure p on P which satisfies:

4.1.3 $p(e) = 1.$

Obviously the restriction of a measure μ on a maximal Boolean σ-algebra B_i in P is a well known concept of a measure on a Boolean σ-algebra. Gleason [3] has given a way to express every probability on $P = L(H)$ and between all these probabilities on L(H) there are such that: $0 < p(a) < 1$ for every $a \in L(H)$ with $a \neq 0$ and $a \neq e$, i.e strictly positive. Hence in the case $L = L(H)$ is every maximal Boolean σ-algebra B_i in L(H), a probability σ-algebra and that p-separable i.e. every B_i belongs to one of the three types of p-separable probability σ-algebra [7b,chII,Theorem 4.1].

4.2 Let μ be a measure on any orthomodular σ-poset P and V (P) be the set of all place functions over P, then one can introduce the integration theory in the meaning of Carathèodory [1β,7β] and define all the subspaces $L^q(P)$ of V (P) by the equations:

$$L^q(P) = \bigcup_{i\in I} L^q(B_i),\ 0 \le q \le \infty.$$

i.e the spaces of all r.v's $X \in V$ such that $|X|^q$ is integrable. In this way we have a measure and integration theory on an orthomodular σ-poset.

4.3 We do not intend to give a detailed presentation of such a theory. We mention moreover that many problems are open in such a generalized theory. For example 1) can one always define on an orthomodular poset a finite additive measure? Note that this is possible on a Boolean algebra*).

2) Let μ be a measure μ on an orthomodular subposet P_o of an orthomodular σ-poset P, can then this measure μ be extented to the smallest orthomodular σ-subposet of P over P_o. Other open problems can be found in our paper [7γ]. We remark, that it is interesting to study a generalized measure and integration theory also in the particular case of an modular σ-ortholattice or of a modular complemented lattice (compare about such problems Riecan [9a bis 9δ]).

References

1α. Carathèodory, C.: Entwurf für eine Algebraisierung des Integralbegriffes, Bayer. Acad. Wiss. Math - Nat. Klasse , München 1938,27-69.

1β. ——————, : Maß und Integral und ihre Algebraisierung. Birkhäuser Verlag 1956. Th same in English: Algebraic theory of measure and integration, chelsea 1963.

2. Bodiou, G.: Thèorie dialèctique des probabilites englopant leur calcul classique et quantique. Paris 1964

3. Finch, P.D.: On the structure of quantum logic Journal symbolic Logic 34(1969) 275 - 281.

*) compare Kappos [7a,sec.4]

4. Gleason, A.: Measures on classes of subspaces of a Hilbert space. J.Rat.Mech.Anal. 6(1957) 885 - 893.

5α. Gudder, S.P.: A generalized probability model for quantum mechanics
Ph. D. Thesis university of Illinois 1964

5β. ———, : Axiomatic quantum mechanics and generalized probability theory
Probabilistic methods in applied Math. Acad.Press Vol. 2(1970) 53 - 129

6. Jauch, I.M.: Foundation of quantum mechanics
Addison Wesley 1968

7α. Kappos, D.A.: Strukturtheorie der Wahrscheinlichkeitsfelder und Räume, Ergebnisse der Math. etc.
Springer Verlag Br. 24(1960)

7β. ———, : Probability algebras and stochastic spaces
Acad. Press 1969

7γ. ———, : Generalized probability with applications to the quantum mechanics
Caratheodory Symposium, September 1973
Greek Math. Society, 253 - 270.

8. Olmsted, J.M.H.: Lebesgue theory on a Boolean algebra, Trans. Amer. Math. Soc. 61(1947) 164 - 193.

9. Ramsey, A.A.: A theorem on two commuting observables, J. Math. Mech. 15(1966) 227 - 234.

10α. Riecan, B.: On the extension of a measure on lattices
Matem. Casopis Slov. Acad. 19(1970) 44 - 49.

10β. ———, : A note on the extension of measures om lattices, Matem. Casopis Slov. Acad. 20(1970) 239 - 244.

10γ. Riecan, B. and Györffy, L.: On the extension of the measures in relatively complemented lattices.
Matem. Casopis, Slov. Acad. 23(1973) 158 - 163.

10δ. Reican, B.: Regularity and approximation theorems for measures and integrals
Matem. Casopis Slov. Acad. 24(1974) 209 - 224

11α. Varadarajan, V.S.: Probability in Physics and a theorem on simultaneous observables
Comm.pure and aplied Math. XV(1962) 189 - 217

11β. Varadarajan, V.S.: Geometry of Quantum Mechanics
Vol 1, van Nostrand 1968

Professor Dr. D.A. Kappos
29 Lykabettou St.
Athens 135, Greece

A NEW APPROACH TO THE THEORY OF PROBABILITY VIA ALGEBRAIC CATEGORIES.

by Victor M. Bogdan

Catholic University of America, Washington, D. C.

1. Statistical Motivation.

In this section will be presented a statistical motivation which will permit later to define categories of the Theory of Probability and to establish isomorphism theorems between such categories. This approach will permit one to find axioms of the expectation spaces and later of a random process. The usual approach to the Theory of Probability starts from considerations of games of chance leading to the notion of probability space. This notion will come out naturally from the isomorphism of the category of expectation spaces with the category of probability spaces.

Let us consider the problem of an insurance company. Let $P = \{p\}$ be the set of all persons insured by the company. To be concrete let us assume that this company insures automobiles. Let $x:P \to R^n$ be a function such that its value $x(p) = (x_1(p),\ldots,x_n(p))$ for a given person p represents such parameters as age of the person, cost of his car, number of accidents in which the person was involved, cost of car repair in the past year due to accidents, etc..

Introduce the collection of all possible insurance policies for the company. By an insurance policy we shall understand a function $f:P \to R$ such that the value of the function $f(p)$ represents the amount the person p is charged for the insurance coverage. This value may be negative if the company pays to the person to cover his cost of repair. Such a policy is fair if for two persons with the same parameters $x(p_1) = x(p_2)$ the values $f(p_1)$, $f(p_2)$ are equal. Denote by

$$L = \{f:P \to R \mid f(p_1) = f(p_2) \ \text{ if } \ x(p_1) = x(p_2)\}$$

the collection of all fair insurance policies.

Let M be the number of persons in the set P . The insurance company is interested in the total amount of money they collect from all insured persons. This amount can be easily found if we know the average

$$E(f) = (\Sigma_{p \in P} f(p)) / M$$

of the amount of money collected when a given policy f is in effect.

Let us consider the properties of the triple (P,L,E) . Notice that $L \subset R^P$ and the collection of functions L is closed under the composition with any function $u: R^k \to R$ (for $k = 1,2$) . This is a very large collection of functions. It is more convenient to narrow it down to functions which might be of interest to the insurance company. The following functions seem to be natural, namely the function

(A) $(r_1, r_2) \to r_1 + r_2$,

and the collection of functions

(B) $r \to tr$ $(t \in R)$.

The first one has obvious interpretation that if in two consecutive years we know the insurance policies the sum of the insurance policies will give us a fair insurance policy from which one can derive the total of the collected money. The second one might be interpreted as a surcharge on the services.

Another important function which comes out naturally is

$$r \to |r| .$$

Indeed the company would be interested to know how much money it has received when a given insurance policy is in effect. This is the same as to assume that the collection of the fair policies L (of all admissable insurance policies) is closed under the composition with the function

$$r \to r^+ = \sup\{r,0\} .$$

Similarly to find the total amount of money the company pays to the insured it is natural to assume that the collection L is closed under the composition with

$$r \to r^- = \sup\{-r, 0\} .$$

This implies that the collection should be closed under the composition with the function

$$|r| = r^+ + r^- .$$

The collection L in the presence of its linearity is closed under the composition with the absolute value if and only if it is closed under the composition with

(C) $\quad (r_1, r_2) \to r_1 \cup r_2 = \sup\{r_1, r_2\}$

and

(D) $\quad (r_1, r_2) \to r_1 \cap r_2 = \inf\{r_1, r_2\}$.

Any collection of functions which is closed under the composition with the functions (A) - (D) is called a <u>linear lattice</u> and for short will be called in the sequel an <u>LL-space</u> . If such a linear lattice is closed under the composition with

(E) $\quad c_R : r \to 1$

it will be called an <u>ULL-space</u> . In the above we have used the following notation to denote the characteristic function c_A of a set A defined by $c_A(r) = 1$ if $r \in A$ and $c_A(r) = 0$ if $r \notin A$. From the identity $c_R \circ f = c_P$ it follows that the space L is closed under the composition with c_R if and only if $c_P \in L$.

Now let us consider the properties of the functional E . Notice that the functional is normalized, that is we have $E(c_P) = 1$, is linear, and is positive, that is if $f \geq 0$ and $f \in L$ then $E(f) \geq 0$.

<u>Definition 1.1.</u>

Consider a triple (X, L, E) such that $L \subset R^X$ and where L is an LL-space and E represents a positive linear functional on L . We shall say that a sequence $f_n \in L$ is a <u>Beppo Levi sequence</u> if f_n is either increasingly or decreasingly convergent everywhere to a finite function f and the sequence $E(f_n)$ is bounded. The function f will be called the <u>Beppo Levi limit</u> of the sequence f_n .

It is easy to see that the collection of fair insurance policies is closed under the Beppo Levi convergence.

Definition 1.2.

A triple (X,L,E) , where L is an ULL-space of R^X and E is a positive linear normalized functional on L is called an expectation space if L is closed under Beppo Levi convergence and E is continuous under Beppo Levi convergence.

It is easy to see that the triple (P,L,E) is an expectation space in the sense of this definition.

Consider the set $Y = x(P) \subset R^n$. Notice that the relation $f \in L$ is equivalent to the existence of a function $g \in R^Y$ such that $f = g \circ x$. Put $L_Y = R^Y$ and notice that we have

$$\frac{1}{M} \Sigma_{p\in P} f(p) = \Sigma_{y\in Y} g(y)r(y)$$

where the function $r(y)$ represents the number of elements in the set

$$x^{-1}(y) = \{p \in P: x(p) = y\}$$

that is the number of people with the same value of the parameter y divided by the number M of all insured. So it is natural to introduce the following functional

$$E_Y(g) = \Sigma_{y\in Y} g(y)\, r(y) \, .$$

Notice that the triple (Y,L_Y,E_Y) also represents an expectation space. The map $\rho: g \to g \circ x$ from the space L_Y into the space L is an example of a morphism in the category of expectation spaces. The precise definition of this notion will be given in the next section. Notice that this map satisfies the equality

$$E(\rho(g)) = E_Y(g) \quad \text{for all} \quad g \in L_Y \, .$$

The above function $r(y_1,\ldots,y_n)$ is known in statistics as n-dimensional contingency table.

From considering the above example one comes to the conclusion that to develop the theory of all possible insurace policies it is enough to work in the category of expectation spaces. For definitions of a category and a functor see MacLane [13] .

2. Category of Expectation and Lebesgue Integral Spaces.

Consider two expectation spaces (X_1,L_1,E_1) and (X_2,L_2,E_2). By a morphism

$$\rho: (X_1,L_1,E_1) \to (X_2,L_2,E_2)$$

in this category we shall understand a map $\rho: L_1 \to L_2$ which preserves the compositions with the functions (A),(B),(C),(D),(E) defined in the previous section, that is an ULL-space morphism ρ, such that ρ preserves the expectation and the Beppo Levi convergence, that is for every sequence $f_n \in L_1$ convergent in the sense of Beppo Levi to a function $f \in L_1$ the images $\rho(f_n) \in L_2$ converge in the Beppo Levi sense to the image $\rho(f) \in L_2$.

To get an equivalent but simple characterization of the morphisms in this category let us introduce the Stone's operation $r \to r \cap 1$. Any linear lattice closed under this operation will be called an <u>SLL-space</u>. Notice that an ULL-space is also an SLL-space.

We shall use the prefix P in front of the notion of a linear lattice (LL-space) to indicate that the linear lattice is closed under convergence everywhere. That is for every sequence of functions $f_n \in L$ which converges pointwise everywhere to a finite function $f_n \to f \in R^X$ we have $f \in L$.

Similarly we shall use the prefix D to indicate that the linear lattice (LL-space) is closed under the dominated convergence, that is for every sequence $f_n \in L$ such that $|f_n| \leq g$ for some function $g \in L$ the condition $f_n \to f$ implies $f \in L$.

<u>Definition 2.1.</u>

(X,L,E) is called a <u>Lebesgue Integral Space</u> if L is an SLL-space of R^X and E is a positive linear functional on L. Moreover L is closed under Beppo Levi's convergence and E is continuous on L under that convergence.

By a *morphism*

$$\rho: (X_1, L_1, E_1) \to (X_2, L_2, E_2)$$

between two Lebesgue integral spaces we shall understand a map $\rho: L_1 \to L_2$ which is an SLL-morphism preserving the integral, that is

$$E_2(\rho(f)) = E_1(f)$$

for every $f \in L_1$, and preserving the Beppo Levi convergence.

In the papers [1], [2], [3] has been presented the theory of locally convex function lattices which include Lebesgue spaces. Following it one obtains the theory of Lebesgue's integral from its axioms. Let (X, L, E) be a Lebesgue integral space. Then the set

$$L_o = \{f \in L : E|f| = 0\}$$

will be called *the null space* of the Lebesgue integral space. The null space L_o is an SPLL-space and the space L is an SDLL-space. Moreover the null set L_o is *solid* in L. This last property means that whenever $f \in L$ and $|f| \leq |g|$, where $g \in L$, we have $f \in L$. Notice that the class of SDLL-spaces contains the classes of SPLL-, UPLL-, UDLL-spaces.

A map $\rho: L_1 \to L_2$ between two SDLL-spaces will be called an SDLL-morphism if it is an SLL-space morphism which preserves the dominated convergence. The basic relations between the DLL-, SDLL-, PLL-, SPLL-spaces have been investigated in [4] - [9].

One can prove that a morphism in the category of Lebesgue Integral Spaces is an SDLL-morphism which preserves the integral and that a morphism in the category of expectation spaces is an UDLL-morphism which preserves the expectation.

3. Category of Delta Rings and Trace Functor.

Consider a delta ring space (X, V), that is a collection V of sets of the space X, which is closed under finite union, the difference, and countable intersection operations. By a morphism

$$\eta: (X_1, V_1) \to (X_2, V_2)$$

between two such spaces we shall understand a map $\eta: V_1 \to V_2$ mapping the first delta ring into the second and preserving all the delta ring operations. The class of delta rings with the class of the delta ring morphisms forms a category.

Now take the category of SDLL-spaces. Let (X,L) be a given SDLL-space. Assign to this space the pair (X,V), where

$$V = \{A \subset X : c_A \in L\}.$$

We remind that $c_A(x) = 1$ on A and 0 on $X \setminus A$.

To a morphism

$$\rho : (X_1,L_1) \to (X_2,L_2)$$

between two SDLL-spaces assign a map $\eta : V_1 \to V_2$ such that $A = \eta(B)$ whenever $c_A = \rho(c_B)$. These assignments define a functor on the category of SDLL-spaces into the category of delta ring spaces. This functor will be called the <u>TRACE functor.</u>

An interval of the form $I = (-\infty, -a)$ or (a, ∞) where $a > 0$ will be called a ray. Take a delta ring space (X,V) and define a pair (X,L) as follows. Let

$$M(V,R) = \{f \in R^X \mid f^{-1}(I) \in V \text{ for every ray } I\}.$$

This space we shall call the <u>space of measurable functions</u> generated by the delta ring space (X,V). Put $L = M(V,R)$.

Now consider a delta ring morphism $\eta : (X_1,V_1) \to (X_2,V_2)$. Assign to it a map $\rho : L_1 \to L_2$ defined by the following conditions: $\rho(f) = g$ where $f \in L_1$ and $g \in L_2$ is equivalent to $g^{-1}(I) = \eta(f^{-1}(I))$ for every ray I. Thus defined assignment establishes a functor from the category of delta-rings into the category of SDLL-spaces. This functor will be called <u>MEAS functor</u>.

<u>Theorem 2.1.</u>

(A) The MEAS functor restricted to the subcategory of sigma rings and the TRACE functor restricted to the subcategory of SPLL-spaces are inverse to each other.

(B) The TRACE functor restricted to the subcategory of UPLL-spaces and the MEAS functor restricted to the subcategory of sigma algebra spaces are inverse to each other.

It is worth-while to mention the following important result concerning SDLL-spaces. Consider two SDLL-spaces $L_1 \subset L_2$ and let V_1 and V_2 denote their traces respectively. By Theorem 4, Section 4, [1](see [9] for extensions of the theorem) the SDLL-space L_1 is solid in the SDLL-space L_2 if and only if the delta ring V_1 is an ideal in the delta ring V_2. Being <u>an ideal</u> means that the conditions

$$A \in V_2 \text{ and } A \subset B \in V_1$$

imply

$$A \in V_1 .$$

Consider a Lebesgue integral space (X,L,E). Assign to it the pair (X,L). If

$$\rho: (X_1,L_1,E_1) \to (X_2,L_2,E_2)$$

represents a Lebesgue integral space morphism then the map $\rho: L_1 \to L_2$ is an SDLL-morphism. Thus these assignments define a forgetful functor of the category of Lebesgue integral spaces into the category of SDLL-spaces.

Now consider a delta ring space (X,V). Assign to it the space (X,V^δ), where

$$V^\delta = \{A = \cup_{m\in N} A_m \mid A_m \in V,\ A_m \text{ being disjoint}\}.$$

To a delta ring morphism

$$\eta_o: (X_1,V_1) \to (X_2,V_2)$$

assign a morphism

$$\eta: (X_1,V_1^\delta) \to (X_2,V_2^\delta)$$

defined by the condition

$$A = \eta(B)$$

if and only if there exist disjoint sets $B_m \in V_1$ such that

$$A = \cup_{m\in N} \eta_o(B_m) \text{ and } B = \cup_{m\in N} B_m .$$

It is easy to prove that this definition is correct, that is the morphism η is well defined. The above assignment defines a functor from the category of delta rings into its subcategory of sigma rings. This functor will be called the SIGMA functor. Using it define the BAIRE functor by means of the composition

$$\text{BAIRE} = \text{MEAS} \circ \text{SIGMA} \circ \text{TRACE} .$$

Theorem 2.2

The BAIRE functor maps the category of SDLL-spaces into the category of SPLL-spaces. The object (X,L^δ) into which an object (X,L) is mapped can be characterized as the smallest SPLL-space containing the space L. The morphism ρ^δ into which the morphism

$$\rho: (X_1,L_1) \to (X_2,L_2)$$

is mapped can be characterized as the unique extension of the morphism ρ to an SDLL-morphism of the SPLL-space $(X_1,L_1{}^\delta)$ into the SPLL-space $(X_2,L_2{}^\delta)$.

Notice the inclusion $V \subset V^\delta$. It is easy to prove that V is an ideal in V^δ. This implies by a previously mentioned theorem that the SDLL-space L is solid in the SPLL-space L^δ.

4. Isomorphism of the Category of Expectation Spaces with the Category of Probability Spaces.

Consider a probability space (X,V,p), that is a sigma algebra V of sets of the space X and a positive normalized measure p on the sigma algebra V. By a morphism between two probability spaces

$$\rho: (X_1,V_1,p_1) \to (X_2,V_2,p_2)$$

we shall understand a map $\rho: V_1 \to V_2$ which is a delta ring morphism such that $p_2 \circ \rho = p_1$.

Given a probability space (X,V,p) define a triple (X,L,E) by the formulas

$$L = \{f \in M(V,R) \mid \int |f| dp < \infty\}$$

and

$$E(f) = \int f dp \quad \text{for all} \quad f \in L$$

where the integral is understood in the Lebesgue sense.

To a morphism $\eta: (X_1,V_1,p_1) \to (X_2,V_2,p_2)$ between two probability spaces assign a map ρ defined by the following conditions

$$\rho(f) = g \text{ , where } f \in L_1 \text{ and } g \in L_2$$

if and only if

$$g^{-1}(I) = \eta(f^{-1}(I)) \text{ for every ray } I .$$

This assignment defines a functor from the category of probability spaces into the category of expectation spaces. It will be called the LEBESGUE functor.

Now consider an expectation space (X,L,E). Assign to it a triple (X,V,p) where $V = \{ A \subset X \mid c_A \in L \}$ and $p(A) = E(c_A)$ for all $A \in V$. To a morphism

$$\rho\colon (X_1,L_1,E_1) \to (X_2,L_2,E_2)$$

between two expectation spaces assign a map $\eta\colon V_1 \to V_2$ of the corresponding sigma algebras defined by the condition

$$A = \eta(B) \text{ if and only if } c_A = \rho(c_B) .$$

These assignments define a functor from the category of expectation spaces into the category of probability spaces. This functor will be called the MTRACE functor.

Theorem 3.1

The MTRACE functor on the category of expectation spaces and the LEBESGUE functor on the category of probability spaces are inverse to each other.

For the case of the Lebesgue integral space category the above theorem can be generalized in several ways. Notice that if (X,L,E) is a Lebesgue integral space then

$$V = \{ A \subset X \mid c_A \in L \}$$

is a delta ring of sets. It can be extended to V^δ , the smallest sigma ring containing V , or to V^a , the smallest sigma algebra containing V , or to V^r , the largest sigma ring in which V is an ideal. This variety gives rise to four categories of measures isomorphic to the category of Lebesgue integral spaces.

5. Compositors for UPLL-spaces.

We shall say that a map $u:R^T \to R$ is a <u>compositor for all UPLL-spaces</u> if for every infinite tuple $(f_t)_{t\in T}$ of functions such that $f_t \in L$ for all $t \in T$ we have

$$v = u \circ (f_t)_{t\in T} \in L$$

for every UPLL-space L. To see that this notion appears in statistics in a natural way, let us return to the insurance problem considered in the first section. Thus consider the representations (Y,L_Y,E_Y) and (P,L,E). As was mentioned a function satisfies the relation $f \in L$ if and only if it is of the form $f = \rho(g) = g \circ x$. This yields $f(p) = g(x_1(p),\ldots,x_n(p))$ for all $p \in P$. This example shows that the statistician is interested in finding expectation of some composition of functions with given functions which constitute the process. Moreover a more careful analysis shows that this can be accomplished in either one of the two representations. Thus to investigate the problem one could use either the object $((P,L,E),(x_j)_{j\in T})$ where $T = \{1,\ldots,n\}$, or the object $((Y,L_Y,E_Y),(g_j)_{j\in T})$. In the last object the functions should satisfy the relation $\rho(g_j) = x_j$ where $g_j = e_j{}^T$ represents the natural projection $e_j{}^T(r_1,\ldots,r_n) = r_j$ onto the j-th coordinate. Now it is easy to see that both expectations $E_Y(g \circ (g_j)_{j\in T})$ and $E(g \circ (x_j)_{j\in T})$ are equal.

Let us denote by <u>Com(T)</u> the collection of all compositors $u:R^T \to R$ for all UPLL-spaces and for a fixed index set T.

Given a set $M \subset R^X$ of functions we shall denote by <u>UPLL(M)</u> the smallest set of functions closed under composition with the functions (A)-(E) of Section 1 and under pointwise convergence everywhere on X and containing the set M. We shall denote by <u>P(M)</u> the smallest set containing the set M and closed under pointwise convergence everywhere on X.

We shall use the notation

$$e_t^T\colon R^T \to R$$

to denote the natural projection onto the t-th coordinate

$$e_t^T\colon (x_s)_{s\in T} \to x_t \quad .$$

Theorem 5.1

The following identity holds

$$\mathrm{Com}(T) = \mathrm{UPLL}\{e_t^T : t \in T\} = P(C(R^T,R)) \quad ,$$

where $C(R^T;R)$ denotes the space of all continuous functions from the product space R^T with its Tikhonov topology into the space R of reals. Thus the collection $\mathrm{Com}(T)$ consists of Baire functions of the product space R^T .

Theorem 5.2

Let (X_j,L_j) for $j = 1,2$ denote an UPLL-space. Let

$$\rho\colon (X_1,L_1) \to (X_2,L_2)$$

be an UPLL-morphism. If $u \in \mathrm{Com}(T)$ and $f_t \in L_1$ for all $t \in T$ then

$$\rho(u \circ (f_t)_{t\in T}) = u \circ (\rho(f_t))_{t\in T} \, .$$

Thus every UPLL-morphism preserves compositions with Baire functions.

These two theorems permit one to introduce several categories isomorphic to the category of expectation spaces. For instance, the <u>category of extended expectation spaces and the category of complex extended expectation spaces</u>. The objects of these categories, roughly speaking, correspond to the Lebesgue integral considered on the collection of all finite measurable functions real or complex, respectively, with the usual restrictions. That is the integral of a real function exists if the function can be decomposed into difference of two nonnegative functions the integral of one of which is finite. The integral is defined in the usual way. In the case of complex functions the integral exists if the absolute value of the function has a finite Lebesgue integral. Using the above theorems one can easily characterize axiomatically such objects. The morphisms in these categories can be defined as maps preserving the composition with the Baire functions and preserving the integral.

6. The Category RARP of Representations of All Random Processes.

Abstraction from the objects obtained in considering the insurance problem leads to the notion of a representation of a random process. By a representation of a random process we shall understand an object

$$A = ((X,L,E),\ T \ni t \to f_t \in L^\delta)$$

where the triple (X,L,E) denotes an expectation space and $t \to f_t$ is a map from the set T into the UPLL-space $L^\delta = \mathrm{BAIRE}(L)$.

A morphism from such an object into the object

$$A_1 = ((X_1,L_1,E_1),\ T_1 \ni s \to g_s \in L_1{}^\delta)$$

can be defined as a pair $\eta = (\rho,\delta)$ where

$$\rho\colon (X\ ,L^\delta\ ) \to (X_1,L_1{}^\delta)$$

is an UPLL-morphism and its restriction $\rho\colon L \to L_1$ yields an expectation space morphism, the second transformation $\delta\colon T \to T_1$ is such that the diagram

$$\begin{array}{ccc} t & \to & f_t \\ \downarrow\delta & & \downarrow\varphi \\ s & \to & g_s \end{array}$$

commutes.

Now take a compositor $u \in \mathrm{Com}(T)$. Notice that the function $f = u \circ (f_t)_{t\in T}$ defined by the formula

$$f(x) = u((f_t(x))_{t\in T}) \quad \text{for all} \quad x \in X$$

belongs to the UPLL-space L^δ. Since every UPLL-morphism preserves compositions with the Baire functions we get

$$g = \rho(u \circ (f_t)_{t\in T}) = u \circ (\rho(f_t))_{t\in T} \in L^\delta .$$

This permits us to define the following triple (R^T,L_A,E_A) by the formulas

$$L_A = \{u \in \mathrm{Com}(T) \mid u \circ (f_t)_{t\in T} \in L\}$$

and

$$E_A(u) = E(u \circ (f_t)_{t\in T}) \quad \text{for all} \quad u \in L_A .$$

This triple represents an expectation space such that

$$\mathrm{BAIRE}(L_A) = L_A{}^{\delta} = \mathrm{Com}(T).$$

Assign to the object A the following object

$$B_A = ((R^T, L_A, E_A),\ T \ni t \to e_t{}^T \in L_A{}^{\delta}) .$$

We remind the reader that $e_t{}^T\colon R^T \to R$ denotes the natural projection onto the t-th coordinate.

Now consider the morphism $\eta = (\rho,\delta)\colon A \to A_1$. Notice that the formula

$$\rho_\eta(f) = f \circ (e_{\delta(t)}{}^S)_{t\in T} \quad \text{for all} \quad f \in \mathrm{Com}(T)$$

defines an UPLL-morphism of the space $\mathrm{Com}(T)$ into the space $\mathrm{Com}(S)$. Moreover the pair $\mu_\eta = (\rho_\eta,\delta)$ represents a morphism from the object B_A into the object B_{A_1} in the category of representations of all random processes.

The assignment $A \to B_A$ on the objects and $\eta \to \mu_\eta$ on the morphisms establishes a functor from the category RARP of representations of all random processes into itself. This functor will be called the <u>statistical functor</u> and will be denoted by STAT .

Let be given an expectation space of the form (R^T,L,E) where $L^\delta = \mathrm{Com}(T)$. The object $((R^T,L,E),\ T \ni t \to e_t{}^T \in L^\delta)$ will be called a Baire object.

<u>Theorem 6.1.</u>

The only objects of the RARP category which are invariant under STAT functor are the Baire objects.

<u>Definition 6.1.</u>

If A is a Baire object of the category RARP then the subcategory $\mathrm{STAT}^{-1}(A)$ will be called a <u>statistical random process</u> and any object $A_1 \in \mathrm{STAT}^{-1}(A)$ will be a representation of the statistical process.

Remark 6.1.

Notice that for every object $A_1 \in STAT^{-1}(A)$ there exists a morphism $\eta: A \to A_1$ defined by the pair $\eta = (\rho, \delta)$ where $\delta: T \to T$ is the identity map and ρ represents a morphism from the Baire object A into the object $A_1 = ((X,L,E), T \ni t \to f_t \in L^\delta)$ given by

$$\rho(f) = f \circ (f_t)_{t \in T} \quad \text{for all} \quad f \in Com(T) .$$

Remark 6.2.

To investigate all statistical random processes one may restrict himself to just Baire objects. Any property which is proven for such objects and is invariant under RARP morphism holds for all representations of the statistical random process.

Remark 6.3.

Since Baire functions depend only on countable number of coordinates, the STAT functor can be well defined in the quotient objects of the category RARP. Using this fact one can prove that the category of the quotient Baire objects is isomorphic with the category of Baire objects.

7. Isomorphism Between the Categories of Baire Expectation Spaces, Baire Measures, Kolmogorov's Distributions and Characteristic Functions.

Consider a Baire object $((R^T,L,E), T \ni t \to e_t{}^T \in L^\delta)$. Notice that this object is uniquely determined if one knows the expectation space $A = (R^T,L,E)$. Thus the forgetful functor from the category of Baire objects to the Baire expectation spaces establishes an isomorphism. A morphism in the category thus derived $\rho: A \to A_1$ is an expectation space morphism such that there exists a map $\delta: T \to T_1$ with the property

$$\rho(c_{(-\infty,a)} \circ e_t{}^T) = c_{(-\infty,a)} \circ e_{\delta(t)}{}^{T_1}$$

for all $a \in R$ and all $t \in T$.

Now consider an object of the form (R^T,V,p) where V is the smallest sigma ring with respect to which all continuous functions

$f:R^T \to R$ are measurable and p is a probability measure on it. A morphism between such objects $\eta:(R^T,V,p) \to (R^{T_1},V_1,p_1)$ is defined as a map $\eta:V \to V_1$ being a probability space morphism for which there exists a map $\delta:T \to T_1$ such that

$\eta((e_t^{\,T})^{-1}(-\infty,a)) = (e_{\delta(t)}^{\,T_1})^{-1}(-\infty,a)$ for all $t \in T$ and $a \in R$.

The LEBESGUE functor and the MTRACE functor establish isomorphism of this category, which will be called the <u>category of Baire measures</u>, with the category of Baire expectation spaces. Again consider a Baire expectation space (R^T,L,E). Take a finite subset $J \subset T$, $J \neq 0$ and define a function $F_J:R^J \to R$ by

$$F_J(a) = E(\prod_{j\in J} c_{(-\infty,a_j)} \circ e_j^{\,T})$$

where $a = (a_j)_{j\in J}$. The object consisting of all such functions

$$\{F_J:J \subset T,\ J \neq 0\ ,\ J \text{ is finite}\}$$

will be called <u>Kolmogorov's distribution</u> (compare [10], Section 3).

The function F_J is a <u>finite dimensional distribution</u>, that is it satisfies the following properties: It is nondecreasing and continuous on the left in each variable and it has the following limit properties:

$\lim\ F_J((a_t)_{t\in J}) = 0$ when $a_s \to -\infty$ for some $s \in J$,

$\lim\ F_J((a_t)_{t\in J}) = 1$ when $a_s \to \infty$ for all $s \in J$.

Thus the Kolmogorov's distribution can be characterized as a collection of finite dimensional distribution functions indexed by non-empty finite subsets of a given index set T satisfying the compatibility condition

$\lim\ F_S((a_t)_{t\in S}) = F_J((a_t)_{t\in J})$ when $a_t \to \infty$ for all $t \in S\backslash J$.

A morphism from Kolmogorov's distribution $\{F_J:J \subset T\ ,\ J\text{-finite}\}$ into the Kolmogorov's distribution $\{G_S:S \subset T_1,\ S\text{-finite}\}$ is a map $\delta:T \to T_1$ such that if we define

$$I(J,\delta):(a_t)_{t\in J} \to (b_s)_{s\in\delta(J)}$$

where

$$b_s = \inf\{a_t : \delta(t) = s\ ,\ t \in J\}\ ,$$

then

$$F_J(a) = G_{\delta(J)}(I(J,\delta)a)$$

for all $a \in R^J$ and all finite nonempty subsets $J \subset T$.

If the map δ is injective then the operator $I(J,\delta)$ is such that $b_{\delta(t)} = a_t$ for all $t \in J$. It is easy to prove that the category of Kolmogorov's distributions is isomorphic with the category of Baire expectation spaces and thus is isomorphic with the category of Baire measures. The isomorphism functors one obtains by restricting the objects or extending them. (Compare [10], Section 3).

Now let us define the category of characteristic functions. Let $(R^T)'$ denote the strong dual (of the locally convex space R^T) equipped with the inductive limit topology generated by its finite dimensional subspaces. Every element $z \in (R^T)'$ of the dual can be uniquely represented in the form $z = \Sigma_{t\in T}\ \mu_t(z)e_t^{\ T}$ where $\mu_t(\cdot)$ is a linear functional such that for every fixed argument z the condition $\mu_t(z) \neq 0$ is satisfied at most for a finite number of $t \in T$. The objects of the category of characteristic functions will consist of pairs $((R^T)', \varphi)$ where $\varphi:(R^T)' \to C$ is a positive definite and continuous function taking values in the space of complex numbers. A morphism $\varepsilon:((R^T)', \varphi) \to ((R^S)', \varphi_1)$ between two such objects is a map ε such that

$$\varphi_1(\varepsilon(z)) = \varphi(z) \quad \text{for all} \quad z \in (R^T)'$$

and there exists a map $\delta: T \to S$ such that

$$\varepsilon(\Sigma_{t\in T}\ \mu_t(z)e_t^{\ T}) = \Sigma_{t\in T}\ \mu_t(z)e_{\delta(t)}^{\ S} \quad \text{for all} \quad z \in (R^T)'.$$

To define the functor establishing the isomorphism of the category of characteristic functions with the category of Baire expectation spaces take a Baire object (R^T,L,E). Notice that the function

$$\varphi(z) = E(\mathrm{Cos}(\Sigma_{t\in T}\ \mu_t(z)\ e_t^{\ T})) + i\ E(\mathrm{Sin}(\Sigma_{t\in T}\ \mu_t(z)\ e_t^{\ T}))$$

is well defined and represents a positive definite and continuous function in the inductive limit topology generated by all finite dimensional subspaces. This assignment yields the object $((R^T)', \varphi)$ of the category of characteristic functions. Since a morphism in both categories is generated by a unique map $\delta: T \to T_1$ the definition of the assignment of the morphisms to obtain a functor establishing the isomorphism between these two categories is evident. The proof that the functor obtained in such a way yields an isomorphism makes use of the Bochner theorem on representation of positive definite functions by measures. (Compare [11]).

8. Kolmogorov's Strong Law of Larger Numbers.

In this section we shall present for illustration how one can derive the Kolmogorov strong law of large numbers within the categories defined in the previous section.

Consider a one dimensional distribution function $F: R \to R$ such that $\int |x| dF(x) < \infty$. Let T denote the set of all positive integers. For any finite nonempty subset $J \subset T$ define

$$F_J(a) = \prod_{j\in J} F(a_j) \quad \text{where} \quad a = (a_j)_{j\in J} .$$

Consider the object $K = \{F_J: J \subset T\}$. It is easy to see that this is a Kolmogorov's distribution. Now consider the injective map given by

$$\delta(n) = n + 1 \quad \text{for all} \quad n \in T .$$

Notice that this map establishes a morphism of the Kolmogorov object K into itself. Indeed we have

$$F_{\delta(J)}(I(J,\delta)a) = F_{\delta(J)}((b_{\delta(t)})_{t\in J}) =$$

$$F_{\delta(J)}((b_s)_{s\in\delta(J)}) = \prod_{s\in\delta(J)} F(b_s) = \prod_{t\in J} F(a_t) = F_J(a)$$

for all $a \in R^J$ and all finite nonempty subsets $J \subset T$.

Let us consider now the isomorphic object $((R^T, L, E), T \ni n \to e_n^T \in L^\delta)$ to this Kolmogorov's distribution K in the category of Baire representations. Let ρ denote the morphism in that category isomorphic to the morphism δ. The morphism ρ represents an UPLL-morphism

$$\rho: L^\delta \to L^\delta = \mathrm{Com}(T) .$$

Using its definition one can derive that it must be of the form

$$\rho(f) = f \circ \delta' \quad \text{for all } f \in L^{\delta} ,$$

where δ' maps R^T into itself and is given by the formula

$$\delta'(x_t)_{t\in T} = (x_{\delta(t)})_{t\in T} \quad \text{for all } x \in R^T .$$

Thus we get

$$E(f \circ \delta') = E(f) \quad \text{for all } f \in L ,$$

which yields that the map $\delta' : R^T \to R^T$ is measure preserving. Thus the conditions of the Birkhoff-Khinchin Pointwise Ergodic Theorem are satisfied (See Halmos [12]). We shall apply this theorem to the projection $e_1{}^T$ onto the first coordinate space. To prove the summability of this projection consider a set $S = \{1\}$ being a singleton and put $F_{\{1\}} = F$. This defines a Kolmogorov's object $K_o = \{F_J : J \subset S,\ J \neq 0\ ,\ J$ is finite$\}$. The identity map $\delta_o(1) = 1$ of S into T establishes a morphism

$$\delta_o : K_o \to K .$$

Now take the isomorphic Baire objects in the category RARP. Put

$$((R^S, L_S, E_S),\quad S \ni t \to e_1{}^S \in L_1^{\delta}\ ) .$$

From the relation

$$F(x) = F_S(x) = E_S\, c_{(-\infty,x)} \circ e_1{}^S = E_S\, c_{(-\infty,x)}$$

for all $x \in R$ one can easily derive the summability of the projection and the equalities

$$E_S(e_1{}^S) = \int x \, d\, F(x), \quad E_S(|e_1{}^S|) = \int |x| \, d\, F(x) .$$

Denote by ρ the morphism isomorphic to δ_o . Since $\rho_o : e_1{}^S \to e_1{}^T$ we get

$$E_S(e_1{}^S) = E(e_1{}^T) \quad \text{and} \quad E_S(|e_1{}^S|) = E(|e_1{}^T|) .$$

Now from the identity

$$e_1{}^T \circ (\delta')^n = e_n{}^T$$

and the pointwise ergodic theorem we derive

$$s_n = \frac{1}{n} \Sigma_{j=1}{}^n e_n{}^T \to \int x \, d\, F(x) \quad \text{a.e.} \qquad \text{(A)}$$

Notice the relation $s_n \in \mathrm{Com}(T)$.

Consider a function $u: R^T \to R$ defined by $u(r) = \lim_n r_n$ if the last limit exists and $u(r) = 0$ otherwise, where $r = (r_n)_{n \in T}$. It is easy to prove that this is a Baire function that is $u \in Com(T)$.

Put $a = \int x \, d\, F(x)$ and

$$v = u \circ (s_n)_{n \in T} - ac_{R^T} .$$

Since the space $Com(T)$ is itself an UPLL-space, it is closed under composition with Baire functions. Thus we get $v \in Com(T)$. Therefore the limit condition (A) can be written now as $E(|v|) = 0$.

Now consider any other representation

$$((X_1, L_1, E_1) , T \ni n \to f_n \in L_1^\delta)$$

of the statistical random process with the Baire object

$$((R^T, L, E) , T \ni n \to e_n^T \in L^\delta) .$$

Let $\eta = (\rho, e)$ be the morphism from the Baire object into that representation. Thus by the definition of the morphism we have

$$\rho(v) = \rho(u \circ (\tfrac{1}{n} \Sigma_{j=1}^n e_j^N) - a) = u \circ (\tfrac{1}{n} \Sigma_{j=1}^n f_j) - a .$$

This yields

$$E_1 |\rho(v)| = E|v| = 0$$

which proves

$$\tfrac{1}{n} \Sigma_{j=1}^n f_j \to a \quad \text{a.e.}$$

where the convergence almost everywhere is with respect to the null sets generated by the expectation space (X_1, L_1, E_1). Thus the statistical character of the strong law of large numbers is established.

9. General Remarks.

It is easy to show that such notions like convergence in probability, almost sure convergence, and convergence in distribution can be expressed in terms of expectation of a Baire function of a process. From such an expression one can easily deduct that these notions do not depend on the representation of a given statistical process. Similarly one can prove that a measurable or summable process is a statistical notion. The same is also true for the notion of a conditional expectation of a random variable with respect to a family of random variables.

Finally it is worthwhile to mention about another natural functor in the category of representations of all random processes. This functor will be called PHYS. To define it consider a representation

$$A = ((X_1, L_1, E_1) , T \ni t \to f_t \in L_1^{\delta})$$

of a random process. The point

$$(f_t(x))_{t \in T} \in R^T$$

for a given $x \in X$ will be called a trajectory of the process. The collection of all possible trajectories of the process $B = Sp(A)$ will be called the spectrum of the process. If

$$((R^T, L, E) , T \ni t \to e_t^T \in Com(T))$$

is a Baire object representing the statistical random process A then by restricting the functions $f \in Com(T)$ to the spectrum and putting

$$E_B(f|_B) = E(f) \quad \text{for all} \quad f \in L$$

one obtains a well defined random process

$$((B, L_B, E_B) , T \ni t \to e_t^T|_B \in L_B^{\delta}) .$$

This assignment to a process A, the process defined on its spectrum together with the assignment of morphisms as in the case of the STAT functor yields the PHYS functor. This functor does not distinguish statistically equivalent processes which have the same trajectories.

10. Calculus of Baire Functions and the New Process Functor.

Let

$$A_o = ((R^T, L, E), T \ni t \to e_t^T \in L^{\delta})$$

be a Baire object in the category RARP. Consider any fixed collection $S \subset Com(T)$ of Baire functions. Take any representation

$$A \in STAT^{-1}(A_o)$$

where

$$A = ((X_1, L_1, E_1), T \ni t \to f_t \in L_1^{\delta})$$

and put

$$S \circ A = ((X_1, L_1, E_1), S \ni s \to g_s \in L_1^{\delta}) ,$$

where

$$g_s = s \circ (f_t)_{t \in T} \quad \text{for all} \quad s \in S .$$

Let

$$B = \text{STAT} \quad (S \circ A_o)$$

and let δ_T denote the identity map on the set T. The assignment $A \to S \circ A$ on objects and $(\rho, \delta_T) \to (\rho, \delta_S)$ on morphisms yields the New Process Functor from the category $\text{STAT}^{-1}(A_o)$ into the category $\text{STAT}^{-1}(B)$.

This functor permits one, roughly speaking, to identify any statistical random process with a family of Baire functions and thus to replace the calculus of random process by the calculus of Baire functions.

References

[1] Bogdanowicz, W. M.,''Theory of a class of locally convex vector lattices which includes the Lebesgue spaces'', Proc. Nat. Acad. Sci. USA, 66(1970), 275-281.

[2] Bogdanowicz, W.M.,''Locally convex lattices of functions in which Lebesgue type theory can be developped'',Bull. de l'Acad. Polon. Sci., 19(1971), 731-735.

[3] Bogdanowicz, W.M.,''Minimal extension of Daniell functionals to Lebesgue and Daniell-Stone integrals'',Bull. l'Acad. Polon. Sci., 19(1971), 1093-1100.

[4] Bogdanowicz, W. M.,''Measurability and linear lattices of real functions closed under convergence everywhere'', Bull. l'Acad. Polon. Sci.,20(1972), 981-986.

[5] Bogdanowicz, W. M.,''Multipliers for a linear lattice of functions closed under pointwise convergence'', Bull. l'Acad. Polon. Sci., 20(1972), 987-989.

[6] Bogdanowicz, W. M.,''The smallest P-linear lattice of functions extending a D-linear lattice'', Bull. l'Acad. Polon. Sci., 21(1973), 9-16.

[7] Bogdanowicz, W. M.,''Characterizations of linear lattices of functions closed under dominated convergence'',(to appear in Bull. l'Acad. Polon. Sci.).

[8] Bogdanowicz, W. M.,''The largest linear lattice of functions in which a D-linear lattice forms a solid subset'',(to appear in Bull. l'Acad. Polon. Sci.).

[9] Bogdanowicz, W. M.,''Necessary and sufficient conditions for one D-linear lattice to be solid in another'',(to appear in Bull. l'Acad. Polon. Sci.).

[10] Kolmogorov, A. N., ''Foundations of the Theory of Probability'', Chelsea Publishing Co., New York, (1950).

[11] Bochner, S., and Chandrasekharan,''Fourier transforms'', Princeton University Press, Princeton (1949).

[12] Halmos, P. R.,''Lectures on Ergodic Theory'', Math. Society of Japan, Tokyo (1956) .

[13] MacLane, S., and Birkhoff, G.,''Algebra'', MacMillan Co.,New York, (1967).

A PROBLEM OF EQUIDISTRIBUTION ON THE UNIT INTERVAL $[0, 1]$

by

Shizuo Kakutani

Yale University

1. Let $P = \{x_0, x_1, \ldots, x_n\}$ be a finite ordered set of real numbers such that $0 = x_0 < x_1 < \ldots < x_n = 1$, where n is a positive integer. P is called a partition of the unit interval $I = [0, 1]$. If $n \geq 2$, then P may be considered as a decomposition of the interval $I = [0, 1]$ into a finite number of subintervals $I_i = [x_{i-1}, x_i]$, $i = 1, 2, \ldots, n$, which are mutually disjoint except possibly the endpoints.

We put

(1) $$\ell(P) = \min_{1 \leq i \leq n} (x_i - x_{i-1}), \qquad L(P) = \max_{1 \leq i \leq n} (x_i - x_{i-1}).$$

$I_i = [x_{i-1}, x_i]$ is called a minimal interval or maximal interval of P according as $x_i - x_{i-1} = \ell(P)$ or $x_i - x_{i-1} = L(P)$. We observe that it is possible that there exist more than one minimal or maximal intervals for a given partition P.

2. Let $P = \{x_0, x_1, \ldots, x_n\}$ be a partition of the unit interval $I = [0, 1]$. Let μ_P be the probability measure defined on $I = [0, 1]$, or, more precisely, defined on the sigma-field $\mathcal{B}$ of all Borel subsets B of $I = [0, 1]$, by

(2) $$\mu_P(B) = \frac{1}{n+1} \sum_{i=0}^{n} \chi_B(x_i),$$

where χ_B is the characteristic function of the set B. μ_P is called the probability measure on $I = [0, 1]$ associated with the partition P. μ_P may be considered as a bounded linear functional defined on the Banach space $C[0, 1]$ of all real-valued continuous functions f defined on $I=[0,1]$ with the norm $\|f\| = \sup_{0\le t\le 1}|f(t)|$ by the relation:

(3) $$\int_0^1 f(t)\mu_P(dt) = \frac{1}{n+1}\sum_{i=0}^{n} f(x_i).$$

3. Let $\{\mu_k : k = 1, 2, \ldots\}$ be a sequence of probability measures on the unit interval $I=[0,1]$. We say that the sequence $\{\mu_k : k = 1, 2, \ldots\}$ <u>converges weakly to</u> a probability measure μ on $I=[0,1]$ if

(4) $$\lim_{k\to\infty}\int_0^1 f(t)\mu_k(dt) = \int_0^1 f(t)\mu(dt)$$

for any $f\in C[0,1]$. We are particularly interested in the case when $\mu_k = \mu_{P_k}$ for some partition P_k of the unit interval $I = [0,1]$, $k = 1, 2, \ldots$, and when μ is the normalized Lebesgue measure λ. A sequence $\{P_k : k = 1, 2, \ldots\}$ of partitions of the unit interval $I=[0,1]$ is said to be <u>equidistributed on</u> $I=[0,1]$ if the corresponding sequence $\{\mu_{P_k} : k = 1, 2, \ldots\}$ of associated probability measures is weakly convergent to the normalized Lebesgue measure λ, i.e. if

(5) $$\lim_{k\to\infty}\int_0^1 f(t)\mu_{P_k}(dt) = \int_0^1 f(t)dt$$

for any $f\in C[0,1]$.

We observe that, in order to prove that the sequence $\{\mu_k : k = 1, 2, \ldots\}$ converges weakly to the normalized Lebesgue measure λ, it is sufficient to show that, if μ is a weak limit of any subsequence of $\{\mu_k : k = 1, 2, \ldots\}$, then $\mu = \lambda$. This follows from the fact that there exists a metric d defined on the set $\mathcal{P}$ of all probability measures μ on the unit interval $I=[0,1]$ with the following two properties: (i) $(\mathcal{P}, d)$ is a compact metric space, (ii) a sequence $\{\mu_k : k = 1, 2, \ldots\}$ of elements of $\mathcal{P}$ converges weakly to an element μ of $\mathcal{P}$ if and only if $\lim_{k\to\infty} d(\mu_k, \mu) = 0$.

4. Let $P = \{x_0, x_1, \ldots, x_n\}$ and $Q = \{y_0, y_1, \ldots, y_m\}$ be two partitions of the unit interval $I = [0, 1]$. Q is said to be a <u>refinement</u> of P if P is a subset of Q, or, more precisely, if every interval $I_i = [x_{i-1}, x_i]$ of P is the union of a finite number $(= m_i)$ of subintervals $J_j = [y_{j-1}, y_j]$ of Q. (Each I_i is decomposed into m_i subintervals J_j if $m_i \geqq 2$, while I_i is not decomposed and $I_i = J_j$ for some j, $1 \leqq j \leqq m$, if $m_i = 1$). We note that $m = \sum_{i=1}^{n} m_i$.

Let α be a real number, $0 < \alpha < 1$. Let Q be a refinement of P. Q is said to be the α-<u>refinement</u> of P if $m_i = 2$ for all i, and if each interval $I_i = [x_{i-1}, x_i]$ of P is decomposed into two subintervals $J_i' = [x_{i-1}, x_i']$ and $J_i'' = [x_i', x_i]$ of Q, where $x_i' = x_{i-1} + \alpha(x_i - x_{i-1})$, $i = 1, 2, \ldots, n$.

Further, Q is said to be the α-<u>maximal</u> <u>refinement</u> of P if the following two conditions are satisfied: (i) if $I_i' = [x_{i-1}, x_i]$ is a maximal interval of P, then $m_i = 2$, and I_i is decomposed into two subintervals J_i' and J_i'' in exactly the same way as in the case of α-refinement, (ii) if $I_i = [x_{i-1}, x_i]$ is not a maximal interval of P, then $m_i = 1$, and $I_i = J_j$ for some j, $1 \leqq j \leqq m$.

5. Let α be a real number, $0 < \alpha < 1$. Consider the sequence $\{P_k^\alpha : k = 0, 1, 2, \ldots\}$ of partitions of the unit interval $I = [0, 1]$ defined inductively as follows: (i) $P_0^\alpha = \{0, 1\}$, (ii) P_{k+1}^α is the α-refinement of P_k^α, $k = 0, 1, 2, \ldots$ It is easy to see that, for $k = 1, 2, \ldots$, P_k^α is a decomposition of the unit interval $I = [0, 1]$ into 2^k subintervals $J_k^\alpha(\varepsilon_1, \varepsilon_2, \ldots, \varepsilon_k)$, $\varepsilon_i = 0$ or 1, $i = 1, 2, \ldots, k$, which are defined in-

ductively as follows: (i) $J_1^{\alpha}(0) = [0, \alpha]$ and $J_1^{\alpha}(1) = [\alpha, 1]$, (ii) if $J_k^{\alpha}(\varepsilon_1, \varepsilon_2, \ldots, \varepsilon_k) = [a, b]$, then $J_{k+1}^{\alpha}(\varepsilon_1, \varepsilon_2, \ldots, \varepsilon_k, 0) = [a, c]$ and $J_{k+1}^{\alpha}(\varepsilon_1, \varepsilon_2, \ldots, \varepsilon_k, 1) = [c, b]$, where $c = a + \alpha(b - a)$, $k = 0, 1, 2, \ldots$ $J_k^{\alpha}(\varepsilon_1, \varepsilon_2, \ldots, \varepsilon_k)$ is called an α-<u>dyadic interval of rank</u> k.

Let $\mu_{P_k^{\alpha}}$ be the probability measure on the unit interval $I = [0, 1]$ associated with the partition P_k^{α}, $k = 1, 2, \ldots$ We observe that it is well known that, for each α, $0 < \alpha < 1$, there exists a probability measure μ^{α} defined on $I = [0, 1]$ such that (i) $\lim_{k\to\infty} \mu_{P_k^{\alpha}} = \mu^{\alpha}$ weakly, (ii) μ^{α} and $\mu^{\alpha'}$ are singular to each other if $0 < \alpha < \alpha' < 1$, (iii) $\mu^{\alpha} = \lambda$ if $\alpha = 1/2$.

6. Let α be a real number, $0 < \alpha < 1$. Consider the sequence $\{Q_k^{\alpha} : k = 0, 1, 2, \ldots\}$ of partitions of the unit interval $I = [0, 1]$ defined inductively as follows: (i) $Q_0^{\alpha} = \{0, 1\}$, (ii) Q_{k+1}^{α} is the α-maximal refinement of Q_k^{α}, $k = 0, 1, 2, \ldots$

The main purpose of this article is to prove the following:

<u>Theorem</u>: <u>For each real number</u> α, $0 < \alpha < 1$, <u>the sequence</u> $\{Q_k^{\alpha} : k = 1, 2, \ldots\}$ <u>of partitions of the unit interval defined above is equidistributed on the unit interval, i.e.</u>

$$\lim_{k\to\infty} \int_0^1 f(t)\, \mu_{Q_k^{\alpha}}(dt) = \int_0^1 f(t)\,dt \tag{6}$$

<u>for any</u> $f \in C[0, 1]$.

7. We first observe that

$$\beta\, L(Q_k^{\alpha}) \leqq \ell(Q_k^{\alpha}) \leqq L(Q_k^{\alpha}), \quad k = 1, 2, \ldots, \tag{7}$$

where $\beta = \min(\alpha, 1 - \alpha)$.

The second inequality is obviously true for $k = 1, 2, \ldots$ The first inequality is also obviously true for $k = 1$. Let us now assume that $\beta\, L(Q_k^{\alpha}) \leqq \ell(Q_k^{\alpha})$ for some positive integer k.

There are two possibilities: either (i) $\ell(Q_{k+1}^{\alpha}) = \ell(Q_k^{\alpha})$ or (ii) $\ell(Q_{k+1}^{\alpha}) < \ell(Q_k^{\alpha})$. In the first case, we have $\beta \cdot L(Q_{k+1}^{\alpha}) \leqq \beta\, L(Q_k^{\alpha}) \leqq \ell(Q_k^{\alpha}) = \ell(Q_{k+1}^{\alpha})$. In the second case, every minimal interval of Q_{k+1}^{α} is obtained from a maximal interval of Q_k^{α} by decomposition into two subintervals, and hence $\beta\, L(Q_{k+1}^{\alpha}) \leqq \beta\, L(Q_k^{\alpha}) = \ell(Q_{k+1}^{\alpha})$. Thus the first inequality is proved for $k = 1, 2, \ldots$ by induction.

From (7) follows that, if μ is a probability measure on the unit interval $I = [0, 1]$ which is a weak limit of a subsequence of $\{\mu_{Q_k^{\alpha}} : k = 1, 2, \ldots\}$, then μ is equivalent with the Lebesgue measure λ, i.e. $\mu(B) = 0$ for a Borel subset B of the unit interval $I = [0, 1]$ if and only if $\lambda(B) = 0$.

8. Let $J = J_{k_0}^{\alpha}(\varepsilon_1, \varepsilon_2, \ldots, \varepsilon_{k_0})$ be any α-dyadic interval of rank k_0, where k_0 is any positive integer. It is easy to see that J appears in Q_k^{α} as one of the intervals for certain integral values of k. To be more precise, there exist two positive integers $k_1(J)$ and $k_2(J)$ with the following two properties: (i) $k_1(J) \leqq k_2(J)$, (ii) J appears in Q_k^{α} as one of the intervals if and only if $k_1(J) \leqq k \leqq k_2(J)$.

Let $J = J_{k_0}^{\alpha}(\varepsilon_1, \varepsilon_2, \ldots, \varepsilon_{k_0})$ and $J' = J_{k_0}^{\alpha}(\varepsilon_1', \varepsilon_2', \ldots, \varepsilon_{k_0}')$ be two α-dyadic intervals of the same rank and of the same length. This means that $(\varepsilon_1, \varepsilon_2, \ldots, \varepsilon_{k_0})$ is a permutation of $(\varepsilon_1', \varepsilon_2', \ldots, \varepsilon_{k_0}')$. We observe that it is quite possible that $k_1(J) \neq k_1(J')$, but we must always have $k_2(J) = k_2(J')$. This means that the intervals J and J' may start to appear in Q_k^{α} at two different stages, but they will disappear (i.e. these intervals J and J' are decomposed into two subintervals) in the same passage from $Q_{k_2}^{\alpha}$ to $Q_{k_2+1}^{\alpha}$, where $k_2 = k_2(J) = k_2(J')$. From this follows that the configuration of Q_k^{α} on J is congruent with the configuration of Q_k^{α} on J' for $k \geqq k_2$, i.e. that

$T(Q_k^\alpha \cap J) = Q_k^\alpha \cap J'$ for $k \geqq k_2$ if we denote by T the translation of the real line which maps J onto J'. This implies that $\mu_{Q_k^\alpha}(T(B)) = \mu_{Q_k^\alpha}(B)$ for any Borel subset B of J and for any integer $k \geqq k_2$, where T is the translation of the real line which maps J onto J'.

We thus conclude: Let J and J' be two α-dyadic intervals of the same rank and of the same length. Let T be the translation of the real line which maps J onto J'. Let further μ be a probability measure on the unit interval $I = [0, 1]$ which is the weak limit of a subsequence of $\{\mu_{Q_k^\alpha} : k = 1, 2, \ldots\}$. Then $\mu(T(B)) = \mu(B)$ for any Borel subset B of J.

9. Let X^α be the set of all real numbers x of the unit interval $I = [0, 1]$ which is not the endpoint of any of the α-dyadic intervals $J_k^\alpha(\varepsilon_1, \varepsilon_2, \ldots, \varepsilon_k)$, $\varepsilon_i = 0$ or 1, $i = 1, 2, \ldots, k$; $k = 1, 2, \ldots$ We observe that $I - X^\alpha$ is a countable set. Let further $\mathcal{B}^\alpha$ be the sigma-field of all subsets B^α of X^α of the form $B^\alpha = B \cap X^\alpha$, where B is a Borel subset of $I = [0, 1]$. We also put $X_k^\alpha(\varepsilon_1, \varepsilon_2, \ldots, \varepsilon_k) = X_\alpha \cap J_k^\alpha(\varepsilon_1, \varepsilon_2, \ldots, \varepsilon_k)$, $\varepsilon_i = 0$ or 1, $i = 1, 2, \ldots, k$; $k = 1, 2, \ldots$

Let us put

$$(8) \qquad Y_{p,q}^\alpha = X_{p+q+2}^\alpha(\overbrace{0, 0, \ldots, 0}^{p}, \overbrace{1, 1, \ldots, 1}^{q}, 1, 0),$$

$$(9) \qquad Z_{p,q}^\alpha = X_{p+q+2}^\alpha(\overbrace{1, 1, \ldots, 1}^{q}, \overbrace{0, 0, \ldots, 0}^{p}, 0, 1),$$

$$p, q = 0, 1, 2, \ldots$$

We observe that (i) $Y_{p,q}^\alpha$ and $Z_{p,q}^\alpha$ have the same Lebesgue measure for $p, q = 0, 1, 2, \ldots$ (ii) $X^\alpha = \bigcup_{p=0}^\infty \bigcup_{q=0}^\infty Y_{p,q}^\alpha$ (disj) $= \bigcup_{p=0}^\infty \bigcup_{q=0}^\infty Z_{p,q}^\alpha$ (disj). We denote by S the one-to-one mapping of X^α onto itself such that

(10) S maps $Y^{\alpha}_{p,q}$ onto $Z^{\alpha}_{p,q}$ by translation,
$p, q = 0, 1, 2, \ldots$

S is uniquely defined on X^{α} by (10), and it is clear that S is a measure preserving transformation of the probability space $(X^{\alpha}, \mathcal{B}^{\alpha}, \lambda)$ onto itself. We observe that S is a transformation which was discussed in detail in [1], where it was proved that S is ergodic as a measure preserving transformation of $(X^{\alpha}, \mathcal{B}^{\alpha}, \lambda)$ onto itself.

Let now μ be a probability measure on the unit interval $I = [0, 1]$ which is the weak limit of a subsequence of $\{\mu_{Q^{\alpha}_k} : k = 1, 2, \ldots\}$. From the observation made in § 7, μ is equivalent with the Lebesgue measure λ. On the other hand, from the result obtained in § 8, it follows that S is a measure preserving transformation of the probability space $(X^{\alpha}, \mathcal{B}^{\alpha}, \mu)$ onto itself. Thus we have $\mu = \lambda$, and, because of the observation made at the end of § 3, this completes the proof of our Theorem.

Reference

[1] Arshag Hajian, Yuji Ito and Shizuo Kakutani, Invariant Measures and Orbits of Dissipative Transformations, Advances in Mathematics, 9(1972), 52-65.

ON ABSOLUTE CONTINUITY OF MEASURES GENERATED BY ITÔ-MCSHANE STOCHASTIC DIFFERENTIAL EQUATIONS

Zoran R. Pop-Stojanovic
University of Florida

The setting. A. Throughout this paper $(\Omega,\mathcal{F},P)$ denotes a probability space; $\{\mathcal{F}_t;\ t\in[0,1]\}$ is a non-decreasing family of σ-subalgebras of $\mathcal{F}$; $z=(z_t,\mathcal{F}_t;\ t\in[0,1])$ is a quasi-martingale, sample continuous a.e.(P), such that $z_0=0$, $E((z_t-z_s)^2|\mathcal{F}_s)\le t-s$, a.e.$(P)$, $s\le t$.

B. Let $(C,\mathcal{C})$ denote a measurable space of sample continuous random functions $x=(x_t,\mathcal{C}_t;\ t\in[0,1])$, $x_0=0$, where $\mathcal{C}=\mathcal{B}(\{x_s;s\le 1\})$, $\mathcal{C}_t=\mathcal{B}(\{x_s;s\le t\})$, and $\mathcal{B}(\{\cdot\})$ denotes the Borel field generated by sets in the question. Finally, $\mathcal{C}_{[0,1]}$ denotes Borel field of subsets of $[0,1]$.

Definition. A continuous random process $y=(y_t,\mathcal{F}_t;\ t\in[0,1])$ is said to be Itô-McShane type with respect to z, if there exists a random process $a=(a_t,\mathcal{F}_t;\ t\in[0,1])$ such that

(i) $\int_0^1|a_t(\omega)|dt<+\infty$ a.e.(P), and

(ii) $\forall t,t\in[0,1]$: $dy_t=a_t(\omega)dt+dz_t, y_0=0$,

where the last equation is Itô-McShane stochastic differential equation [3].

Let μ_y,μ_z be measures in $(C,\mathcal{C})$ corresponding to $y(\omega)=(y_t(\omega);\ t\in[0,1])$ and $z(\omega)=(z_t(\omega);\ t\in[0,1])$, respectively, such that $\forall B, B\in\mathcal{C}$: $\mu_y(B)=P(\{\omega;y(\omega)\in B\})$, $\mu_z(B)=P(\{\omega;z(\omega)\in B\})$. Finally, put $\mathcal{F}_t^y=\mathcal{B}(\{\omega;y_s(\omega),s\le t\})$, $\mathcal{F}_t^z=\mathcal{B}(\{\omega;z_s(\omega),s\le t\})$, $s,t\in[0,1]$.

Now we can state the following

Theorem. If $\int_0^1 a_t^2(\omega)dt < +\infty$ a.e.(P) then $\mu_y << \mu_z$.

Proof. Let n be a positive integer and put

$$I_s^{(n)} = I_{\{\omega;\int_0^s a_u^2 du \le n\}},\quad a_s^{(n)}(\omega) = a_s(\omega)I_s^{(n)},\quad s \in [0,1].$$

Set

$$y_t^{(n)} = \int_0^t a_s^{(n)}(\omega)ds + z_t,\quad t \in [0,1].$$

Now, one can show that

$$(1)\qquad \mu_y(n) << \mu_z.$$

Having in mind this fact let us now introduce a sequence of stopping times $\{\tau^{(n)}\}_{n\in N}$, where

$$\tau^{(n)} = \inf\{t;\ t\le 1,\ \int_0^t a_s^2\,ds \ge n\}.$$

It is clear that $y_t^{(n)} = y_t$ on $\{\omega;\ \tau^{(n)}=1\}$ for all $t\le 1$; therefore

$$(2)\quad \mu_y(A) = P(\{\omega)\in A\}) = P(\{y(\omega)\in A,\tau^{(n)}=1\})+P(\{y(\omega)\in A,\tau^{(n)}<1\})$$
$$= P(\{y^{(n)}(\omega)\in A,\tau^{(n)}=1\})+P(\{y(\omega)\in A,\tau^{(n)}<1\}).$$

Now let us assume that $\mu_z(A) = 0$. Since $\mu_y(n) << \mu_z$ by (1), it follows that $P(\{y^{(n)}(\omega)\in A,\tau^{(n)}=1\}) = 0$, so (2) implies

$$(3)\quad \mu_y(A) = P(\{y(\omega)\in A,\tau^{(n)}<1\}) \le P(\{\tau^{(n)}<1\}).$$

The assumption $\int_0^1 a_s^2(\omega)ds < +\infty$ a.e.(P) implies that $P(\{\tau^{(n)}<1\}) = P(\{\int_0^1 a_s^2(\omega)ds > n\}) \to 0$ as $n \to +\infty$, which combined with (3) gives $\mu_y(A) = 0$.□

In order to show that (1) holds, one should use Riesz decomposition theorem for quasi-martingale z, [1], [4], i.e., $\forall t,\ t \in [0,1],\ z_t = \tilde{z}_t+\hat{z}_t$, a.e.($P$), where $(\tilde{z}_t,\mathcal{F}^{\tilde{z}}_t;\ t \in [0,1])$ is a martingale and $(\tilde{z}_t,\mathcal{F}^{\hat{z}}_t;\ t \in [0,1])$ is a process whose trajectories are a.e.(P) of bounded variation. Then, using a similar method as in [2], p. 862, one shows that $\mu_y(n) << \mu_{\tilde{z}}$ and $\mu_y(n) << \mu_{\hat{z}}$, which in turn implies (1).

REFERENCES

[1] D. L. Fisk, Quasi-Martingales, Trans. Amer. Math. Soc., 120(1965), pp. 369-389.

[2] P. Ŝ. Lipcer, A. N. Ŝiraev, On absolute continuity of measures corresponding to the processes of diffusion type, Proc. of the Academy of Sciences of the USSR, 36(1972), pp. 847-885 (in Russian).

[3] J. E. McShane, Stochastic Integration, Vector and Operator valued measures and Applications, Academic Press, New York (1973), pp. 247-281.

[4] P. A. Meyer, Probabilities and Potentials, Blaisdell, 1966.

A PROBLEM IN L^p-SPACES

by

Alexandra Bellow

Let $(\Omega,\mathcal{F},\mu)$ be a measure space. For $1 < p < \infty$ let $L^p = L^p(\Omega,\mathcal{F},\mu)$ and $L^p_+ = \{f \in L^p \mid f \geq 0\}$.

Recently the following inequality came up in the context of ergodic theory:

L^p-Inequality. Let $1 < p < \infty$. Let $f \in L^p_+$, $g \in L^p_+$. Then for any $0 < \varepsilon < 1$ we have with $\alpha = (p-1) + \frac{1}{(p-1)}$

$$(1) \qquad \int f^{p-1} g \leq \varepsilon\|f\|_p^p + \varepsilon\|g\|_p^p + \frac{1}{\varepsilon^\alpha} \int f\, g^{p-1}$$

A proof of this elementary inequality may be found in [6].

By minimizing the right hand side of (1) over ε we obtain the homogeneous form of this inequality.

This turns out ot be:

$$(2) \qquad \left(\int f^{p-1} g\right)^{\alpha+1} \leq \|f\|_p^{p(p-1)} \|g\|_p^{\frac{p}{p-1}} \left(\int f\, g^{p-1}\right)$$

This is none other than Hölder's inequality for three functions

$$\int f_1 f_2 f_3 \leq \left(\int f_1^{r_1}\right)^{\frac{1}{r_1}} \left(\int f_2^{r_2}\right)^{\frac{1}{r_2}} \left(\int f_3^{r_3}\right)^{\frac{1}{r_3}}$$

$\left(\frac{1}{r_1} + \frac{1}{r_2} + \frac{1}{r_3} = 1\right)$ with the identifications:

$$r_1 = \frac{\alpha+1}{p-1}, \quad r_2 = (\alpha+1)(p-1), \quad r_3 = \alpha+1$$

$$f_1 = f^{\frac{p(p-1)}{\alpha+1}}, \quad f_2 = g^{\frac{p}{(\alpha+1)(p-1)}}, \quad f_3 = f^{\frac{1}{\alpha+1}}\, g^{\frac{p-1}{\alpha+1}}$$

However, for the specific problem in ergodic theory that we shall discuss later, it turns out that the non-homogeneous form (1) of the L^p-Inequality is the useful one.

There are several reasons why this inequality is of interest:

I) Canonical duality between L^p and L^q. Let p be fixed, $1 < p < \infty$ and let $\Phi : L^p \to L^q$ be the "canonical duality map" (here $\frac{1}{p} + \frac{1}{q} = 1$) ; it is given by

$$\Phi(u) = \operatorname{sgn} u.\ |u|^{p-1}$$

We recall that for every $u \in L^p$

$$(u, \Phi(u)) = \|u\|_p \, \|\Phi(u)\|_q$$

and

$$\|\Phi(u)\|_q = \|u\|_p^{p-1} .$$

When $p = 2$, Φ is simply the identity mapping.

For $p \neq 2$, Φ provides a very good example of the pathology of behavior of non-linear mappings.

We note the following facts about Φ :

A) $\Phi : L^p \to L^q$ is strongly continuous. Infact Φ is uniformly continuous on any bounded part of L^p (see [8], p.221).

B) For $p \neq 2$, $\Phi : L^p \to L^q$ is nowhere weakly continuous if the measure space is non-atomic.

Proof of B) using the notion of strongly mixing transformation. We assume for simplicity that $(\Omega, \mathfrak{F}, \mu)$ is the unit interval with Lebesgue measure. Let $\tau : \Omega \to \Omega$ be a measurable, measure-preserving transformation. Assume that τ is strongly mixing, that is (see [13])

$$\mu(\tau^{-n}(A) \cap B) \to \mu(A)\,\mu(B)$$

for all $A \in \mathfrak{F}$, $B \in \mathfrak{F}$. For each $1 \leq r < \infty$, let $T = T_\tau$ be the operator induced by τ in L^r : $Tf = f \circ \tau$, for $f \in L^r$.

To say that τ is strongly mixing is equivalent with the statement

$$T^n f \xrightarrow{\text{weakly } L^r} Pf = \int f \, d\mu .$$

for each $f \in L^r$.

Note also that $T \circ \Phi = \Phi \circ T$.

Case $f \neq 0$, $f \in L^p$. Let $A \in \mathfrak{F}$ with $\mu(A) = \frac{1}{2}$ and define the function u to be 2 on A and 0 on $\Omega - A$.

Note that

$$0 \leq u \leq 2, \quad \int u = 1, \quad \int \Phi(u) = 2^{p-2}.$$

Let now

$$f_n = f(T^n u)$$

Then, since u and $\Phi(u)$ are bounded, we have for any $E \in \mathfrak{F}$

$$(3) \quad \int_E f_n = \int (1_E f)\, T^n u \to \int (1_E f)\left(\int u\right) = \int_E f,$$

and

(4) $$\int_E \Phi(f_n) = \int_E \Phi(f) \;\; \Phi(T^n u) =$$

$$= \int (1_E\, \Phi(f)).\, T^n(\Phi(u)) \;\; \rightarrow \;\; \int (1_E\, \Phi(f)) \,(\int \Phi(u)) = \int_E \Phi(f).\, 2^{p-2}$$

The relations (3) and (4) above show that $(f_n)_{n\geq 1}$ converges weakly in L^p to f, but $(\Phi(f_n))_{n\geq 1}$ converges weakly in L^q to $2^{p-2}\,\Phi(f)$.

Case $f = 0 \in L^p$. Let g be a function in L^p such that

$$\int g = 0, \;\; \int \Phi(g) = c \neq 0.$$

Then, as was noticed in [6],

$$T^n g \xrightarrow{\text{weakly } L^p} 0$$

but

$$\Phi(T^n g) \xrightarrow{\text{weakly } L^q} c \neq 0.$$

C) The L^p-Inequality shows that

$$\Phi | L^p_+ : \; L^p_+ \; \rightarrow \; L^q_+$$

is weakly continuous at 0, even though $\Phi | L^p_+$ fails to be weakly continuous at any other $f \in L^p_+$, $f \neq 0$ (see the proof of B), case $f \neq 0$). It also shows that if $f_n \in L^p_+$ and $(f_n)_{n\geq 1}$ converges to 0 weakly in L^p, then the rate of weak convergence to 0 of $\Phi(f_n) = f_n^{p-1}$ is comparable to that of f_n.

II) Application to Ergodic Theory: "Weak Convergence of Iterates implies Strong Convergence of Averages".

We now consider a matrix $(a_{ni})_{1 \leq n < \infty,\, 1 \leq i < \infty}$ satisfying the following two conditions

(a) $$\sup_n \sum_i |a_{ni}| = M < \infty$$

(b) $$\sup_i |a_{ni}| = m_n \rightarrow 0 \text{ as } n \rightarrow \infty .$$

Lemma 1: Suppose that $a_{n,i} \geq 0$ for all (n,i). Let $(g_n)_{n\geq 1}$ be a bounded sequence of elements of L^p_+. The following assertions are then equivalent:

(j) $\lim_n \| \sum_i a_{ni} \, g_i \|_p = 0$

(jj) $\lim_n \sum_{(i,j)} a_{ni} \, a_{nj} \int g_i^{p-1} \, g_j = 0.$

Proof: We may assume without loss of generality that

$$\sum_i a_{ni} \leq 1 \quad \text{for all} \quad n \geq 1$$

and that

$$\|g_n\|_p \leq 1 \quad \text{for all} \quad n \geq 1.$$

(jj) ⇒ (j). Let $0 < \varepsilon < 1$. By inequality (1) we may write

$$\| \sum_i a_{ni} \, g_i \|_p^p = \int (\sum_i a_{ni} \, g_i) \, (\sum_j a_{nj} \, g_j)^{p-1}$$

$$= \sum_i a_{ni} \int g_i \, (\sum_j a_{nj} \, g_j)^{p-1}$$

$$\leq \sum_i a_{ni} \, \{\varepsilon + \varepsilon + \frac{1}{\varepsilon^{\alpha}} \int g_i^{p-1} \, (\sum_j a_{nj} \, g_j) \}$$

$$\leq 2\varepsilon + \frac{1}{\varepsilon^{\alpha}} \{ \sum_i \sum_j a_{ni} \, a_{nj} \int g_i^{p-1} \, g_j \}$$

(j) → (jj) Let again $0 < \varepsilon < 1$. By inequality (1) again we have:

$$\sum_{(i,j)} a_{ni} \, a_{nj} \int g_i^{p-1} g_j = \sum_i a_{ni} \int g_i^{p-1} \, (\sum_j a_{nj} \, g_j) \leq$$

$$\leq \sum_i a_{ni} \{ \varepsilon + \varepsilon + \frac{1}{\varepsilon^{\alpha}} \int g_i \, (\sum_j a_{nj} \, g_j)^{p-1} \} \leq$$

$$\leq \varepsilon + \varepsilon + \frac{1}{\varepsilon^{\alpha}} \int (\sum_i a_{ni} \, g_i) \, (\sum_j a_{nj} \, g_j)^{p-1} =$$

$$= 2\varepsilon + \frac{1}{\varepsilon^{\alpha}} \| \sum_i a_{ni} \, g_i \|_p^p$$

This completes the proof of Lemma 1.

Corollary: Let $(g_n)_{n\geq 1}$ be a bounded sequence of elements of L^p_+ satisfying

(5) $$\lim_{|i-j| \to \infty} \int g_i^{p-1} \, g_j = 0$$

<u>Then</u>

$$\lim_n \| \sum_i a_{ni}\, g_i \|_p = 0.$$

<u>Proof</u>: By Lemma 1 it suffices to show that

$$\lim_n \sum_{(i,j)} |a_{ni}||a_{nj}| \int g_i^{p-1} \; g_j = 0.$$

We may assume that $\|g_n\|_p \leq M$ for all $n \geq 1$.

Let $\varepsilon > 0$ be given. By assumption, there is N_0 such that

$$i,j \geq 1, \quad |i-j| \geq N_o \Rightarrow \int g_i^{p-1} \; g_j \leq \frac{\varepsilon}{M^2}.$$

We deduce:

$$\sum_{(i,j)} |a_{ni}||a_{nj}| \int g_i^{p-1} \; g_j = \sum_{\substack{(i,j)\\ |i-j|< N_0}} + \sum_{\substack{(i,j)\\ |i-j|\geq N_0}}$$

$$\leq \sum_i |a_{ni}| \, (2N_0 \, m_n) \; M + \frac{\varepsilon}{M^2} \sum_{\substack{(i,j)\\ |i-j|\geq N_0}} |a_{ni}||a_{nj}|$$

$$\leq (2N_0 \, m_n) \, M + \frac{\varepsilon}{M^2} \; (\sum_i |a_{ni}|) \, (\sum_j |a_{nj}|)$$

$$\leq (2N_0 \, M) \, m_n + \frac{\varepsilon}{M^2} \; M^2 = (2N_0 \, M) \, m_n + \varepsilon .$$

This last sum becomes $< 2\varepsilon$ if n is large enough, since $m_n \to 0$ as $n \to \infty$. Thus the Corollary is proved.

<u>Remarks</u>: 1) If a sequence $(g_n)_{n\geq 1}$ in L_+^p converges to 0 weakly in L^p, it need not necessarily satisfy condition (5) in the above Corollary (take for example a sequence (h_n) in L_+^p converging to 0 weakly but not strongly in L^p, and form a new sequence by repeating the h_n's for longer and longer blocks)

2) From any sequence $(f_n)_{n\geq 1}$ in L_+^p that converges to 0 weakly in L^p, one can extract a subsequence $(g_k)_{k\geq 1}$, $g_k = f_{n_k}$ (using the fact that $(f_n^{p-1})_{n\geq 1}$ also converges weakly to 0 in L^q), satisfying for each $k \geq 2$

$$\int g_k^{p-1} \; g_\ell < \frac{1}{2^k} \quad \text{for } 1 \leq \ell \leq k-1$$

and

$$\int g_\ell^{p-1} \; g_k < \frac{1}{2^k} \quad \text{for} \quad 1 \leq \ell \leq k-1$$

and therefore satisfying condition (5) above.

Using an interplay between the L^p-Inequality and an estimate derived from the uniform convexity of L^p one may prove the following:

Lemma 2: Let $T: L^p \to L^p$ be a positive contraction (in the case $p = 2$, $T: L^2 \to L^2$ an arbitrary not necessary positive, contraction . Suppose that for some $f \in L^p_+$ (in the case $p = 2$, $f \in L^2$) the sequence $(T^n f)_{n \geq 1}$ converges to 0 weakly in L^p. Then

$$\lim_{|i-j| \to \infty} (T^i f, \; \Phi(T^j f)) = 0$$

For details of proof see [6].

We recall that a matrix $(a_{ni})_{1 \leq n < \infty \; , 1 \leq i < \infty}$ is called "uniformly regular" if it satisfies conditions a) and b) preceding lemma 1 and in addition the following condition

c) $$\lim_n \; \sum_i \; a_{ni} = 1.$$

From Lemma 2 and the Corollary of Lemma 1, one contains an extremely simple proof of the following Theorem due to Akcoglu and Sucheston (see [6]):

Theorem: Let $T = L^p \to L^p$ be a positive contraction (in the case $p = 2$, $T = L^2 \to L^2$ an arbitrary contraction). Then the following assertions are equivalent:

1) $\lim_n \; T^n$ exists in the weak operator topology

2) If (a_{ni}) is any uniformly regular matrix, then $\lim_n \sum_i a_{ni} \; T^i$ exists in the strong operator topology.

Final Remarks: The oldest and probably best known result in this area of Mean Ergodic Theory goes back to 1960, to Blum and Hanson ([7]):

Theorem: (Blum-Hanson) Let τ be a measurable measure preserving transformation on the probability space $(\Omega, \mathfrak{F}, \mu)$. Then τ is strongly mixing if and only if for each $f \in L^2$,

$$\frac{1}{n} \sum_{i=1}^{n} T^{k_i} f \xrightarrow{\text{strongly } L^2} \int f d\mu$$

for any increasing infinite sequence (k_i).

Subsequently there was frenzy and furor to extend the Blum-Hanson Theorem to more general situations (spaces and operators). In the years that followed there was a proliferation of papers along this line of research [9],[18],[16],[10],[15],[17],[19],[12], and [2],[3],[4],[5].

The final version of the theorem, stated earlier, was given by Akcoglu and Sucheston ([2],[3],[4],[5]). The Akcoglu -Sucheston approach for the L^p case $1 < p < \infty$, $p \neq 2$, was complicated: their proof was based on approximation by finite-dimensional operators, reduction of the contraction case to the invertible isometry case, and ultimately application of Akcoglu's "Dilation Theorem". Our proof, based on the L^p-Inequality is straightforward and extremely elementary (for details of proof see [6]).

<u>Problem</u>: Let $T: L^p \to L^p$ be a contraction. Let $f \in L^p$ be such that the sequence $(T^n f)_{n\geq 1}$ converges weakly in L^p to a limit. Let (a_{ni}) be a uniformly regular matrix. Does the sequence $(\sum_i a_{ni} T^i f)_{n\geq 1}$ converge strongly in L^p to a limit? In other words is the local ergodic theorem true?

REFERENCES

[1] M. A. Akcoglu, A pointwise Ergodic Theorem in L_p-spaces, Canadian J. Math (to appear)

[2] M. A. Akcoglu and L. Sucheston, On Operator Convergence in Hilbert Space and in Lebesgue Space, Periodica Math. Hungarica 2, (1972) 235-244.

[3] M. A. Akcoglu and L. Sucheston, On convergence of iterates of Positive Contractions in L^p-Spaces, to appear in J. Approximation Theory.

[4] M. A. Akcoglu and L. Sucheston, On weak and strong convergence of positive contractions in L^p-spaces, Bull. Amer. Math. Soc, vol 81 (1975), No. 1, 105-106.

[5] M. A. Akcoglu and L. Sucheston, Weak convergence of positive contractions implies strong convergence of averages, Zeitschrift Wahrscheinlichkeits theorie verw. Gebiete 32(1975), 139-145.

[6] A. Bellow. An L^p-Inequality with application to Ergodic Theory, to appear in the new (forthcoming) Houston J.Mathematics.

[7] J. R. Blum and D. L. Hanson, On the mean ergodic theorem for subsequences, Bull Amer. Math. Soc. 66(1960), 308-311.

[8] N. Bourbaki, <u>Intégration</u>, Chap IV, Hermann, Paris, 1952.

[9] G. Bray, A propos de la généralization d'un résultat de la théorie ergodique a des espaces uniformement convexes (unpublished manuscript)

[10] A. Brunel and M. Keane, Ergodic Theorems for operator sequences, Zeit. Wahrscheinlichkeits theorie verw. Gebiete 12(1969),231-240.

REFERENCES CONTD.

[11] N. Dunford and J. T. Schwartz, Linear Operators Part I, Interscience Publ. New York, 1966.

[12] H. Fong and L. Sucheston, On a mixing property of operators in L_p-spaces, Zeit. Wahrscheinlichkeits theorie verw. Gebiete 28 (1974), 165-171.

[13] P. R. Halmos, Lectures on Ergodic Theory, New York, 1956.

[14] D. L. Hanson and G. Pledger, On the mean ergodic theorem for weighted averages, Zeit. Wahrscheinlichkeits theorie verw. Gebiete 13(1969), 141-149.

[15] L.K. Jones and V. Kuftinec, A note on the Blum-Hanson theorem, Proc. Amer.Math. Soc. 30(1971), 202-203.

[16] U. Krengel and L. Sucheston, On mixing in infinite measure spaces, Zeit. Wahrscheinlichkeits theorie verw. Gebiete 13(1969), 150-164.

[17] M. Lin, Mixing for Markov operators, Zeit. Wahrscheinlichkeits theorie verw. Gebiete 19(1971), 231-242.

[18] A. Renyi. On stable sequences of events, Sankhya, Ser. A 25(1963), 293-302.

[19] R. Sato, On Ackoglu and Sucheston's Operator Convergence theorem in Lebesgue Spaces, Proc. Amer.Math. Soc. 40 (1973), 513-516.

On Positive Dilations to Isometries in L_p Spaces

Mustafa A. Akcoglu and Louis Sucheston[1]

A positive contraction operator T on an L_p space is said to have a positive dilation to an isometry if T admits the representation: $T^n = PQ^n$, $n = 0, 1, \ldots,$ where Q and P act on a larger L_p space, Q is an isometry, P is a positive projection. (A more precise definition is given below.) We consider here a positive operator T acting on measurable functions, and give simple conditions for T to be an L_p contraction admitting of a positive dilation to a positive invertible isometry.

Let (J,β) be the unit interval with the σ-algebra of Borel sets. m is a finite measure on β; real-valued measurable functions on J are considered modulo sets of m measure zero. Let M be the vector space of all such functions $\overline{M}$ the set of all extended real-valued measurable functions, M^+ the class of non-negative members of M. Let N be a subspace of M such that for each $f \in M^+$ there is a sequence of functions f_n in N^+ with $f_n \uparrow f$. Let T be a linear operator mapping N to M, (i) positive, i.e. such that $TN^+ \subset M^+$ and (ii) monotone, i.e. such that if f, $f_n \in N^+$ and $f_n \uparrow f$, the $Tf_n \uparrow Tf$. Then T naturally extends so that it maps M^+ to $\overline{M}^+$. There exists also the adjoint operator $T^*: M^+ \to \overline{M}^+$, defined as follows: For each $g \subset M^+$ there exists a unique $g' \in \overline{M}^+$ such that for each $f \in M^+$, $\int (Tf)\, g \, dm = \int f g' \, dm$. Let $T^* g = g'$. It is easy to see that $(T^*)^* = T$.

Theorem 1. Let T be a linear positive monotone operator mapping N to M and p a positive number. Assume that there exists a $u \subset M^+$ such that $u > 0$ and either

$$\text{(A)} \quad T^*(Tu)^{p-1} \le u^{p-1} \quad \text{or} \quad \text{(B)} \quad T[(T^*u)^{\frac{1}{p-1}}] \le u^{\frac{1}{p-1}} .$$

[1]Research supported by the National Science Foundation (USA), grant MPS 72-04752A03

Then T is a contraction on $L_p = L_p(J,\beta,m)$ that has a positive dilation to an invertible isometry. More explicitly, there exist:

(i) A σ-finite countably generated measure space $(\tilde{J},\tilde{\beta},\tilde{m})$;

(ii) A positive and invertible isometry R on $\tilde{L}_p = L_p(\tilde{J},\tilde{\beta},\tilde{m})$;

(iii) A positive self-adjoint idempotent operator $P: \tilde{L}_p \to \tilde{L}_p$. This operator has the form $P = E\chi$, where E is a conditional expectation with respect to a σ-algebra $\mathcal{B} \subset \tilde{\beta}$ and χ is multiplication by the characteristic function of an $\mathcal{B}$-measurable set;

(iv) A positive linear isometry $\xi : L_p \to P\tilde{L}_p$, such that

$$\xi T^n f = PR^n \xi f$$

for each $f \in L_p$ and each $n = 0, 1, \ldots$. ξ is such that if $f_n \in L_p$ and ξf_n coverges pointwise on $\tilde{X}$, then f_n converges pointwise on X .

The proof of this result, which extends the dilation theorems in [1] and [2], is given below. Here we only mention that, for the purpose of application, N can be an L_r of a measure space (r need not be equal to the number p appearing in the theorem), an Orlicz function space, or a Lorentz function space (see [5], in particular Section 3.7). In most cases a positive bounded operator will be automatically monotone, a notable exception being $N = L_\infty$ where the monotonicity of T is equivalent with T being the adjoint of an L_1 operator. If the measure m is atomic and T is a contraction on $L_p(X,\beta,m)$, then it can be shown that there exists a positive u such that $T^*(Tu)^{p-1} \leq u^{p-1}$; thus every positive contraction on ℓ_p admits of a positive dilation. This is theorem 2 below.

<u>Proof of Theorem 1</u>. Let $h^{p-1} = T^*u^{p-1}$ and $Th = ru$ where r is a function with $0 \leq r \leq 1$. First we define a subset Z of J_0^1 as follows

$$Z = \{(x_0,x_1) \mid (x_0,x_1) \in J_0^1 , 0 \le x_1 \le r(x_0)\} .$$

We consider the measure $d\nu = u^p\, dm$ on J_0 and the measure $\nu \times \lambda$ on J_0^1, λ being the standard Borel measure on (J,β). The restrictions of β_0^1 and $\nu \times \lambda$ to Z make Z a measure space which will be denoted as $(Z,\mathfrak{Z},\tilde{\nu})$.

<u>Lemma 1</u>: Let $d\mu = h^p\, dm$. There exists a conditioned family $\{\alpha\}_{J_0}$ on (J_{-1},β_{-1}) and an equivalence $\Pi: (J_{-1}^0,\beta_{-1}^0) \to (Z,\mathfrak{Z})$ that transports $\{\alpha\} \times \mu$ to $\tilde{\nu}$ so that

$$(Tf)(x_0) = u\,(x_0) \int_0^{r(x_0)} \frac{f(\Pi_0^{-1}(x_0,x_1))}{h(\Pi_0^{-1}(x_0,x_1))}\, dx_1$$

for each $f \in M^+$ with support $f \subset \operatorname{supp} h$. Here $\Pi_0^{-1}(x_0,x_1)$ denotes the J_0-coordinate of the point $\Pi^{-1}(x_0,x_1) \in J_{-1}^0$ with $(x_0,x_1) \in Z$.

<u>Proof of Lemma 1</u>: First note that $h < \infty$ m a.e. on J. In fact, if $h = \infty$ on a set $A \in \beta$ with $m(A) > 0$, then $\infty = \int_J h\, T^* f\, dm = \int_J Th\, f\, dm \le \int_J u\, f\, dm = \infty$ whenever $f \in M^+$ and $T^* f > 0$ on a positive subset of A. In particular, let $f = u^{p-1} \chi_{B_n}$ where B_n is an increasing sequence of sets so that $\bigcup_{n=1}^{\infty} B_n = J$ and so that $m(B_n) < \infty$ and u is bounded on B_n. Then $T^*(u^{p-1} \chi_{B_n}) \uparrow T^* u^{p-1} = h^{p-1}$ and so there exists an n with $T^*(u^{p-1} \chi_{B_n}) > 0$ on a positive subset of A. This means that $\int u^p \chi_{B_n}\, dm = \infty$ which contradicts the definition of B_n. Hence $h < \infty$ m.a.e. on J.

Then, without loss of generality, we may assume that h is bounded on $[0,\alpha]$ for each α with $0 \le \alpha < 1$; in addition assume that $m([0,\alpha]) < \infty$ for such α.

We define j_s, χ_s, G_s, ψ_s as in [2], p. 356, starting with $p_s(x_0) = \dfrac{T(\chi_s h)(x_0)}{u(x_0)}$. Then everything is as in [2], except that instead of

$p_0 + p_1 = 1$ we now have $p_0 + p_1 = r$. Hence G_s's form a partition of Z, instead of full square J_0^1. Then we have

$$T(\chi_s h)(x_0) = u(x_0)p_s(x_0) = u(x_0)\int_0^{r(x_0)} \psi_s(x_0,x_1)dx_1 .$$

We also define $g: Z \to J$ as in [2], p. 357. We now show that g transports $\tilde{\nu}$ on Z to μ on J. In fact for each binary interval $j_s \subset J$

$$\tilde{\nu}(g^{-1}j_s) = \tilde{\nu}(G_s) = (\nu \times \lambda)(G_s)$$

$$= \int_{J_0} \frac{T\chi_s h}{u} u^p \, dm = \int_{J_0} \chi_s h \, T^* u^{p-1} \, dm$$

$$= \int_{j_s} h^p \, dm = \mu(j_s) .$$

Therefore by Rohlin - Maharam theorem [6, 7], one can again get a conditioned family $\{\alpha\}_{J_0}$ on J_{-1} and an equivalence $\Pi: J_{-1}^0 \to Z$ that transports $\{\alpha\} \times \mu$ to $\tilde{\nu}$; i.e. the restriction of $\nu \times \lambda$ to Z. This mapping is such that $g\Pi: J_{-1}^0 \to J_0$ is the projection of J_{-1}^0 to its J_0-component, $\{\alpha\} \times \mu$ - a.e.

Remark. The form of the Rohlin-Maharam theorem used is the following: Let (Ω, Σ, σ) be a Borel space and let $g: \Omega \to J$ be a measurable function transporting σ to a measure ν on (J,β). Then there exists an isomorphism $\Pi: (J^2 \to \Omega$ between $(J^2, \beta^2, \nu \times \{\eta\})$ and (Ω, Σ, σ), for some choice of the conditioned family $\{\eta\}$, such that $g\Pi : J^2 \to J$ is the projection of J^2 to its first component J, $\nu \times \{\eta\}$ - a.e. Since (Ω, Σ) is equivalent to (J,β), Π can be chosen as an equivalence. This form can be immediately derived from [6] or [7] using a standard theorem due to Doob, asserting the existence of 'regular conditional distributions' (see [4], p. 29 ff.).

We now show that

$$(Tf)(x_0) = u(x_0)\int_0^{\Pi(x_0)} \frac{f(\Pi_0^{-1}(x_0,x_1))}{h(\Pi_0^{-1}(x_0,x_1))} \, dx_1 .$$

If $f = h\chi_S$ then the integrand becomes $\psi_S(x_0,x_1)$ and the equation holds. Let $\mathcal{C}$ be the subset of β consisting of sets C such that this equation holds if $f = h\chi_C$. Then it is clear that $\mathcal{C}$ is a monotone class and contains the algebra of binary intervals. Hence $\mathcal{C} = \beta$. It is clear that this equation holds when $f = h\phi$ with $\phi \in M^+$. $\square$

The Proof of the Theorem 1 (continued)

We construct $\tau : J_{-\infty}^{\infty} \to J_{-\infty}^{\infty}$ as in [2]. This map is again one-to-one, but not onto. In fact, the image by τ is exactly $\ldots \times J_{-2} \times J_{-1} \times Z \times J_2 \times J_3 \times \ldots = \tau J_{-\infty}^{\infty}$ where Z replaces $J_0 \times J_1$. Now τ transports $y_{-\infty}^{\infty}$ not to

$$\nu_{-\infty}^{\infty} = \ldots \times \{\alpha_{-2}\} \times \{\alpha_{-1}\} \times \nu \times \lambda \times \lambda \times \ldots$$

but to $\tilde{\nu}_{-\infty}^{\infty}$, which can be defined either as the restriction of $\nu_{-\infty}^{\infty}$ to $\tau J_{-\infty}^{\infty}$ or as

$$\tilde{\nu}_{-\infty}^{\infty} = \ldots \times \{\alpha_{-2}\} \times \{\alpha_{-1}\} \times \tilde{\nu} \times \lambda \times \lambda \times \ldots \quad .$$

Hence $\rho = \dfrac{d\mu_{-\infty}^{\infty}}{d\nu_{-\infty}^{\infty}}$ is not a function of x_0 alone, but depends on x_0 and x_1, namely

$$p(\ldots,x_{-1},x_0,x_1,\ldots) = \frac{d\tilde{\nu}}{d\mu\times\lambda}(x_0,x_1) = \begin{cases} \dfrac{u^p(x_0)}{h^p(x_0)} & \text{if } 0 \le x_1 \le r(x_0) \\ 0 & \text{otherwise.} \end{cases}$$

Let Q be the positive isometry of $L_p(J_{-\infty}^{\infty}, \beta_{-\infty}^{\infty}, \mu_{-\infty}^{\infty})$ induced by τ. Then

$$(QF)(x) = \begin{cases} \dfrac{u(x_0)}{h(x_0)} F(\tau^{-1}x) & \text{if } x \in \tau J_{-\infty}^{\infty} \\ 0 & \text{if } x \notin \tau J_{-\infty}^{\infty} \end{cases}$$

Lemma 2. Let $F \in L_p(J_{-\infty}^{\infty}, \beta_{-\infty}^{\infty}, y_{-\infty}^{\infty})$ be a function depending only on finitely many coordinates $(x_0,x_1,\ldots,x_n)$, $n \ge 0$. Then $EQF = EQEF = \frac{1}{h} ThEF$.

Proof of Lemma 2. If $F = F(x_0,x_1,\dots,x_n)$ then QF depends only on $(x_0,\dots,x_{n+1})$ and

$$(QF)(x_0,x_1,\dots,x_{n+1}) = \begin{cases} \frac{u(x_0)}{h(x_0)} F(\pi_0^{-1}(x_0,x_1),x_2,\dots,x_{n+1}) & \text{if } (x_0,x_1,) \in Z \\ 0 & \text{if } (x_0,x_1) \notin Z \end{cases}$$

Hence

$$(EQF)(x_0) = \frac{u(x_0)}{h(x_0)} \int_{[0,r(x_0)]\times J_2\times\dots\times J_{n+1}} F(\pi_0^{-1}(x_0,x_1),x_2,\dots,x_{n+1})dx_1\dots dx_{n+1}$$

$$= \frac{u(x_0)}{h(x_0)} \int_0^{r(x_0)} dx_1 \int_{J_2^{n+1}} F(\pi_0^{-1}(x_0,x_1),x_2,\dots,x_{n+1})\, dx_2\dots dx_{n+1}$$

$$= \frac{u(x_0)}{h(x_0)} \int_0^{r(x_0)} (EF)(\pi_0^{-1}(x_0,x_1))dx_1$$

$$= \frac{1}{h(x_0)}\, T(hEF)(x_0) \qquad \square$$

Completion of the proof of the Theorem 1. From the last lemma it follows that if $f \in L_p(J,\beta,m)$ and if the support of f is contained in the support of h, then $T^n f = hEQ^n \frac{f}{h}$. Hence, if $\tilde{Q}$ is the isometry of $L_p(J_{-\infty}^{\infty}, \beta_{-\infty}^{\infty}, \eta_{-\infty}^{\infty})$ with $d\eta_{-\infty}^{\infty} = \frac{1}{h} d\mu_{-\infty}^{\infty} = \dots \times \{\alpha_{-1}\} \times \eta \times \lambda \times \dots$ where $d\eta = \frac{1}{h} d\mu$ is the measure that is the restriction of m to support of h, then

$$T^n f = E\,\tilde{Q}^n f .$$

Hence T is a contraction on $L_p(J,\beta,\eta)$. Now any $f \in L_p(J,\beta,m)$ can be written as $f = f_1 + f_2$ where $f_1 \in L_p(J,\beta,\eta)$ and the support of f_2 is disjoint from the support of h. We show that $Tf_2 = 0$, thus completing the proof of the fact that T is a contraction on $L_p(J,\beta,m)$ and that it has a dilation to a (not necessarily invertible) positive isometry.

Lemma 3: If $f \in M^+$ and if $fh = 0$ m.a.e. then $Tf = 0$.

Proof: $0 = \int f\, h^{p-1} dm = \int f\, T^* u^{p-1} dm$

$= \int T f\, u^{p-1} dm = 0$.

Hence $Tf = 0$ since $u^{p-1} > 0$ a.e.

Dilation to an Invertible Isometry

We now have $T^n = E\widetilde{Q}^n$, where $\widetilde{Q}$ is induced by a one-to-one point transformation of $J_{-\infty}^{\infty}$ onto $\tau J_{-\infty}^{\infty}$. If $\tau J_{-\infty}^{\infty} = J_{-\infty}^{\infty}$ (i.e. $r = 1$ m.a.e.), then we are done: $\widetilde{Q}$ is invertible. Otherwise let $Y_0 = J_{-\infty}^{\infty} - \tau J_{-\infty}^{\infty}$ with the σ-algebra and measure inherited from $(J_{-\infty}^{\infty}, \beta_{-\infty}^{\infty}, \eta_{-\infty}^{\infty})$, let Y_n $n = 1,2,\ldots$ be countably many disjoint copies of Y_0 (as a measure space), and let $X = J_{-\infty}^{\infty} \cup \bigcup_{n=1}^{\infty} Y_n$. Define $\tau' : X \to X$ as τ on $J_{-\infty}^{\infty}$ and as the identity: $Y_n \to Y_{n-1}$ on the remaining part. Then τ' is an equivalence of X that induces an invertible isometry R of $L_p(X)$. Then $T^n f = E\chi R^n \tilde{f}$

$= \chi\, ER^n \tilde{f}$

for all $f \in L_p(J,\beta,m)$, identifying f with a function $\tilde{f}$ on X such that $\tilde{f} = 0$ on $\bigcup_{n=1}^{\infty} Y_n$ and $\tilde{f}(x) = f(x_0)$ if $x = (\ldots,x_{-1},x_0,x_1,\ldots) \in J_{-\infty}^{\infty}$. Here χ is the multiplication by the characteristic function of $J_{-\infty}^{\infty}$ in X, and E is the conditional expectation operator, corresponding to the σ-algebra β' on X which coincides with β_0 on $J_{-\infty}^{\infty}$ and has $\bigcup_{n=1}^{\infty} Y_n$ as an atom.

Existence of $u > 0$ with $T^*(Tu)^{p-1} \le \|T\|^p u^{p-1}$ in the Discrete Case.

We will be dealing with countable sets and sequences defined on them. If D is a (finite or) countable set then $M(D)$ and $\overline{M}(D)$ denote, respectively, the set of all real valued and the set of all extended real valued sequences $x = (x_i)_{i \in D}$ on D. If $1 \le p \le \infty$ then $\ell_p(D) \subset M(D)$ is the usual Banach space, where D is now considered as a measure space with the counting measure.

Classes of non-negative members of these sequence spaces are denoted, correspondingly, by $M^+(D)$, $\overline{M}^+(D)$ and $\ell_p^+(D)$.

Let D and R be two countable sets. A non-negative matrix $T = (T_{ik})$, $i \in D$, $k \in R$, defines two operators $T: \overline{M}^+(D) \to \overline{M}^+(R)$ and $T^*: \overline{M}^+(R) \to \overline{M}^+(D)$ as $(Tx)_k = \Sigma_{i \in D} T_{ik} x_i$, $x \in \overline{M}^+(D)$, $k \in R$ and $(T^*y)_i = \Sigma_{k \in R} T_{ik} y_k$, $y \in \overline{M}^+(R)$, $i \in D$.

In what follows we fix a number p , $1 < p < \infty$, and assume that (the linear extension of) the operator T defined above is a bounded operator from $\ell_p(D)$ into $\ell_p(R)$. We denote this operator by the same letter, $T : \ell_p(D) \to \ell_p(R)$ and its norm by $\|T\| = \lambda$. Similarly, $T^*: \ell_q(R) \to \ell_q(D)$, with $q = \frac{p}{p-1}$, is a bounded operator and $\lambda = \|T^*\|$. Our purpose is to prove the following result.

<u>Theorem 2</u>. There exists a sequence $u \in M^+(D)$, with strictly positive terms, such that $T^*(T u)^{p-1} \leq \lambda^p u^{p-1}$.

<u>Proof</u>. We will first prove this result for a special type of T and then show that the general case can be reduced to this special case.

<u>Definition</u>. The operator T will be called <u>simple</u> if for any $i, j \in D$ there exist finitely many $i_1, \ldots, i_{n+1} \in D$ and $k_1, \ldots, k_n \in R$ so that $i_1 = i$, $i_{n+1} = j$ and so that $T_{i_m k_m} \cdot T_{i_{m+1} k_m} > 0$ for each $m = 1, \ldots, n$. Such a finite sequence $(i_1, \ldots, i_{n+1})$ will be called a <u>chain</u> connecting i to j . Until further notice we will now assume that T is a simple operator, in addition to our previous hypotheses on T .

Let $D(n) \subset D$ be a non-decreasing sequence of finite subsets of D with $\bigcup_{n=0}^{\infty} D(n) = D$, so that if $T(n): \ell_p(D(n)) \to \ell_p(R)$ is the restriction of T to $D(n)$, then $T(n)$ is also a simple operator for each $n = 0, 1, \ldots$. The existence of such a sequence is clear: If D is finite, let $D(n) = D$ for each n . Otherwise order the elements of D as $(0, 1, 2, \ldots)$ and let $D(0) = \{0\}$ and for $n = 1, 2, \ldots$, let $D(n)$ contain, in addition to the elements of $D(n-1)$, also a chain connecting n to 0.

We denote the norm of $T(n)$ by $\lambda(n)$. Since $\ell_p(D(n))$ is a finite-dimensional space, it is clear that there exists a non-zero vector $u(n) \in \ell_p^+(D(n))$ so that $\|T(n)u(n)\|_p = \lambda(n)\|u(n)\|_p$. Using the uniqueness part of the Hölder's Inequality, one can then easily show that

$$T(n)^*(T(n)u(n))^{p-1} = \lambda(n)^p \, u(n)^{p-1} \quad ,$$

or, more explicitly, that

$$\sum_{k \in R} T_{ik} \Big(\sum_{j \in D(n)} T_{jk} \, u(n)_j\Big)^{p-1} = \lambda(n)^p \, u(n)_i^{p-1}$$

for all $i \in D(n)$.

If D is finite then $u = u(n)$ and $\lambda = \lambda(n)$ are independent of n ; the proof of Theorem 2 follows. Hence we assume that D is infinite and denote the elements of D as $0, 1, 2, \ldots$ as before, with the definition of $D(n)$ as given above. Since $u(n)$'s are determined only up to a positive multiplicative constant, we normalize them by the condition

$$u(n)_0 = 1 \quad , \qquad n = 0, 1, 2, \ldots \quad .$$

In what follows, $u(n)$ will always be considered as satisfying this condition.

<u>Lemma 1</u>. Let $i_0, j_0 \in D$ and $k_0 \in R$. Then

$$\frac{u(n)_{j_0}}{u(n)_{i_0}} \geq \frac{1}{\lambda^{p/p-1}} \left(T_{j_0 k_0}\right)^{\frac{1}{p-1}} T_{i_0 k_0} \quad ,$$

whenever $i_0, j_0 \in D(n)$.

<u>Proof</u>. This follows directly from

$$\lambda^p u(n)_{j_o}^{p-1} \geq \lambda(n)^p u(n)_{j_o}^{p-1} = \sum_{k \in R} T_{j_o k}(\sum_{i \in D(n)} T_{ik} u(n)_i)$$

$$\geq T_{j_o k_o} T_{i_o k_o}^{p-1} u(n)_{i_o}^{p-1} .$$

Corollary. For each $i_o, j_o \in D$ there exists a constant $K(i_o,j_o) < \infty$ such that

$$\frac{u(n)_{i_o}}{u(n)_{j_o}} \leq K(i_o,j_o)$$

whenever $i_o, j_o \in D(n)$. In particular,

$$(0 <) \quad \frac{1}{K(0,i_o)} \leq u(n)_{i_o} \leq K(i_o,0) \quad (< \infty)$$

whenever $i_o \in D(n)$.

Proof. If there is a $k_o \in R$ such that $T_{i_o k_o} T_{j_o k_o} > 0$, then we may let $K(i_o,j_o) = \lambda^{\frac{p}{p-1}} (T_{j_o k_o})^{-\frac{1}{p-1}} T_{i_o k_o}^{-1}$. Otherwise we let n_o to be the smallest integer with $i_o, j_o \in D(N_o)$. Fix a chain $(i_o, i_1, \ldots, i_r, j_o)$ in $D(n_o)$ connecting i_o to j_o, and let

$$K(i_o,j_o) = K(i_o,i_1)\, K(i_1,i_2) \cdots K(i_r,j_o) .$$

The final part follows immediately, since $u(n)_o = 1$ for each n.

Hence there is a subsequence n_m such that $\lim_{m \to \infty} u(n_m)_i = u_i < \infty$ exists and is non-zero for each $i \in D$. Without loss of generality, we will assume that $\lim_{n \to \infty} u(n)_i = u_i$ for each $i \in D$.

Similarly, if we let $v(n) = T(n)u(n)$ then $v(n)_k$ is bounded in n for each fixed $k \in R$. This follows from the fact that $\Sigma_{k \in R} T_{ik} v(n)_k^{p-1} = \lambda(n)^p u(n)_i^{p-1}$ for each $i \in D(n)$. Hence without loss of generality we may also assume that

$$\lim_{n\to\infty} v(n)_k = \lim_{n\to\infty} \sum_{i\in D(n)} T_{ik}\, u_i(n) = v_k$$

exists (and is finite) for each $k \in R$.

Then two applications of Fatou's Lemma give the following for each fixed $i \in D$:

$$\lim_{n\to\infty} \lambda(n)^p\, u(n)_i^{p-1} = \lambda^p\, u_i^{p-1}$$

$$= \lim_{n\to\infty} \Sigma_{k\in R}\, T_{ik}\, (\Sigma_{j\in D^0}\, T_{jk}\, u(n)_j)^{p-1}$$

$$\geq \sum_{k\in R} T_{ik} \lim_{n\to\infty} (\sum_{j\in D(n)} T_{jk}\, u(n)_j)^{p-1}$$

$$= \sum_{k\in R} T_{ik} (\lim_{n\to\infty} \sum_{j\in D(n)} T_{jk}\, u(n)_j)^{p-1}$$

$$\geq \sum_{k\in R} T_{ik} (\sum_{j\in D} T_{jk} \lim \bar{u}(n)_j)^{p-1}$$

$$= \sum_{k\in R} T_{ik} (\sum_{j\in D} T_{jk}\, u_j)^{p-1} ,$$

where $\bar{u}(n)_j$ is defined as $u(n)_j$ if $j \in D(n)$ and as 0 otherwise.

This gives the proof of Theorem 1 in the special case where T is a simple operator.

In the general case we first let

$$D^0 = \{i \mid i \in D\ ,\ T_{ik} = 0\ \forall\ \ k \in R\}$$

On $D - D^0$, we will say that $i,j \in D - D^0$ are <u>related</u>, and write $i \sim j$, if there is a (finite) chain connecting i to j in the sense of the Definition given above. It is easy to see that this is an equivalence relation, which defines a partition of $D - D^0$ as, say, $D^1, D^2, \ldots$. The restriction of T to each D^n, $n = 1,2,\ldots$ is now a simple operator and we have a strictly positive vector $u^n \in \ell_p^+ (D^n)$ for each $n = 1,2,\ldots$ so that if $i \in D^n$ then

$$\sum_{k \in R} T_{ik} \left(\sum_{j \in D^n} T_{jk} u_j^n\right)^{p-1} \leq \lambda^p (u_i^n)^{p-1}$$

Now let $u_i = 1$ if $i \in D^0$ and $u_i = u_i^n$ if $i \in D^n$. Then we claim that

$$\sum_{k \in R} T_{ik} \left(\sum_{j \in D} T_{jk} u_j\right)^{p-1} \leq \lambda^p u_i^{p-1} ,$$

for each $i \in D$.

In fact, this is trivial if $i \in D^0$, since in this case $T_{ik} = 0$ for each $k \in R$. Let $i \in D^n$. Then, if $T_{ik} > 0$, we have $T_{jk} = 0$ whenever $j \notin D^n$. Hence, if $i \in D^n$ and if $T_{ik} > 0$, then

$$\sum_{j \in D} T_{jk} u_j = \sum_{j \in D^n} T_{jk} u_j$$

$$= \sum_{j \in D^n} T_{jk} u_j^n$$

and the required inequality follows. □

REFERENCES

1. M. A. Akcoglu, Positive contractions on L_1-spaces, Math. Z. 143, 1975, 1-13.
2. M. A. Akcoglu and L. Sucheston, On convergence of iterates of positive contractions in L_p-spaces, J. Approximation Theory 13 (1975), 348-362.
3. M. A. Akcoglu and L. Sucheston, Remarks on dilations in L_p-spaces, Proc. Amer. Math. Soc., 53, November 1975.
4. J. L. Doob, Stochastic Processes, Wiley, New York 1953.
5. G. G. Lorentz, "Bernstein Polynomials," University of Toronto Press, Toronto, 1953.
6. D. Maharam, Decompositions of measure algebras and spaces, Trans. Amer. Math. Soc. 69 (1950), 142-160.
7. V. A. Rohlin, On the fundamental ideas of measure theory, Mat. Sb. 25 (1949), 107-150.

MEASURE-FINE UNIFORM SPACES I

Zdeněk Frolík
Matematický ústav ČSAV
Žitná 25, 110 00 Praha 1, Československo

At the beginning the basic results on uniform measures are stated (§ 2). The aim of this note is to introduce a large class of mappings of uniform spaces such that the images of uniform measures are uniform measures, denote $\mathcal{M}(X,Y)$ the set of all these mappings from X into Y, and show that there exists a functor $\mathcal{M}_f$ of uniform spaces (in fact, a coreflection, hence $\mathcal{M}_f X$ is a uniform space finer than X) such that

1) $\mathcal{M}(X,Y) = U(\mathcal{M}_f X, Y)$.

2) X and $\mathcal{M}_f X$ have the same uniform measures.

3) Each uniform measure on X is σ-additive iff $f_n \in U_b(X)$, $f_n \downarrow 0$ imply $\{f_n\}$ is equi-uniformly continuous (and this is equivalent to a much stronger condition).

4) $\mathcal{M}_f X$ is locally fine.

5) $U_b(\mathcal{M}_f X)$ is exactly the set of all Riemann-measurable functions wrt all uniform measures on X.

Another description of the main purpose of this note: a study of the relationship between uniform and Radon measures.

§ 1. Basics. By a space we shall mean a uniform space. The set of all uniformly continuous maps of X into Y is denoted by $U(X,Y)$. Denote by $U_b(X)$ the set of all bounded uniformly continuous functions (real-valued!) on X.

For a space X we denote by $\hat{X}$ the completion of X, and by $\check{X}$ the so called Samual compactification of X (i.e. the completion of the precompact reflection pX of X, which is the structure space of the Banach algebra $U_b(X)$). Recall that pX is projectively generated by $U_b(X)$. We may and shall assume that $X \subset \hat{X} \subset \check{X}$, the first inclusion is uniform, the second is just proximal, hence both are topological.

Denote by $\mathfrak{M}(X)$ the linear space of all continuous functionals on $U_b(X)$ (with the sup-norm); the elements of $\mathfrak{M}(X)$ are called measures on X. For a compact space X, $C(X) = U_b(X)$, and each measure on X can be uniquely represented by a regular Borel measure on X. Now for any space X the restriction operator

$U_b(\check{X}) \longrightarrow U_b(X)$ defines a bijection of measures on X onto the measures on $\check{X}$.

For any measure μ on X denote by $\check{\mu}$ the regular Borel measure on $\check{X}$ which represent the measure on $\check{X}$ which corresponds to μ.

Recall that a measure μ on a space X is called:

σ-additive if for each sequence $\{f_n\}$ in $U_b(X)$ which point-wise decreases to 0 (in symbols: $\{f_n\} \downarrow 0$) $\mu(f_n) \longrightarrow 0$; the set of all σ-additive measures on X is denoted by $\mathfrak{M}^{\sigma}(X)$;

τ-additive if for each net $\{f_a\}$ in $U_b(X)$ decreasing to 0 (in symbols $\{f_a\} \downarrow 0$), $\mu(f_a) \longrightarrow 0$. The set of all τ-additive measures is denoted by $\mathfrak{M}^{\downarrow}(X)$;

tight or a Radon measure if μ is continuous on the unit ball of $U_b(X)$ in the topology of uniform convergence on compact subsets of X (i.e. in the compact-open topology). The set of all tight measures on X is denoted by $\mathfrak{M}^{t}(X)$. Note:

Lemma 1. $\mathfrak{M}^{t}(X) \subset \mathfrak{M}^{\downarrow}(X)$.

We shall need the following well-known theorem (which shows that τ-additivity and tightness are topological concepts). It should be remarked that σ-additivity is not a topological concept. For a uniform space X denote by $t_f X$ the finest uniform space topologically equivalent to X.

Theorem 1. Each τ-additive (Radon) measure on X extends to a unique τ-additive (Radon, respectively) measure on $t_f X$.

Proof. For f in $U_b(t_f X)$, the net $\{\bar{f} - \underline{f}\}$, $\bar{f}$, $\underline{f} \in U_b(X)$, $\underline{f} \leq f \leq \bar{f}$, decreases to 0.

Recall the author's characterization of σ-additive and tight measures on X by means of the corresponding regular Borel measures on $\check{X}$ ([4]) which generalizes the results for topological spaces.

Theorem 2. $\mu \in \mathfrak{M}^{\sigma}(X)$ iff $|\check{\mu}| Z = 0$ for each zero set Z contained in $\check{X} - X$. $\mu \in \mathfrak{M}^{\downarrow}(X)$ iff $|\check{\mu}| C = 0$ for each compact set $C \subset \check{X} - X$.

Recall that a zero set in X is the preimage of a closed set under a uniformly continuous function. Cozero sets are complements of the zero sets.

§ 2. Uniform measures. Following L. Le Cam, for a uniform space X denote by UEB(X) the collection of all subsets of $U_b(X)$ which are uniformly bounded, and equi-uniformly continuous. Denote by

Mol(X) the set of all molecular measures on X , i.e. the set of all finite linear combinations of Dirac measures (evaluation at points). Denote by $\mathfrak{M}_{\mu}(X)$ the set of all measures μ on X which are continuous in pointwise topology on each UEB-set.

Recall that on UEB-sets the topology of uniform convergence on precompact sets, the topology of pointwise convergence, and the topology of pointwise convergence on any dense set coincide. Hence:

$$\mathfrak{M}_{\mu}(X) = \mathfrak{M}_{\mu}(\hat{X}) \ , \qquad \mathfrak{M}^{R}(\hat{X}) \subset \mathfrak{M}_{\mu}(X) \ .$$

One can show easily:

Lemma 2. $\mathfrak{M}^{\downarrow}(X) \subset \mathfrak{M}_{\mu}(X)$.

All what follows depends on the following result from [3]. If $\mathcal{U}$ is a uniform cover of X , we denote by $K(\mathcal{U})$ the union of all int $\bar{U}$, $U \in \mathcal{U}$, where the closure and the interior operator are taken in $\check{X}$. Note that any $f: X \longrightarrow Y \in U$ extends to a continuous mapping $\check{f}: \check{X} \rightarrow \check{Y}$.

Theorem 3 (Frolík [3]). Each of the following two conditions is necessary and sufficient for a measure μ on X to be uniform:

(1) For each uniform cover $\mathcal{U}$, $|\check{\mu}|(\check{X} - K(\mathcal{U})) = 0$.

(2) For each uniformly continuous mapping $f: X \longrightarrow M$,

$$|\check{\mu}|(\check{X} - \check{f}^{-1}[\check{M}]) = 0 \ .$$

Proof. Since each $\check{\mu}$ is σ-additive, the conditions (1) and (2) are equivalent. Assuming (2) we shall prove that μ is uniform. Let $f: X \longrightarrow M$ be a uniformly continuous mapping; we must show that $f[\mu]$ is a uniform measure on M . It is enough to show that $f[\mu]$ is a Radon measure on $\hat{M}$, but this is obvious because $f[\mu] = \check{f}[\check{\mu}]$. Finally, we show that Condition 1 is necessary. We may and shall assume that X is metric, and $\mathcal{U}$ is the cover by all δ-spheres. For each finite set $F \subset X$ let

$$f_F = \{x \longrightarrow \text{dist}\ (x,F) \mid x \in X\} : X \longrightarrow R \ .$$

Since

$$f_F\, x = \min_{y \in F} \{ d \langle x,y \rangle \} \ ,$$

$\check{f}_F$ is $\geqq \delta$ on $\check{X} - K(\mathcal{U})$. On the other hand, $\{f_F\}$ is a Lipschitz family with constant 1 , and $\{f_F\}$ converges to zero in the pointwise topology. Hence for each uniform measure μ

$$\mu(f_F) \longrightarrow 0 ,$$

hence $\check{\mu}(\check{X} - K(\mathcal{U})) = 0$.

Corollary 1. (Le Cam [2], Berezovskij [1].) If X is a complete metric space then

$$\mathfrak{M}^R(X) = \mathfrak{M}^t(X) = \mathfrak{M}_u(X) .$$

Corollary 2. If $X = t_f(X)$ is paracompact then

$$\mathfrak{M}^{\downarrow}(X) = \mathfrak{M}_u(X) .$$

Proof. If $C \subset \check{X} - X$ is compact then there exists a uniform cover $\mathcal{U}$ of X ($= t_f X$) such that $K(\mathcal{U}) \cap C = \emptyset$, Hence each uniform measure is τ-additive.

Corollary 3. If X is a metric space then

$$\mathfrak{M}_u^{\sigma}(X) = \mathfrak{M}^{\downarrow}(X) = \mathfrak{M}_u(t_f X) .$$

Proof. By Theorem 1 and Corollary 2 the second equality holds. It is enough to show that

$$\mathfrak{M}_u^{\sigma}(X) \subset \mathfrak{M}_u^{\downarrow}(X) .$$

Assume that $\mu = \mu^+$ is uniform, and μ is not τ-additive. Hence there exists a compact set $C_1 \subset \check{X} - X$ with $\mu C_1 = 2\varepsilon > 0$, and there exists a compact set $C_2 \subset \hat{X}$ which carries $\check{\mu}$ up to $\varepsilon > 0$. Hence $C = C_1 \cap C_2 \subset \hat{X} - X$, $\check{\mu} C \geq \varepsilon$, and C is a zero set because $\hat{X}$ is metric. By Theorem 1 , μ is not σ-additive.

The following results will not be needed:

The space $\mathfrak{M}_u(X)$ has the topology of uniform convergence on the UEB-sets in $U_b(X)$. It is a complete LCVS, and the set $\mathrm{Mol}(X)$ of molecular measures (free real VS over X) is dense in $\mathfrak{M}_u(X)$. The dual of $\mathfrak{M}$ (X) is $U_b(X)$, and the weak topology coincides with the topology of $\mathfrak{M}_u(X)$ (called uniform topology on $\mathfrak{M}_u(X)$) on the cone $\mathfrak{M}^+(X)$ of non-negative measures. A recent important result of J. Pachl says that relatively weakly countably compact sets are relatively compact in the uniform topology.

There is a canonical uniform embedding of X into $\mathfrak{M}_u(X)$ which assigns to each $x \in X$ the Dirac measure at x , i.e. the evaluation of x . We shall write $X \hookrightarrow \mathfrak{M}_u(X)$. The map $\{X \longrightarrow \mathfrak{M}_u(X)\}$ of spaces into complete LCVS is functorial, and

$$\begin{array}{ccc} X & \longrightarrow & Y \\ \downarrow & & \downarrow \\ \mathfrak{M}_u(X) & \longrightarrow & \mathfrak{M}_u(Y) \end{array}$$

commutes. The space $\mathfrak{M}_u(X)$ can be characterized by the following property:

if f is a uniformly continuous mapping of X into a complete LCVS E, and if the range of f is a bounded subset of E, then there exists a unique continuous linear map $\tilde{f}: \mathfrak{M}_u(X) \longrightarrow E$ such that

$$\begin{array}{ccc} X & \hookrightarrow & \mathfrak{M}_u(X) \\ & {\scriptstyle f} \searrow & \downarrow {\scriptstyle \tilde{f}} \\ & & E \end{array}$$

commutes (see [3]). Stated in other words, μ is uniform iff the weak integral $\int f \, d\mu$ exists for every uniformly continuous mapping into a LCVS such that the range is bounded.

It should be remarked that the most of the theory carries over to vector-valued measures (sometimes with some assumptions on the range space). For the general theory we refer to J. Pachl's papers. There is a lot of unpublished material presented in the Seminar of Abstract Analysis 1973-5. For example, if the projective limit of uniform measures (the bonding maps are uniformly continuous) is a measure, it is a uniform measure; any indirect product of positive uniform measures is a uniform measure.

§ 3. Measure-fine spaces. The proofs are given in § 4. We shall need the following result: if X is a dense subspace of a compact space C, and if f is a continuous mapping of X into a complete metric space S, then there exists a continuous extension g of f to a subspace X_f of C such that: if $h: X \longrightarrow S$ is a continuous extension of f, $X' \subset C$, then $X' \subset X_f$. Moreover, X_f is a G_δ-set in C. Roughly speaking, there exists the largest extension, and the domain of this largest extension is a G_δ.

Definition 1. An $\mathcal{M}$-mapping of a uniform space X into a metric space S is a continuous mapping $f: X \longrightarrow S$ such that the domain of the largest continuous extension g of $f: X \longrightarrow \hat{S}$ to a subspace of $\check{X}$ carries each $\check{\mu}$, $\mu \in \mathfrak{M}_u(X)$, as a Radon measure.

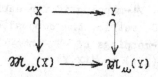

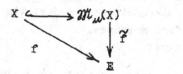

A mapping $f: X \longrightarrow Y$ is an $\mathcal{M}$-mapping if for each uniformly continuous mapping $g: Y \longrightarrow S$, S metric, the composite $g \circ f$ is an $\mathcal{M}$-mapping. The set of all $\mathcal{M}$-mappings of X into Y is denoted by $\mathcal{M}(X,Y)$. By Theorem 3, $U(X,Y) \subset \mathcal{M}(X,Y)$.

Definition 2. A space X is called measure-fine if $\mathcal{M}(X,Y) = U(X,Y)$ for each Y.

Theorem 4. A bounded function f on X is an $\mathcal{M}$-function on X iff the following condition is fulfilled:

for each $\varepsilon > 0$, and each $\mu \in \mathfrak{M}_u(X)$, there exist $\underline{f}$, $\overline{f} \in U_b(X)$ such that $\underline{f} \leq f \leq \overline{f}$, and $|\mu|(\overline{f} - \underline{f}) < \varepsilon$.

Thus the bounded $\mathcal{M}$-functions are just the Riemann-measurable functions w.r.t. all uniform measures.

The main result says:

Theorem 5. For each space X let $\mathcal{M}_f X$ be projectively generated by all $\mathcal{M}$-mappings of X into metric spaces. Then $\mathcal{M}_f$ is a coreflection of uniform spaces into measure-fine spaces, and

$$U(\mathcal{M}_f X, Y) = \mathcal{M}(X,Y)$$

for each X and Y. Moreover, the extension of the identity $\mathcal{M}_f X \longrightarrow X$ to a linear continuous mapping $\mathfrak{M}_u(\mathcal{M}_f X) \longrightarrow \mathfrak{M}_u(X)$ is a bijection.

In addition:

Theorem 6. For each space X, $\mathcal{M}_f X$ is projectively generated by the identity maps

$$i : \mathcal{M}_f X \longrightarrow t_f G ,$$

where G runs over all subspaces of $\check{X}$ with the property that each $\check{\mu}$, $\mu \in \mathfrak{M}_u(X)$, is a Radon measure on G. We obtain a projectively generating family if G's are restricted to paracompact G_δ's.

Theorem 7. Each measure-fine space is locally fine (in the sense of J. Isbell [6]).

Theorem 8. The following conditions on a space X are equivalent:

(a) $\mathfrak{M}_u(X) = \mathfrak{M}_u^{\sigma}(X)$.

(b) $\mathfrak{M}_u^{\sigma}(X) = \mathfrak{M}_u(\mathcal{M}_f X)$.

(c) $\mathcal{M}_f X$ is metric-fine.

(d) $\mathcal{M}_f X$ has the property: if $\{f_n\} \downarrow 0$, and if $\{f_n\}$ ranges in $U_b(\mathcal{M}_f X)$, then $\{f_n\}$ is equi-uniformly continuous.

(e) $\mathcal{M}_f X$ is inversion-closed.

For the proof of Theorem 8 we need elementary:

Lemma 3. If $\mu \in \mathfrak{M}_u^{\sigma}(X)$, then the unique extension of μ to $\mathcal{M}_b(X,R)$ (= the bounded functions with the property in Theorem 4), is σ-additive.

Corollary. $\mathfrak{M}_u^{\sigma}(\mathcal{M}_f X) \longrightarrow \mathfrak{M}_u^{\sigma}(X)$ is bijective.

Remark. Theorem 7 says that $\mathcal{M}_f X$ is quite fine. In particular, $\mathcal{M}_f X$ has a basis for uniform covers consisting of point-finite covers, each uniform cover is refined by cozero-sets of an ℓ_1-uniformly continuous partition of unity, etc.

Theorem 8 says that for a measure-fine space X the following statements are equivalent:

a) every uniform measure is σ-additive;

b) X is metric-fine.

Always (b) implies (a) (Frolík [3]), and the implication (a) $\Longrightarrow$ (b) "usually" does not hold, e.g. on a complete metric space S every uniform measure is tight, hence σ-additive, however, usually $S \neq t_f S$.

Corollary to Theorem 7 [13]. If X is a super-complete (i.e. the space of all closed subsets of X with the Hausdorff uniformity is complete), then each uniform measure on X can be represented by a regular Borel measure on X.

Proof. By Isbell theorem, if X is super-complete then the locally fine coreflection of X is topologically fine, and X is a paracompact topological space. Hence $\mathcal{M}_f X = t_f X$ is paracompact, and the result follows.

§ 4. Proofs of results in § 3.

A. Proof of Theorem 4. Let f be a bounded function on X. Define two functions f^* and f_* on X as follows:

$$f^* = \inf \{\check{g} \mid g \in U_b(X)\ ,\ g \geq f\}\ ,$$

$$f_* = \sup \{\check{h} \mid h \in U_b(X)\ ,\ h \leq f\}\ .$$

Put:

$$G = \{x \mid x \in X\ ,\ f^* x = f_* x\}.$$

First let $f \in \mathcal{M}(X)$. Then f is continuous, and $f^*|G$ is the largest continuous extension of f . Hence each $\check{\mu}$ with $\mu \in \mathfrak{M}_u(X)$, is carried by G as a Radon measure, and by Theorem 1 f has the property in Theorem 4.

Now let f have the property in Theorem 4. Then f is continuous (Dirac function at each $x \in X$ is uniform), and again $f^*|G$ is the largest continuous extension of f . It follows immediately from the condition in Theorem 4 that G carries all $\check{\mu}$, $\mu \in \mathfrak{M}_u(X)$, as Radon measures. By Definition 1 necessarily $f \in \mathcal{M}(X)$.

B. Proof of Theorems 5 and 6. For each space X let $\mathcal{M}_f X$ be projectively generated by $\mathcal{M}$-mappings of X into metric spaces. Hence $\mathcal{M}_f X$ has the meaning given in Theorem 5.

(α) We shall show that Theorem 6 holds. Assume that $G \subset X$ carries all $\check{\mu}$, $\mu \in \mathfrak{M}_u(X)$, as Radon measures, and prove that $i: \mathcal{M}_f X \longrightarrow t_f G$ is uniformly continuous. If $g: t_f G \longrightarrow S \in U$, S being a complete metric space, then the restriction t of g to X is continuous, and the largest continuous extension h of f to a subspace of $\check{X}$ is an extension of g . Since G carries all $\check{\mu}$, $\mu \in \mathfrak{M}_u(X)$, necessarily the domain of h carries all $\check{\mu}$, $\mu \in \mathfrak{M}_u(X)$, and hence f is an $\mathcal{M}$-mapping. This shows that the identity in question is uniformly continuous. Conversely, if $f: X \longrightarrow S \in \mathcal{M}$, S complete metric, and if $h: G \longrightarrow S$ is the largest continuous extension of f to a subspace of $\check{X}$, then by Definition 1, G carries all $\check{\mu}$, $\mu \in \mathfrak{M}_u(X)$, and hence

$$f = h \circ (i: \mathcal{M}_f X \longrightarrow t_f G)$$

factorizes through one of the identity mappings in question.

(β) $\mathcal{M}(X,Y) = U(\mathcal{M}_f X, Y)$. By definition, the inclusion holds. Conversely, if $f: \mathcal{M}_f X \longrightarrow Y \in U$, Y being complete metric, then f factorizes through an $i: \mathcal{M}_f X \longrightarrow t_f G$ because these identities form a projective generating family by (α) which is obviously closed under countable infimas.

(γ) $\mathfrak{M}_u(\mathcal{M}_f X) \longrightarrow \mathfrak{M}_u(X)$ is bijective. The mapping is one-to-one e.g. because of Theorem 4. It is onto because the unique extension of any $\mu \in \mathfrak{M}_u(X)$ to a $\nu \in \mathfrak{M}({}_f X)$ has the property that the image under every uniformly continuous mapping into a complete metric space is a Radon measure, hence a uniform measure by Lemma 2 , and this implies that ν is uniform.

(δ) $U_b(\mathcal{M}_f\ \mathcal{M}_f X) \subset U_b(\mathcal{M}_f X)$. This follows from Theorem 4, and (β). Let f be an element of the left-hand side, $\varepsilon > 0$, $\mu = \mu^+ \in \mathfrak{M}_u(X)$, μ_1 the unique extension of μ to a measure on $\mathcal{M}_f X$. There exist $\overline{f}, \underline{f} \in U_b(\mathcal{M}_f X) = \mathcal{M}_b(X)$, $\underline{f} \leq f \leq \overline{f}$, such that $\mu_1(\overline{f} - \underline{f}) < \varepsilon$. Then we choose $\underline{h}$, $\overline{h} \in U_b(X)$ such that $\underline{h} \leq \underline{f}$, $\overline{h} \geq \overline{f}$, and $\mu_1(\overline{h} - \overline{f}) < \varepsilon$, $\mu_1(\underline{f} - \underline{h}) < \varepsilon$. Now clearly $\underline{h} \leq f \leq \overline{h}$, and $\mu_1(\overline{h} - \underline{h}) = \mu(\overline{h} - \underline{h}) < 3\varepsilon$. This proves that $f \in \mathcal{M}(X)$.

(ε) $\mathcal{M}_f \mathcal{M}_f X = \mathcal{M}_f X$. By ($\delta$) we have

$$U_b(\mathcal{M}_f \mathcal{M}_f X) = U_b(\mathcal{M}_f X),$$

and this implies immediately the relation.

C. Proof of Theorem 7. Recall that a uniform space X is called locally fine if the following condition is fulfilled:

if $\mathcal{U}$ is a uniform cover of X, and if $\{\mathcal{V}_U \mid U \subset \mathcal{U}\}$ is a family of uniform covers, then

$$(*) \qquad \{V \cap U \mid U \in \mathcal{U}, \ V \in \mathcal{V}_U\}$$

is a uniform cover of X. Theorem 5 will be proved just for those who are familiar with basics about locally fine spaces as given in Isbell [6], or know an exposition by the author in Seminar Uniform Spaces, 1973-74.

It is enough to show that if $\mathcal{W}$ is the cover $(*)$ of X, then $K(\mathcal{U})$ carries each $\check{\mu}$, $\mu \in \mathfrak{M}_u(X)$. Let $\mu = \mu^+ \in \mathfrak{M}_u(X)$, and $\varepsilon > 0$. Choose a compact set $C \subset K(\mathcal{W})$ such that $\check{\mu}(\check{X} - C) < \varepsilon$. Since C is compact, C is covered by a finite family

$$\{\operatorname{int} \overline{U} \mid U \in \mathcal{F}\}, \ \mathcal{F} \subset \mathcal{U}.$$

Choose a common uniform star-refinement $\mathcal{T}$ of $\mathcal{U}$ and all $\mathcal{V}_U$, $U \in \mathcal{F}$. The set $K(\mathcal{T})$ carries all $\check{\mu}$, $\mu \in \mathfrak{M}_u(X)$, and hence there exists a compact set K of $K(\mathcal{T})$ such that $\check{\mu}(\check{X} - K) < \varepsilon$. Hence

$$\check{\mu}(\check{X} - (C \cap K)) < 2\varepsilon .$$

It is easy to check that

$$C \cap K \subset K(\mathcal{W}),$$

which concludes the proof.

D. Proof of Lemma 3. Assume $\{f_n\} \downarrow 0$, $\{f_n\}$ ranges in

$\mathcal{M}(X)$, and $\mu = \mu^+ \in \mathfrak{M}_u^\sigma(X)$. Given $\sigma > 0$, choose $\underline{f}_n$ and $\overline{f}_n$ in $U_b(X)$ such that $0 \leq \underline{f}_n \leq f \leq \overline{f}_n$, and $\mu(\overline{f}_n - \underline{f}_n) < \sigma$. Then $\underline{f}_n \to 0$ pointwise, and hence by an elementary theory of σ-additive measures,

$$\mu(\underline{f}_n) \longrightarrow 0 ,$$

hence

$$\lim \mu(\overline{f}_n) \leq \sigma ,$$

hence $\mu(f_n) \to 0$.

E. Proof of Theorem 8. Since (a) is equivalent to (b) by Corollary to Lemma 3, it is enough to show (b) $\Longrightarrow$ (c) $\Longrightarrow$ (d) $\Longrightarrow$ (e) $\Longrightarrow$ (b).

For those who are familiar with the terms in Conditions (c), (d), and (e), it is clear that

$$(c) \Longrightarrow (d) \Longrightarrow (e) .$$

Also (e) implies (b) for any Y (Frolík [5]), and in fact (d) is equivalent to (e) for any Y (a result of Preiss and Zahradník generalizing the result referred to), and self-evidently (d) implies (b) for any Y . The only surprizing implication is (b) $\Longrightarrow$ (c). We must show that if

$$f: \mathcal{M}_f X \longrightarrow S \in U , \quad S \text{ metric},$$

then

$$f: \mathcal{M}_f X \longrightarrow t_f S \in U .$$

It is enough to show that for each $\mu \in \mathfrak{M}_u(\mathcal{M}_f X)$, the image $f[\mu]$ on S is τ-additive, and this follows from Corollary 4 to Theorem 3 because $f[\mu]$ is a σ-additive uniform measure on S .

§ 5. Prospects. The set $\mathcal{M}(X,Y)$ defined in § 3 can be described as follows (this does not seem to be easy):

$f \in \mathcal{M}(X,Y)$ iff the obvious extension $\mathrm{Mol}^+(f): \mathrm{Mol}^+(X) \longrightarrow \mathrm{Mol}^+(Y)$ continuously extends to a continuous map from $\mathfrak{M}_u^+(X)$ into $\mathfrak{M}_u^+(Y)$.

This shows that $\mathcal{M}$ is very natural. Thus the bijection $\mathfrak{M}^+(\mathcal{M}_f X) \longrightarrow \mathfrak{M}^+(X)$ is a homeomorphism, and $\mathcal{M}_f X$ is the finest space with this property.

A simple example shows that if Y_1 and Y_2 are two uniformities finer than X , and if $\mathfrak{M}_u(Y_i) \longrightarrow \mathfrak{M}_u(X)$ are bijections, then $\mathfrak{M}_u(Y) \longrightarrow \mathfrak{M}_u(X)$ does not need to be a bijection, where

Y is the infimum of Y_1 and Y_2. E.g., let X be the set N of natural numbers with the uniformity projectively generated by c (convergent sequences of reals).

For any x in $\beta N - N$ let Y_x be N with the relativization of the fine uniformity on $N \cup (x)$. Then $\mathfrak{M}_\mu(Y_x) \longrightarrow \mathfrak{M}_\mu(X)$ is a bijection for each x, but the infimum of two such distinct spaces is N with the discrete uniformity.

References:

[1] Berezanskij I.A.: Measures on uniform spaces and molecular measures (Russian), Trudy Moskov. mat. obšč. 19(1968), 3-40; English translation: Trans. Moscow Math. Soc. 19(1968), 1 - 40; MR 38 # 4634.

[2] LeCam L.: Note on a certain class of measures (unpublished).

[3] Frolík Z.: Mesures uniformes, C.R. Acad. Sci. Paris 277(1973), A105-108; MR 48 # 2336.

[4] Frolík Z.: Représentation de Riesz des mesures uniformes, C.R. Acad.Sci. Paris 277(1973), A163-166; MR 48 # 2337.

[5] Frolík Z.: A note on metric-fine spaces, Proc. Amer. Math. Soc. 46(1974), 111-119.

[6] Isbell J.R.: Uniform spaces; Math. Surveys of A.M.S., Providence 1964; MR 30 # 561.

[7] Pachl J.: Free uniform measures, Comment. Math. Univ. Carolinae 15(1974), 541-553.

[8] Pachl J.: Free uniform measures on sub-inversion-closed spaces (submitted).

[9] Pachl J.: Compactness in spaces of uniform measures (submitted).

[10] Rajkov D.A.: Free locally convex spaces of uniform spaces (Russian), Mat. Sb. 63(105)(1964), 582-590; MR 28 # 5320.

[11] Zahradník M.: Projective limits of uniform measures (submitted).

[12] Zahradník M.: Inversion-closed spaces have the Daniell property, Seminar Uniform Spaces 1973-4, Matematický ústav ČSAV, Praha, 1975, 233-4.

[13] Fedorova V.P.: On a problem about measures on uniform space (Russian), Uspechi mat. nauk XXIX-5 (179)(1974), 238.

ON A MEASURE THEORETICAL PROBLEM IN MATHEMATICAL ECONOMICS

Dieter Sondermann
University of Hamburg
Department of Economics
D - 2000 Hamburg 13

1. Some Concepts of Mathematical Economics.

In economic theory an economic agent, who participates in an exchange economy with ℓ commodities, is characterized by his consumption possibilities, his tastes and his endowments. In mathematical economics these concepts are given the following precise and restricted formulation: The *endowments* of an economic agent are described by a point e in the commodity space R^{ℓ}, his consumption possibilities by a nonempty subset X of R^{ℓ}, called his *consumption set*, and his tastes by a reflexive, transitive and complete binary relation, $\precsim$, defined on the consumption set X, called his *preference ordering*.

Let P denote the set of all such binary relations with the additional property that their graph is closed in $R^{\ell} \times R^{\ell}$ and that their domain $X = \{ x \in R^{\ell} : (x,x) \in \precsim \}$ is bounded from below. In Hausdorff's topology of "closed convergence" P becomes a separable metrizable space (see e.g. HILDENBRAND [8].

An economic agent with the characteristics $(X, \precsim, e)$ is then described by a point in the catesian product $A = P \times R^{\ell}$, called the *space of agents' characteristics*. For the generic point a in A we denote the corresponding consumption set, preference ordering and endowments by $X(a)$, $\precsim_a$, $e(a)$, respectively.

An exchange economy is a finite set of economics agents, i.e., a finite family of points in A. Since in the study of the consumption sector of an economy economists are mainly interested in a large economy with many (different) agents, such an economy is most appropriately described by an atomless distribution on the space of agents' characteristics, i.e., by a positive nonatomic probability measure μ on the metrizable space A. It is assumed:

(i) $e(a) \in X(a)$ for μ - a.a. $a \in \mathrm{supp}(\mu)$

(ii) $\int_A \mathrm{pr}_e \, d\mu < \infty$, where pr_e denotes the projection of $A = P \times R^\ell$ onto R^ℓ.

Given the positive price vector $p \in P := \mathrm{int}\, R^\ell_+$, an economic agent a with the characteristics $(X(a), \precsim_a, e(a))$ chooses a maximal element for $\precsim_a$ in his *budget set* $\{x \in X(a) : px \le pe\}$. The nonempty set $\varphi(a,p)$ of all such maximal elements is called his *demand set*. Since the demand correspondence $\varphi(a,\cdot)$ from P into R^ℓ is homogeneous of degree zero, prices can be restricted to the open *price simplex* $S := \{p \in P : \sum_{i=1}^{\ell} p_i = 1\}$. For any $p \in S$, the *mean demand* of the economy with distribution μ is then given by the (set-valued) integral[1)]

$$\Phi(p) = \int_A \varphi(a,p)\, d\mu .$$

A price vector $p^* \in S$ is called an *equilibrium price*, if

$$(1) \qquad \int_A \mathrm{pr}_e \, d\mu \in \Phi(p^*),$$

i.e., mean supply can equate mean demand.[2)] For more details and the economic motivitation of these concepts we refer the interested reader to the monograph by HILDENBRAND [8].

1) For measurability and integrability concepts for correspondences see AUMANN [2] or HILDENBRAND [8].

2) The existence of equilibrium prices can be shown under very general conditions. The main tool is Kakutani's fixed point theorem. For existence proofs see e.g. DEBREU [4] or HILDENBRAND [8]. For the computation of economic equilibria see SCARF [9] or SMALE [1o].

2. Statement of the Problem.

The use of demand correspondences in economic equilibrium analysis is hardly satisfactory for two reasons: (1°) The flavour of the economic concept of a price equilibrium is the idea of decentralization. Every individual economic agent chooses a maximal element in his budget set without knowing the actions of the other agents and without knowledge of the total supply. Then, in equilibrium, the total demand should just be equal to the total supply. But this is only guaranteed by condition(1) if the mean demand is a (single-valued) function; (2°) In economic theory one is not only interested in the existence of economic equilibria, but also in their properties, like local uniqueness or stability. The study of such questions involves methods of differential topology[3)]. However, the application of these methods requires continuously differentiable total demand functions. On the other hand, demand correspondences are the appropriate tool to describe individual demand behavior, since the individual consumer typically shows switching behavior in his demand pattern. The question is whether these two aspects can be reconciled. One has the intuitive feeling that in a large economy, where every individual consumer has only a negligible influence on the total demand, the individual switches in demand may smooth out in the aggregate, i.e., a "smooth" mean demand *function* may well be compatible with rather unsmooth individual demand *correspondences*. This was first conjectured, somewhat vaguely, by DEBREU [5] as follows:

Debreu's Conjecture (1971): If the distribution μ is suitably diffused over the space of agents' characteristics A, integration over A of the demand *correspondences* of the agents will yield a (continuous) total demand *function*, possibly even a total demand function of class C^1.

This is, first of all, a measure theoretical problem, since it requires the specification of suitable measures on the space A. But its solution also involves methods of differential topology and of catastrophe theory.

In this paper we can only give a partial answer to this problem. But we hope that this partial answer reveals the structure behind the problem and thus may stimulate further research on an important and difficult field of mathematical economics.

3) See e.g. the monograph by DIERKER [6] or several articles on "Global Analysis and Economics" by SMALE in the Journal of Mathematical Economics (1974 and later).

We shall treat the Debreu conjecture under the following additional assumptions:

(i) There exists a subset $\mathcal{P}^n \subset \mathcal{P}$, which is a finite dimensional differentiable manifold, such that $\mu^*(\mathcal{P}^n \times R^\ell_+) = \mu(A)$.

(ii) $X(a) = R^\ell_+$ for all $a \in \mathcal{P}^n \times R^\ell_+$

(iii) There exists a map

$$u : \mathcal{P}^n \to C^0(R^\ell_+, R)$$
$$\precsim_\rho \to u_\rho$$

such that, for any $\precsim_\rho \in \mathcal{P}^n$ with the local parameter $\rho = (\rho_1, \ldots, \rho_n)$, the following properties hold:

(α) The evaluation map

$$ev : \mathcal{P}^n \times P \to R^\ell_+$$
$$(\precsim_\rho, x) \to u_\rho(x)$$

is of class C^2, where $P := \text{int } R^\ell_+$.

(u is a C^2 *representation* for $\mathcal{P}^n$, cf. [1,p.46]).

(β) $x \precsim_\rho y$ iff $u_\rho(x) \leq u_\rho(y)$ for all $x, y \in R^\ell_+$.

(u_ρ is a *utility representation* for $\precsim_\rho$).

(γ) $Du_\rho(x) > 0$ for all $x \in P$.

(*Monotonicity* of $\precsim_\rho$)

(δ) $u_\rho(x) = 0$ for all $x \in R^\ell_+ \setminus P$.

(*Boundary condition*)

Example: For any $\rho = (\rho_1, \ldots, \rho_\ell) \in S$, the interior of the unit simplex in R^ℓ, and any $x \in R^\ell_+$ define

$$u_\rho(x) := x_1^{\rho_1} \cdot \ldots \cdot x_\ell^{\rho_\ell}$$

and the preference ordering $\precsim_\rho$ by $x \precsim_\rho y$ iff $u_\rho(x) \leq u_\rho(y)$. Then the set $\mathcal{P}^{\ell-1}$ of all these preference relations forms a $(\ell-1)$-dimensional differentiable manifold, satisfying the assumption (ii) and (iii).

Conditions (ii) and (iii) are of technical nature. They impose almost no restriction on the individual demand behavior. E.g., for $\ell = 2$, an individual demand set may look as follows:

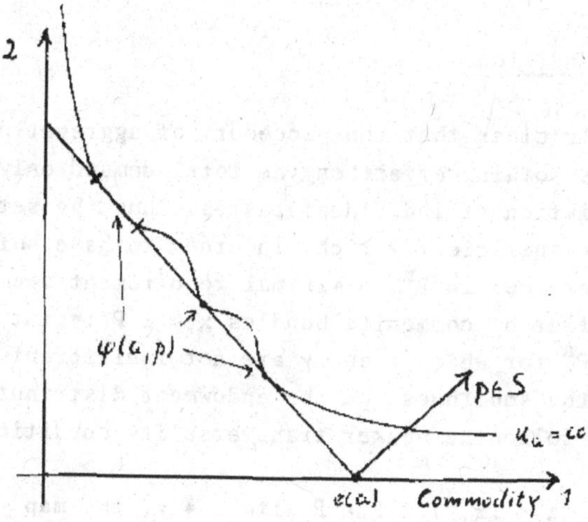

Figure 1

In the terminology of Halmos, condition (i) says that the set $P^n \times R_+^\ell$ is μ-*thick* in A, and thus can be regarded as a measure space itself (HALMOS [7, pp. 74-75]). For simplicity we shall keep the notation $A = P^n \times R_+^\ell$ and μ for this new measure space resp. measure.

The reason why we only consider distributions concentrated on the $(n+\ell)$-dimensional manifold $A = P^n \times R_+^\ell$ is mainly, that it allows the study of specific classes of distributions of demand characteristics, namely distributions which possess local densities with respect to the Lebesgue measure on the underlying parameter space.

A set $N \subset A$ is called a *null set* if, for any $a \in N$ and any chart (U,h) with $a \in U$, the set $h(U \cap N)$ has Lebesgue measure zero in $R^{n+\ell}$. A measure μ on A is called *dispersed* if, for any null set $N \subset A$, $\mu(N)$ is zero. Dispersed measures on A possess local densities with respect to the $(n+\ell)$-dimensional Lebesgue measure. A dispersed measure on A is called *smooth* if all these local densities are continuous functions. Of course, these densities depend on the coordinate maps. Let (U,h) be a chart on A and d_h a local density of $\mu|U$ with respect to the coordinate map h. If g is a different coordinate map on U, then, according to the transformation formula for integrals, one has $d_g = d_h \circ k|\det Dk|$ where det Dk is the Jacobian of the coordinate transformation $k = h \circ g^{-1}$. This shows that the definition of smoothness does not depend on the chart.

3. Transversality Condition.

It is intuitively clear that the procedure of aggregating individual demand will have a smoothing effect on the total demand only if there is a sufficient variation of individual tastes. Thus the set of preferences in P^n must be sufficiently rich. In order to have sufficiently many different preferences in P^n, a minimal requirement seems to be, that for any two different commodity bundles $x,y \in P := \text{int } R_+^{\ell}$ there are some tasted in P^n for which x and y are not indifferent. In fact, by exploiting also the smoothness of the endowment distribution, we shall only need the following weaker transversality condition:

(TC): For any $z = (x,y) \in P \times P$ with $x \neq y$, the map

$$v_z : P^n \times (0,\infty) \to R$$

defined by

$$(\rho,\alpha) \to u_\rho(\alpha x) - u_\rho(\alpha y)$$

has 0 as a regular value.

The interpretation of (TC) is the following: Choose, without loss of generality, $\alpha = 1$. Then $v_\rho(z) = 0$ is equivalent to $u_\rho(x) = u_\rho(y)$, or x and y lie on the same indifference surface of the preference ordering. (TC) says: either the indifference surface will be twisted by moving on the rays through x and y (i.e. $\precsim_\rho$ is non-homothetic), or there is at least one parametric change of the consumption characteristics that will twist the indifference surface such that x and y are no longer indifferen (see Figure 2).

Figure 2

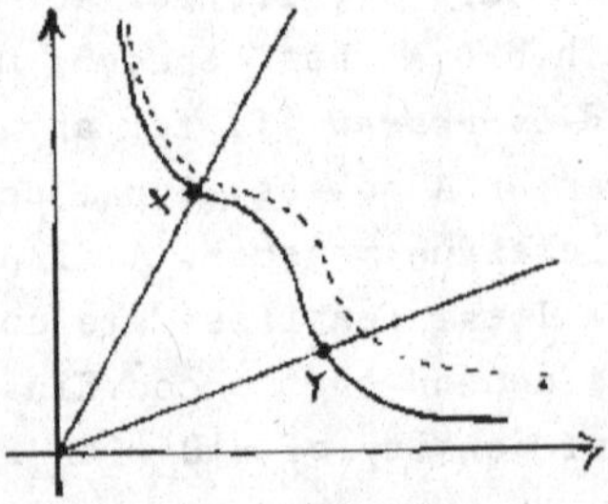

4. Continuity of Mean Demand.

Theorem I: Let P^n satisfy (TC). Then, for every dispersed preference-endowment distribution μ on $A = P^n \times R^\ell_+$, the mean demand $\Phi: S \to P$ given by

$$p \to \Phi(p) = \int_A \varphi(a,p)\, d\mu$$

is a continuous function.

Corollary: For every $p \in S$, there exists a null set $N_p \subset A$ such that, for all agents $a \in A \setminus N_p$, the demand set $p \to \varphi(a,p)$ is a C^1 function in a neighborhood of p.

In [11] we proved Theorem I under stronger conditions with methods of differential topology. Later on (see outline of Proof of Theorem II) we shall show how this Theorem can be proved by methods of catastrophe theory, which also shows, that our Hypothesis (H2) in [11] is superfluous.

5. Differentiability of Mean Demand.

The last Corollary seems to indicate that an economy with a dispersed distribution of demand characteristics will not only have a continuous, but also a differentiable mean demand. Since, for any $p \in S$, the individual demand sets are C^1 functions in a neighborhood of p, except for a null set of individuals, the integral

$$\int_A D_p\varphi(a,p)\, d\mu$$

is well defined. Does this integral give the derivative $D_p\Phi(p)$ of the mean demand, i.e. can one interchange differentiation and integration? The answer is NO, as the following example shows.

Example: For $\rho \in (0,\infty)$ consider the 1-parametric family P^1 of preference relations $\precsim_\rho$ on R^2_+ given by the utility functions $u_\rho(x,y) = \max(x,\rho y)$. Let ν be an absolutely continuous measure on $(0,\infty)$, and consider the distribution $\mu = \nu \times \delta_{(1,0)}$ on $A = P^1 \times R^2_+$, where $\delta_{\tilde{x}}$ denotes the Dirac measure concentrated on $\tilde{x}$. Let $p = \frac{p_y}{p_x}$ be the relative price. Thus, for any ρ and p, one has

$$\varphi_x(\rho,p) \underset{\text{a.e.}}{=} H_\rho(p) = \begin{cases} 0 & p < \rho \\ 1 & p \geq \rho \end{cases}$$

I.e., the demand of the ρ-th consumer for commodity x is equal to the Heavyside function with jump in ρ. Integration yields

$$\Phi_x(p) = \int_A \varphi_x(a,p)\mu(da) = \int_0^\infty H_\rho(p)\nu(d\rho) = \int_0^p d\nu = F(p),$$

where F is the distribution function of ν. For any p, the individual demand is a C^1 function, except for the null set $N_p = \{a \in A : a = (p,e)\}$. But all these demand functions have vanishing derivative. Thus

$$\int_A D_p\varphi_x(a,p)d\mu = 0.$$

However, one has $D\Phi_x(p) = F'(p)$. Consequently the mean demand is a C^1 function if, and only if, the distribution function F is differentiable or, equivalently, if ν possesses a continuous density g. Since then $D\Phi_x(p) = g(p)$, this example shows that the derivative of Φ_x at p is determined completely by the null set N_p, for g(p) is the density of the number of people in N_p. This example also explains why the method of differentiating under the integral sign must fail. □

In the above example one could differentiate under the integral sign by using the distributional derivative ∂_p instead of the functional derivative D_p. As is well-known, the distributional derivative of the Heavyside function is the Dirac measure. Thus one would obtain:

$$\partial_p\Phi = \int_0^\infty \partial_p\varphi(\rho,p)\nu(d\rho) = \int_0^\infty \delta_\rho(p)g(\rho)d\rho = g(p).$$

This leads to the same result, since, if g is a continuous function, then the distributional derivative coincides with the functional derivative and we have $D\Phi(p) = g(p)$. This indicates a different method of attacking the problem of differentiable mean demand. Since the distributional derivatives of the individual demand functions are in general distributions or "generalized" functions, the problem is equivalent to the following question: When is the average of a family of "generalized" functions a continuous function? Observe the analogy to the problem: When is the average of a family of demand correspondences a C^1 function? It may well be, that also the distributional problem has to be attacked by methods of catastrophe theory.

Next we reproduce some concepts from catastrophe theory (cf. THOM [12], [13] or BRÖCKER [3]). Consider the space $C^2(R^m,R)$ of all twice continuously differentiable real functions on R^m. Let Q be the subset of all functions which attain their absolute minimum in a unique non-degenerate critical point of R^m or at infinity. The complement $Mxw := C^2 \setminus Q$ is called the *Maxwell set* of C^2. Furthermore let M^n be an n-dimensional C^2 manifold of *external variables* and $f : M^n \to C^2(R^m,R)$ a C^2 representation. Then the *Maxwell convention* assigns to each external variable $u \in M^n$ an internal variable $x \in R^m \cup \{\infty\}$, such that $f(u)(x)$ is at its absolute minimum. A catastrophe occurs when $f(u)$ attains its absolute minimum in two places or at a degenerate critical point. Thus the *catastrophe set* $C \subset M^n$ is the pre-image of the Maxwell set under the representation f, i.e. $C = f^{-1}(Mxw)$.

In the present problem a catastrophe point is a tripel $(\rho,e,p) \in P^n \times R_+^\ell \times S$ such that the utility function u_ρ restricted to the budget hyperplane $H_{p,e} := p^{-1}(pe)$ attains its absolute maximum at two (or more) different points (see Figure 3). In the (α,p)-Diagram, if we cross the catastrophe set C from below, the demand will jump from the lower region II to the upper region I.

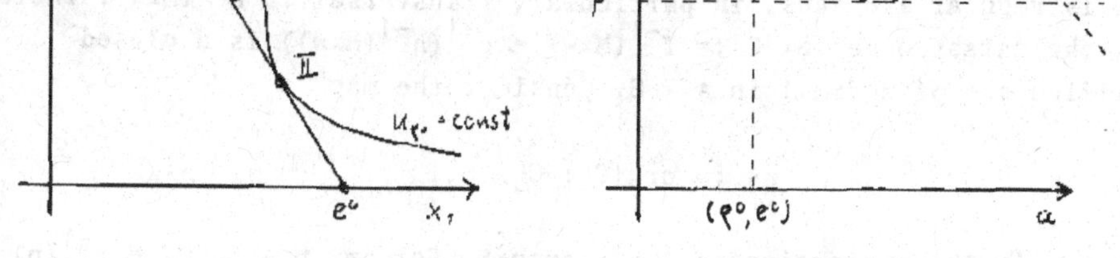

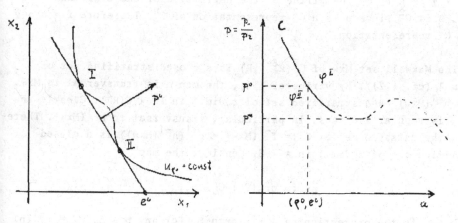

Figure 3

Differentiability of mean demand is closely related to the shape of the catastrophe set C. E.g. if C would continue as indicated by the dotted line, then Φ has a *cusp* at p^*.

Theorem II (Sondermann - Thom [14]): Let P^n satisfy (TC). Then, for every smooth preference-endowment distribution μ on $A = P^n \times R^{\ell}_{+}$ with compact support, the mean demand $\Phi : S \to P$ given by

$$p \to \int_A \varphi(a,p)\,d\mu$$

is a C^1 function on $S \setminus N$, where N is a closed subset of S of Lebesgue measure zero.

Outline of Proof: For the purpose of our problem we shall modify the Maxwell convention by replacing "absolute minimum" through "absolute maximum". Thus what we call Maxwell set in the following is in reality -Mxw. The external variables in our problem are the elements of the $(n+2\ell-1)$-dimensional manifold $A \times S$, the internal variables belong to R^{ℓ}. The representation f is the composition of the two maps

$$A \times S \xrightarrow{g} P^n \times G(R^{\ell}) \xrightarrow{h} C^2(R^{\ell-1},R)$$

$$(\rho,e,p) \longrightarrow (\rho,H_{p,e}) \longrightarrow u_\rho | H_{p,e}$$

where $G(R^{\ell})$ is the Grassmann manifold of hyperplanes in R^{ℓ} and $u_\rho | H_{p,e}$ is the restriction of the utility function u_ρ to the budget hyperplane $H_{p,e}$. g is a C^2 map by virtue of the definition of the Grassmann manifold; h is C^2 since u is a C^2 representation of P^n. Therefore $f = h \circ g$ is a C^2 representation.

The Maxwell set Mxw of $C^2(R^{\ell-1},R)$ is a closed stratified set of codim 1 (cf. [12]). By virtue of (TC), the map h is transversal to Mxw. Thus h^{-1}(Mxw) is a stratified set of codim 1 in $P^n \times G(R^{\ell})$. Clearly the map g is regular and thus, in particular, transversal to h^{-1}(Mxw). Therefore, the catastrophe set $C := f^{-1}(\text{Mxw}) = g^{-1}(h^{-1}(\text{Mxw}))$ is a closed stratified set of codim 1 in $A \times S$. Consider the map

$$\pi := pr_2 | C : C \to S,$$

where pr_2 is the projection of $A \times S$ onto S. For any $p \in S$, $C_p = \pi^{-1}(p)$ is a closed stratified set of codim 1 in $A \times \{p\}$. On $(A \times \{p\}) \setminus C_p$, $\varphi(a,p)$ is a C^1 function. Therefore the mean demand $\Phi(p) = \int \varphi(a,p)d\mu$ is unique, since C_p has μ-measure zero. By upper hemi-continuity of Φ (cf. HILDENBRAND [8]), this implies that $\Phi : S \to P$ is a C^0 function. This proves Theorem I (Observe, that so far we did not use the continuity of the local densities and the compactness of the support of μ).

To prove differentiability under the conditions of Theorem II, we first remark that the set of bifurcation values of p (cf. Figure 4) is a closed null set N in S. Consider $p \in S \setminus N$. Since supp(μ) is compact, by partition of unity it suffices to study $\varphi(a,p)$ on a neighborhood $W = V \times U$ of (a,p). Since C is transversal to $A \times \{p\}$, there exists a diffeormorphism h of W onto $(-1,+1)^{n+\ell} \times U$ taking $C \cap W$ onto the set $[\tau = o]$ (see Figure 4).

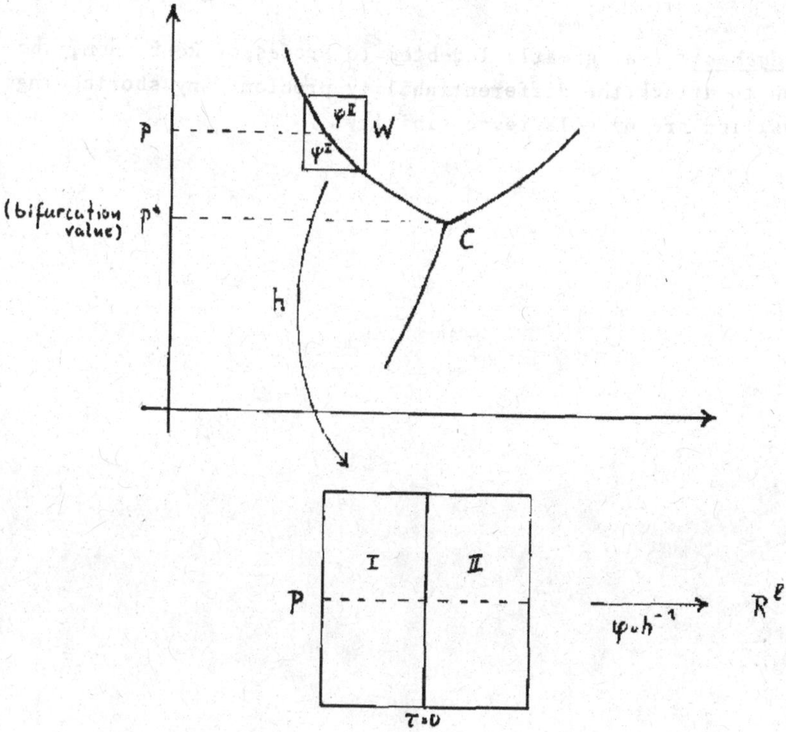

Figure 4

This means that we can suppose C fixed and the distribution $h(\mu)$ varying smoothly with p. Since $\varphi^I \circ h^{-1}$ is a C^1-function on the compact region $I = [\tau \leq o]$ (resp. $\varphi^{II} \circ h^{-1}$ on $II = [\tau \geq o]$), we can now differentiate under the integral sign to obtain that Φ is C^1.

6. Open Problems.

The restriction to a finite-dimensional manifold of preferences is an unnatural assumption. The problem would become easier if the support of the distribution μ on the space $A = P \times R^{\ell}$ is so large, that all necessary transversality condition are automatically satisfied. Then one could even expect differentiability of the mean demand for *all* prices. This would, however, require to consider "suitably diffused" distributions on infinite-dimensional manifolds in A. (It seems to us that abstract Wiener measures may provide the appropriate tool).

Acknowledgement: I am greatly indebted to Professor René Thom, who showed me how to attack the differentiability problem. Any shortcomings of this exposition are my sole responsibility.

References

1. R. ABRAHAM and J. Robbin, Transversal Mappings and Flows, Benjamin 1967.

2. R.J. AUMANN, Integrals of Set-Valued Functions, Journal of Mathematical Analysis Appl., 12, (1965), 1-12.

3. BRÖCKER, Differenzierbare Abbildungen, Lecture Notes, University of Regensburg, 1972.

4. G. DEBREU, Theory of Value, Wiley, 1959.

5. G. DEBREU, Smooth Preferences, Econometrica, 4o (1972), 6o3-615.

6. E. DIERKER, Topological Methods in Walrasian Economics, Springer, 1974.

7. P. HALMOS, Measure Theory, van Nostrand, 195o.

8. W. HILDENBRAND, Core and Equilibria of a Large Economy, Princeton University Press (1974).

9. H. SCARF, The Computation of Economic Equilibria, Yale University Press, 1973.

1o. S. SMALE, Convergent Process of Price Adjustment and Global Newton Methods, to appear in Journal of Mathematical Economics.

11. D. SONDERMANN, Smoothing Demand by Aggregation, Journal of Mathematical Economics, 2(1975), 2o1-223.

12. R. THOM, Sur le cut-locus d'une variété plongée, J. Differential Geometry, 6, (1972), 577-586.

13. R. THOM, Stabilité structurelle et morphogenèse, Benjamin, 1972.

14. R. THOM, Private Communication, Berkeley, August 1974.

Concluding Remarks

by Alexandra Bellow
(formerly A. Ionescu Tulcea)

This has probably been the most "measure-theoretical" conference I ever attended. I learnt a great deal. Here are some of the highlights of the conference, as seen from my vantage point.

I) General Measure Theory. As far as general measure theory is concerned, I learnt of some very nice results - some already known, others of very recent date:

1. Christensen's theorem: Every finitely additive probability measure on $\mathcal{P}(\mathbb{N})$, with values in an Abelian topological group G, which is Borel measurable, is necessarily countably additive.
2. Musial's example of a perfect measure space which is not compact.
3. Fremlin's very pretty theorem about the dichotomy occurring for sequences of measurable functions in a perfect measure space: either (f_n) has a subsequence converging almost everywhere, or (f_n) has a subsequence with no measurable $\mathcal{T}_p$- cluster point ($\mathcal{T}_p$ = the topology of pointwise convergence).

Here several comments are in order.

To begin with, I disagree with Fremlin's opening statement in his lecture that it is somewhat artificial to consider compact sets of measurable functions for the topology $\mathcal{T}_p$.

Compact metrizable sets of measurable functions are the natural context for the Egorov Theorem: P.A. Meyer gave an elegant proof of this beautiful form of the Generalized Egorov Theorem (Séminaire de Probabilités V (1971), Springer-Verlag Lecture Notes); he attributed the theorem to G. Mokobodzki. This was in fact my motivation for looking at compact sets of measurable functions in the topology $\mathcal{T}_p$. (For interesting applications, see my article "On measurability, pointwise convergence and compactness", Bull. A.M.S., Vol. 80, March 1974). The following important question arose then as a consequence:

Let $(E,\mathcal{E},\mu)$ be a probability space. Denote by $\mathcal{L} = \mathcal{L}(E,\mathcal{E})$ the set of all $f: E \to R$ which are $\mathcal{E}$-measurable. On $\mathcal{L}$ we consider the topology $\mathcal{T}_p$ of pointwise convergence and the (non-Hausdorff) topology $\mathcal{T}_\mu$ of convergence in μ-measure.

Problem. Let $H \subset \mathcal{L}$ be compact for the topology $\mathcal{T}_p$. Suppose that $\mathcal{T}_\mu$ separates the points of H (in the terminology of the above article H satisfies the «separation property»). Is it true that $\mathcal{T}_p|H = \mathcal{T}_\mu|H$, or equivalently that H is metrizable for the topology $\mathcal{T}_p$?

Hitherto it was known that: 1) If H is convex the answer is yes. I had talked about this at the previous Lifting Theory Conference at Oberwolfach. For a proof see my article "On pointwise convergence, compactness and equicontinuity II", Advances in

Math., vol. 12, February 1974. ii) With the assumption that $(E,\mathcal{E},\mu)$ is a perfect measure space, the answer is always yes. This was proved by D.H. Fremlin in "Pointwise compact sets of measurable functions", Manuscripta Math., vol. 15 (1975).

Very recently (June 1976), M. Talagrand gave an ingenious affirmative solution to the problem in the general case ("Solution d'un problème de A. Ionescu Tulcea", to appear in Comptes Rendus Acad. Sci. Paris).

II) Finitely Additive Measures emerge as an important concept:

1. Dubin's example. Start out with (countably additive) Lebesgue measure on the torus Z; take the equivalence relation modulo the group Q of all $z \in Z$ having finite period. This leads to a disintegration $\omega \to \lambda_\omega$ in which λ_ω is purely finitely additive for almost all ω.

2. The notion of lifting is really a finitely additive notion. Let us recall that if $(E,\mathcal{E},\mu)$ is non-atomic (= diffuse) then a lifting ρ fails to be countably additive, i.e. in general

$$\rho(\bigcup_n A_n) \neq \bigcup_n \rho(A_n).$$

3. D. Maharam's talk on finitely additive probabilities on $\mathcal{P}(N)$ was very illuminating. It was fascinating to see that in this context there are cases when a lifting exists and there are also cases when a lifting does not exist.

III) Ergodic Theory. We had very little ergodic theory at this conference. Only Prof. Kakutani's talk on equidistributed sequences of partitions of (0,1); and a beautiful talk it was indeed.

Perhaps at the next conference somebody will come up with an answer to the following puzzling question: For $\alpha = \frac{3}{2}$ is it true that the sequence α^n (mod 1) is equidistributed?

There are a number of important topics discussed at this conference that I did not have a chance to touch upon in this brief review: Differentiation, the Radon-Nikodym property, Vector-valued measures, Logarithmic Sobolev inequalities (quantum field theory), Measure theory and equilibrium analysis (economy), Lifting theory in the classical sense, etc. All in all this conference has been a rich and rewarding mathematical experience.